AN INTRODUCTION TO SALINE LAKES
ON THE QINGHAI–TIBET PLATEAU

MONOGRAPHIAE BIOLOGICAE

VOLUME 76

Series Editors
H.J. Dumont and M.J.A. Werger

The titles published in this series are listed at the end of this volume.

A C.I.P. Catalogue record for this book is available from the Library of Congress.

ISBN 978-94-010-6295-4   ISBN 978-94-011-5458-1 (eBook)
DOI 10.1007/978-94-011-5458-1

Printed on acid-free paper

All Rights Reserved
© 1997 Springer Science+Business Media Dordrecht
Originally published by Kluwer Academic Publishers in 1997
No part of the material protected by this copyright notice may be reproduced or
utilized in any form or by any means, electronic or mechanical,
including photocopying, recording or by any information storage and
retrieval system, without written permission from the copyright owner.

The Author
—Research Professor Zheng Mianping

Zheng Mianping Male; born in Zhangzhou, Fujian Province, China in November 1934; graduated from the Department of Geology, Nanjing University, in 1956. Now he is chief of the R & D Center of Saline Lakes and Epithermal Deposit of the Chinese Academy of Geological Sciences, research professor of the Institute of Mineral Deposits and Chinese Academy of Engineering. He serves as member of the Standing Committee of the Society of the Qinghai–Tibet Plateau Research and Vice–Chairman of the International Association of Saline Lakes.

He has studied the geology of saline deposits, total utilization of saline lake resources and hydrothermal water deposits for 40 years, published 64 academic papers in Chinese and foreign journals and 4 monographs, made systematic and pioneering studies on saline lakes of the Qinghai–Tibet Plateau and the exploration and evaluation of saline deposits, and made outstanding contributions to the exploration, exploitation and utilization of K, B, Li and Cs resources. He has conducted an intensive study and prediction of the saline lake resources on the Qinghai–Tibet Plateau and thus directed mineral prospecting, exploitation and mining.

Contents

Preface xi
Foreword xiii
Part 1 General Overviews (1)

Chapter 1 Introduction (1)
Chapter 2 Saline Lakes and Lake Districts (18)
Chapter 3 Evolution of Cenozoic Lake Basins and the Formation of Saline Lakes (23)
   3.1. Cenozoic Salt-bearing Strata and Salt-Forming Periods (23)
      3.1.1. Cenozoic Salt-bearing Strata (23)
      3.1.2. Quaternary Salt Forming Periods and Periodicity (23)
   3.2. Genesis and Classification of Lake Basins (32)
      3.2.1. Glacier Lakes (32)
      3.2.2. Silted-river Valley Lakes (32)
      3.2.3. Interbank Lakes (33)
      3.2.4. Dissolved-Salt Lakes (34)
      3.2.5. Hot Water Lakes (34)
      3.2.6. Volcanic Lakes (35)
      3.2.7. Tectonic Lakes (35)
      3.2.8. Meteorite Crater Lakes (41)
   3.3. Evolution of Lake Basins and Formation of Saline Lakes (42)
      3.3.1. Eogene-Miocene or Early Pliocene (42)
      3.3.2. Pliocene-early-middle Early Pleistocene (44)
      3.3.3. Middle-late Early Pleistocene to Early Late Pleistocene (45)
      3.3.4. Middle-Late Pleistocene to Holocene (48)
Chapter 4 Hydrochemistry and Mineral Associations of Saline Lakes (55)
   4.1. Salinity and pH Values (55)
   4.2. Hydrochemical Types and Their Zoning (59)
   4.3. Chemical Compositions (63)
      4.3.1. Compositions of Lake Waters (64)
      4.3.2. Dominant Elements (69)
      4.3.3. Features of Distribution of B, Li and K (69)
      4.3.4. Correlation of Elements (72)
      4.3.5. The Rare Earth Elements (75)
   4.4. Hydrochemical Types and Their Corresponding Saline Mineral Associations (75)
Chapter 5 Classification of Saline Lakes and Types of Mineral Deposit (79)
Chapter 6 Tectonogeochemistry and Regional Geochemistry (85)
   6.1. Tectonogeochemical Characteristics (85)

- 6.2. B, Li, Cs and Rb Contents .... (86)
  - 6.2.1. Boron Content in Rocks (Table 1.13) .... (87)
  - 6.2.2. Lithium Content in Rocks .... (92)
  - 6.2.3. Cesium Content in Rocks .... (95)
  - 6.2.4. Rubidium Content in Rocks .... (97)
- 6.3. Boron Content in Plants .... (98)
- 6.4. Regional Hydrochemistry .... (100)
  - 6.4.1. Hydrochemical Features of River Waters .... (100)
  - 6.4.2. B, Li and K Contents .... (103)
  - 6.4.3. Hydrogeochemistry of Groundwater .... (104)
- 6.5. Origins of Hydrochemical Types and Their Zonation .... (114)
  - 6.5.1. Factors Controlling the Formation of Different Hydrochemical Lake Types .... (114)
  - 6.5.2. Formation of Various Water Types and the Hydrochemical Zones .... (116)
  - 6.5.3. Formation of the Chloride Type and Its Hydrochemical Zone .... (121)

Chapter 7 Material Sources and Model of Salt Formation of Saline Lakes .... (123)
- 7.1. Material Sources of Saline Lakes .... (123)
  - 7.1.1. Weathering of Rocks .... (123)
  - 7.1.2. Inherited Salt-bearing Rock Series .... (124)
  - 7.1.3. Deep-seated Recharge .... (125)
  - 7.1.4. Affinity between Dispersed Components in Special Saline Lakes and Geothermal Waters .... (130)
- 7.2. Quatitative Estimate of Material Sources .... (131)
  - 7.2.1. Output of B, Li etc. from Geothermal Waters .... (131)
  - 7.2.2. Comparison between the Input and Geochemical Reserves of Lakes .... (134)
  - 7.2.3. Sources of Special Components .... (136)
- 7.3. Salt Formation Models and Minerogenetic Series .... (139)
  - 7.3.1. Tectono-geomorphological Features and Genetic Models .... (139)
  - 7.3.2. Salt Formation Models and Minerogenetic Series .... (141)

Chapter 8 Shoreline Deposits and Palaeoclimatic Analysis .... (147)
- 8.1. The Paleoclimatic Evolution of the Qinghai-Tibet Plateau since the Late Part of the Late Pleistocene .... (147)
  - 8.1.1. Climatic Evolution since 40 Ka .... (148)
  - 8.1.2. Some Understandings of the Features of Climatic Evolution .... (160)
  - 8.1.3. Genetic Mechanism .... (160)
- 8.2. Shoreline Deposits of the Zabuye Lake Area and Its

    Paleoclimatic Analysis ·· (164)
        8.2.1. Differences between the Paleo-lake Shoreline and
               Structural Terraces ·· (164)
        8.2.2. Deposits of the Zhaduixiong Spit Swarm
               and Their Ages ·· (169)
        8.2.3. Evolution Characteristics of the Zhaduixiong ············ (174)
        8.2.4. Paleoclimatic Analysis ·· (175)

Part 2 Study of Halophilic Organisms ······························· (181)

**Chapter 9 Discovery and Study of Halobacteria and Halophilic Algae
            in the Zabuye Caka** ·· (181)
9.1. Introduction ·· (181)
9.2. Physiogeographic Conditions ·· (182)
9.3. Composition of Halobacteria and Halophilic Algae ·················· (187)
9.4. Geological Conditions and Their Implications························· (189)
9.5. Preliminary Study on the Practical Application and Culture
     Conditions of Halophilic Algae ··· (192)
        9.5.1. Experimental Culture of Algae in Synthetic
               Culture Fluid ·· (194)
        9.5.2. Experimental Culture of Algae in Culture Fluids with
               Different NaCl Concentrations ································ (195)
        9.5.3. Experimental Culture of Algae with N- and
               P-Nutrients ·· (196)
        9.5.4. Experimental Culture in Fluids with Different pH ·········· (197)
        9.5.5. Experimental Culture in the Concentrated Sea
               Water ··· (198)
9.6. Some Ideas of the Halobacteria and Halophilic Algae
     in Saline Lakes ·· (198)

**Chapter 10 Biomineralization of Boron and Other Halotolerant
             Organisms** ·· (200)
10.1. Diversity of Boron Mineralization ·· (200)
10.2. Basic Facts ·· (200)
10.3. Preliminary Experimental Studies on Biomineralization
      of Boron ··· (202)
        10.3.1. Experiments on the Solubility of Boron (Based on the
                Experiments Carried Out by Cheng Peng) ················ (202)
        10.3.2. Preliminary Discussion on the Mechanism of
                Biomineralization of Boron ···································· (203)
10.4. Halotolerant Organisms in Other Lakes ································· (205)
        10.4.1. *Daphniopsis tibetana* Sars ·································· (205)
        10.4.2. *Artemia salina Linneans* ···································· (208)

Part 3 Division of Minerogenic Zones, Prediction of Resource Potential and Prospects of Development of Saline Lakes on the Qinghai–Tibet Plateau (211)

Chapter 11 Division of Minerogenic Zones and Prediction of Resource Potential (211)
    11.1. First–Class Minerogenic Zone of B, Li, K, Cs (Rb, Halite, Soda and Mirabilite)–bearing Carbonate–type Saline Lakes (A) (212)
        11.1.1. Gangbei (North of the Gangdisê) Principal Minerogenic Subzone of B, Li, Cs, Br and Rb–Bearing Brine–Borax (Halite, Soda and Mirabilite) (A_1) (214)
        11.1.2. Gangnan (South of the Gangdisê) Subordinate Minerogenic Subzone of Borax–B and Li–bearing Brine (A_2) (218)
    11.2. First–Class Minerogenic Zone of B, Li, K, (Cs Rb, Halite and Mirabilite)–Yielding Saline Lakes of Sodium (Magnesium) Sulfate Type (B) (218)
    11.3. Second–Class Minerogenic Zone of Li, B, K, Halite and Mirabilite–bearing Saline Lakes of Magnesium Sulfate Subtype (C) (222)
    11.4. First–Class Minerogenic Zone of K, Mg, (B and Li)–bearing Saline Lakes of Sulfate–Chloride Type (223)
    11.5. Third–Class Minerogenic Area of Halite and Mirabilite –bearing Saline Lakes of Sodium Sulfate Subtype (E) (230)

Chapter 12 Prospects of Development of Resources of Saline Lakes and Their Environmental Protection (231)
    12.1. Problems in Exploitation of Saline Lakes (232)
    12.2. Suggestions about the Utilization of Resources of Saline Lakes and Environmental Protection (234)

Concluding Remarks (237)
References (243)
    Appendix 1 Rules for Translating Geographic Names Related to Saline Lakes on the Qinghai–Tibet Plateau (251)
    Appendix 2 List of Lakes on the Qinghai–Tibet Plateau (253)
    Appendix 3 Index of the Place Names (273)
    Appendix 4 Explanation of Plates (290)
    Appendix 5 Map of Hydrochemical Zones of Lakes on Qinghai–Xizang (Tibet) Plateau

Preface

As the senior author, I wrote the book "Saline Lakes on the Qinghai−Tibet Plateau" in Chinese in 1989. The present book has been written on the basis of that book, with all the case histories of saline lakes omitted and the latest research results supplemented. In fact, it is a summary of my long−term study on saline lakes of the Qinghai−Tibet Plateau undertaken since 1956 and appears in the English language. The main purpose in publishing this book is to introduce my colleagues and other scholars who are interested in saline lakes to saline lakes on the world's roof and promote the exchange between the Chinese scientists studying saline lakes and their colleagues of other countries. This book was translated into English by Mr. Jiang Minxi, Mr. Fei Zhenbi and I, and the whole English version was checked, revised and proofread by Fei Zhenbi and me. Mrs. Liu Yuxia carefully prepared the camera−ready typescript, and Mrs. Fu Ziji and Mrs. Wang Qingxin fair−drew all the text−figures. During the 39 years of field investigations of the Qinghai−Tibet Plateau, the author has been accorded great support and provided with various conveniences by the Geological Bureaus and Parties of Qinghai and Tibet as well as local governments at all levels and local people. The Institute of Rock and Mineral Analysis, Institute of Hydrogeology and Engineering Geology and Institute of Mineral Deposits under the Ministry of Geology and Mineral Resources and the Institute of Saline Lakes of Qinghai of the Chinese Academy of Sciences assisted in analyzing the chemical composition of solid and liquid phases, the Institute of Geology of the State Seismological Bureau and the Institute of Vertebrate Paleontology and Paleoanthropology of the Chinese Academy of Sciences assisted in determining the ^{14}C ages of the samples, the Beijing Institute of Botany of the Chinese Academy of Sciences and the Institute of Geology of the Chinese Academy of Geological Sciences assisted in sporopollen analysis, and related divisions and groups of the Institute of Mineral Deposits assisted in stable isotope study, section grinding and chemical analysis. Zhang Fasheng and Liu Shuqin helped the author with some work in different periods of time; especially Xiang Jun, Liu Wengao, Wei Xinjun, Xu Guowen, Zheng Yuan, Zhou Taiguang, Shen Jinming, Min Linsheng, Lu Defu, Gao Shiyang, Chen Bingmu, Jing Xuqing and Qian Jun had conducted field surveys or participated in laboratory work together with the author in the fifties, sixties and eighties. My grateful thanks are due to all of them. I would also like to express my heartfet thanks to the older generation of geologists such as the late academians Liu Dagang, Yuan Jianqi and Meng Xianming, Chief Engineer Li Yueyan and Academians Guo Wenkui,

Song Shuhe, Ye Lianjun, Zhang Bingxi and Zhang Zonghu for their help. I thank especially, Dr Porman for revising Chapters 1-8 of this book in English rhetoric. I also wish to extend my appreciation to Prof. W.D. Williams for writing a foreword for the book and recommending it to the publisher so that the book can be publish abroad and more colleagues of mine can gain knowledge of saline lakes on the "world's roof". Finally, I am especially grateful to my wife Liu Junying, associate research fellow of the Chinese Academy of Geological Sciences, who made a lot of sacrifices for the successful completion of this book.

Zheng Mianping
June 1995

Foreword

Until recent decades, the secrets of the Qinghai–Tibet plateau's natural environment, the wonders of its landscapes, remained mostly beyond the reach of science and scientists. Located at high altitude, inaccessible, and with a forbidding climate, the plateau was visited by few with a scientific interest in its environment. Even mountaineers were generally confined to regions of very high altitude in the southern periphery of the area. The situation has changed rapidly in the recent past and continues to evolve. Now, thanks largely to several expeditions involving Chinese scientists (and in a few cases also, non–Chinese scientists), supported by an increasingly important and active regional scientific community, we are beginning to tease out the secrets and more fully appreciate the wonders.

Amongst the most noteworthy and significant features of the Qinghai–Tibet plateau are its lakes. But these are no ordinary lakes: contrary to what cursory and unknowing glances at any atlas may suggest, the lakes on the plateau are – for the most part – saline, with salinities from less than 3 to over 500 parts per thousand (the latter value, 14 times greater than that of the sea!), include some of the world's largest lakes (with almost 40 lakes of area greater than 50 km^2), and certainly include some to the world's highest lakes (e.g. Nam Co, area 1,961 km^2, lies nearly 4,600 m above sea–level). Without doubt, the lakes of the plateau are unique.

My friend and distinguished colleague, Professor Zheng Mianping, has played a significant role in the scientific investigation of these lakes. He has participated in many expeditions, promoted the development of regional scientific efforts, and written widely on scientific matters. Thanks largely to Professor Zheng and his colleagues, we are now in a much stronger position than we were with regard to our knowledge of the limnology of the Qinghai–Tibet plateau. In this connection, it is salutary to place their endeavours within familiar perspectives. Note, then, that the plateau is larger than the whole of western Europe, that access to most lakes involves many hundreds of kilometres of remote travel on unsealed roads, and that prolonged exposure to low oxygen concentrations at high altitudes is not a healthy pastime for those who live normally at altitudes not far above sea–level. Winter temperatures of −40℃, intense solar radiation, and high winds must also be contended with. The sheer physical efforts of Professor Zheng and his colleagues in their limnological investigations are surely orders of magnitude greater than those needed by most limnologists. I speak from slight experience in this matter because after the Sixth International Symposium on Salt Lakes, hosted by Professor Zheng and his colleagues in Beijing in 1994, I and several other colleagues were able to join him on a short expedition to some of his more accessible lakes in Tibet. Even to an Australian, the expedition was a salutary lesson in practical limnology!

Unfortunately, much of the work of Professor Zheng and his colleagues so far remains largely unknown to many limnologists who do not read Chinese (and of whom I am one). This is indeed a great pity. I am glad therefore that the pleas of many of these limnologists, including my own pleas, have at last been acceded to: behind this foreword lies summarised in English the results of many years of Professor Zheng's work, which, together with that of others, represents a masterly review of the Qinghai–Tibetan limnology so far as present data permit. Of course, much remains for discovery, but the present text provides a magnificent basis for future labours. Professor Zheng deserves not only our congratulations on his efforts and achievements but also our thanks. I am pleased to write this foreword and so by a small shaft of reflected illumination be associated with this volume.

W.D. Williams
Adelaide, Australia, September 1995

PART 1: GENERAL OVERVIEW
CHAPTER 1 INTRODUCTION

The Qinghai—Tibet (Xizang) Plateau saline lake region lies between 29 and 39 ° N in western China. It extends from the western part of the Karakorum—Himalaya mountains in the west to a line connecting Qinghai Lake and Yamzho Yumco in the east. This region belongs to the eastern sector of the saline—lake belt in the northern hemisphere and has the highest elevation of any salt lake region on the earth. Its unique and abundant saline—lake mineral resources have long made it the focus of scientific and economic exploration.

Exploitation and utilization of the saline lakes on the Qinghai—Tibet Plateau began as early as the 6th century A.D. when borax was first used in Tibet. Indeed it is thought that this represents the world's first discovery and practical use of this resource. The use of borax in medical treatment is recorded in the "Sibu Yidian", a pharmocopoeia written in Tibetan in the year 720 A.D. (Plate 1.1). There are also records concerning Tibetan in the works of Arabic chemists in the 8th century. As far back as the 13th century borax from Tibet was being sold as far afield as Europe (Gorbov, 1976). According to the "Rihuazi Zhujia Bencao", a pharmocopoeia on Chinese herbal medicine written around 970 A.D., borax was "introduced from the areas of its natural occurrence around Qinghai Lake into the Central Plains of China" (Zhang, 1954). Throughout the period from the reign of Emperor Qianlong of the Qing Dynasty to the years of the Chinese Republic the borax deposits have been utilised by the peoples of Tibet (Lee, 1976). After 1956 the first Chemical Plant was established in Tibet with the support of the Government of the People's Republic of China. Since then the borax production has shown a steady annual increase, such that by the end of July 1961, the cumulative production of crude borax had reached about 100,000 tons in the Dujiali Lake area.

Exploitation of rock salt on the Qinghai—Tibet Plateau began even earlier than that of borax. There are records describing rock salt production in the area in the ancient book "Note to Water Sutra" (Hayden, 1921). In addition to local consumption, rock salt from Tibet was sold to the neighbouring countries, including Nepal, Bhutan and India.

The Qinghai—Tibet Plateau has long attracted the interest of both Chinese and foreign geologists and explorers. In the late 19th century explorers, geographers, and geologists from Europe, India and elsewhere began to set foot on the plateau. Their investigations of some of the saline and fresh water lakes of the area are mentioned in books of the period (Hedin, 1901, 1909; Coulson, 1933; Hutchinson, 1937; Xu, 1960), and some examples of the findings of these early explorations are given below.

In the winter of 1879, N.M. Predevalsky, from Tsarist Russia, made the first recorded investigations of Qinghai Lake and he noted that the waters had a relatively high salinity. He was followed by Sven Hedin who, during 1899−1908, carried out a series of geographical surveys in the western and northern parts of Tibet, and made records concerning the nature and geographical settings of some of the lakes. These notes included details concerning the extent and distribution of the lakes, together with their surface area, elevation above sea level and water depths. He also recorded the salinity of large number of lakes along with details of recessional shorelines and terraces, and water erosion features. With respect to salt deposits, Hedin recorded the nature of sediments in the central and southern parts of the plateau and marked in detail the localities of saline lakes, playas and borax occurrences (Hedin, 1901, 1909). Despite some errors in his notes (e.g. abundant magnesium borates in some lakes were mistaken for gypsum), his records are still of reference value, especially in the relatively unexplored northern parts of the plateau.

Between 1931 to 1935, E. Norin made several visits to the western Ngari area and travelled to the Kunlun Mountains where he examined the geology of the area. In his report he referred to a small lake in the northwestern Mangrik Basin (northwest of Ngari near Qitai Daban, in the southwest corner of Xinjiang). According his analyses of salt from this location, the sediments of the lake "consist mainly of halite, with some thenardite, Na_2SO_4, and minor columnar pinnoite $MgB_2O_4 \cdot 3H_2O$ (Ne = 1.573 and No = 1.565)" (Norin, 1946). Norin's geological reports were very brief and have not attracted a great deal of attention by geologists, but he was the first to discover magnesium borates within the saline lakes of Tibet. From his descriptions, it is apparent that the lake Norin described belongs to the same sulphate type as the Chagcam Caka, which also contains abundant magnesium borate. G.E. Hutchinson also made some measurements of the water chemistry of selected lakes from western Tibet, which he reported in his works published in 1937. He did not analyze such specific elements as B and Li.

In addition, there have been more recent investigations concerning the geomorphology, sedimentation and evolution of freshwater and saline lakes in Tibet, including records describing the Yamzho Yumco, Bangong Co and Mapam Yumco (Xu, 1960). In their works on the glaciers and rivers of Tibet, Burrard and Huyden (1934) recognised that the Yamzho Yumco originally had an outlet in the north of the lake, through which it originally into the Yarlung Zangbo River. This outlet was later closed by a rise in the river bed which blocked the valley. In describing the geomorphology and sediments of the Bangong Co, Huntington (1966) considered the lake to be glacial in origin and catalogued a sedimentary succession from older alluvial sediments to younger lacustrine sediments. This study is still of value in studies of the evolutionary characteristics of lakes in Northern Tibet.

Hedin (1917) and Kaskyap et al. described the hydrodynamic regimes of the Mapam Yumco and Raka Lake, and also noted the the occurrence of a number of geothermal springs in the area. Kaskyap also carried out hydrological observation of lakes during the period 1924–1929.

These reconnaissance surveys were carried out under relatively primitive conditions in these pioneering years, so that the early records and descriptions inevitably appear to be relatively simple and superficial. However, they remain a valuable reference resource concerning the geology of saline lakes, this is especially true of the relatively unexplored areas of the plateau.

There are only limited data available for investigations of saline lakes on the plateau carried out by Chinese scientists in the years before the founding of the People's Republic of China, but records indicate that officials were assigned to make surveys of the Nam Co during the Qing Dynasty. In 1937 the geographer, Xu Jinzhi, made a personal investigation of Nam Co and produced a detailed report (Xu, 1937). Sun Jianchu carried out a geological traverse in the area of Qinghai Lake in 1938 and developed some hypotheses concerning the origin of the lake. In 1944, Yuan Jianqi, a senior salt scientist during these early years, travelled to the Caka salt lake in Qinghai and carried out the first study of its geology and genesis (Yuan, 1946).

After the founding of the People's Republic of China in 1949, working conditions were provided for extensive geological investigations of the saline lakes on the virtually unexplored Qingtai–Tibet Plateau. Since the early 1950s valuable data concerning the mineral resources of saline lakes on the plateau have been obtained by the Ministry of Geology, Ministries of Petroleum, Chemical and Light Industries, and the Chinese Academy of Sciences (Academia Sinica) through their exploration and scientific investigations of this area. At the end of 1952 the first scientific reconnaissance survey of the lakes of the Bangkog Co area was carried out (Li et al., 1959). As a result of surface pitting explorations this expedition determined the areal extent of surficial borax deposits at the Bangkog Co, and they also recognised that the saline lakes of Tibet show geographical differences in the chemical composition of their sediments and waters, and proposed that the lakes of northern Tibet are glacial in origin.

Between 1955 and 1956 a series of major new general surveys were carried out on the Qaidam Basin. They described the Qarhan Playa as "a giant salt storehouse" and indicated that "its salt bed has a K content of 0.40% and it is estimated that the K content may reach 10% or more".

In 1956, the author, Zheng Mianping, and his co-workers carried out studies in the Da Qaidam, Mahai and Qarhan areas. They recorded that "the Da Qaidam boron deposit is related to thermal springs, and that chemical analyses indicate a relatively high content of boron in the water of the Da Qaidam Lake" (Zheng et al., 1957). This report also noted that brine from the Qarhan Playa contains a relatively high potassium content.

CHAPTER 1

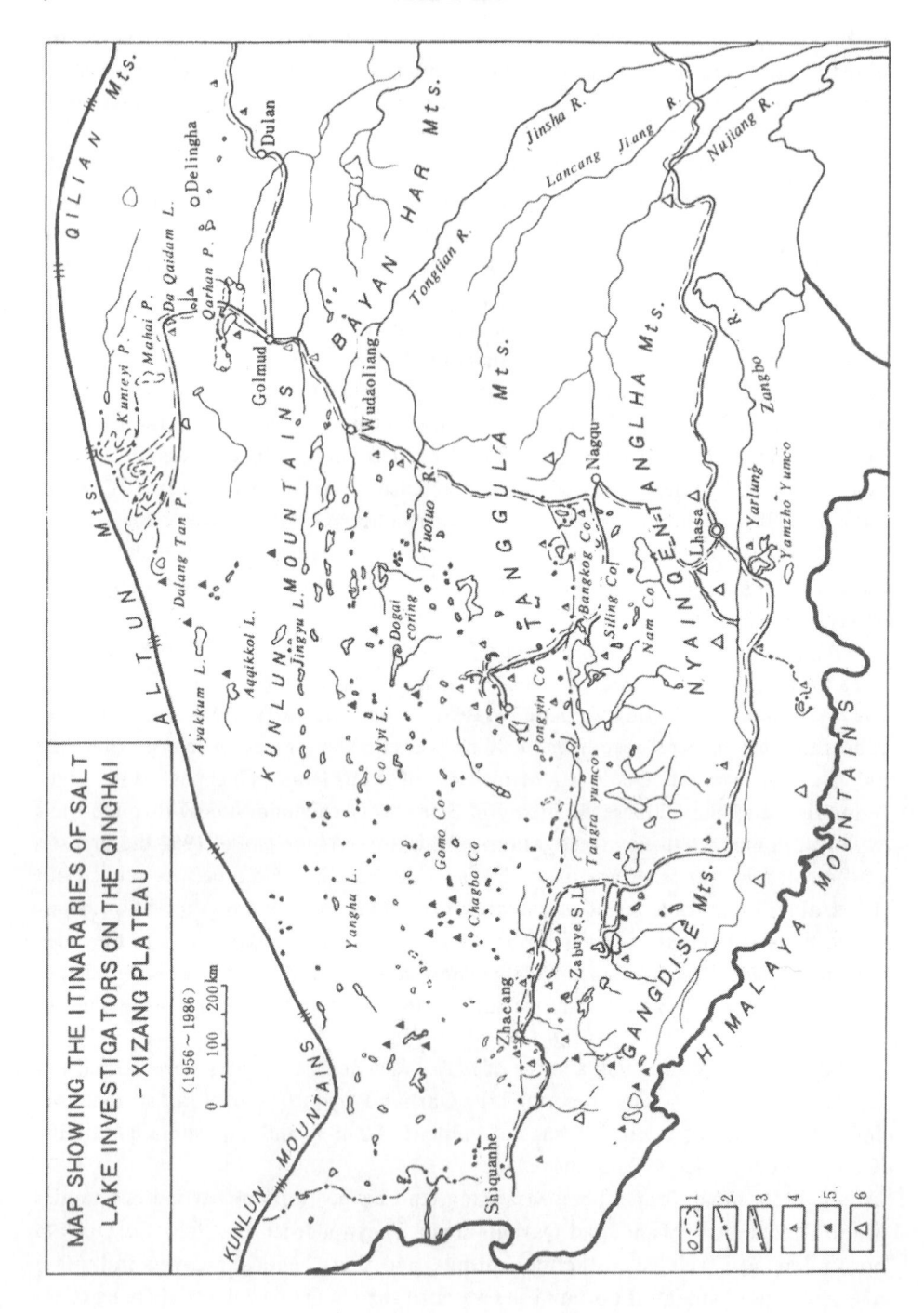

MAP SHOWING THE ITINARARIES OF SALT LAKE INVESTIGATORS ON THE QINGHAI - XIZANG PLATEAU (1956~1986)

INTRODUCTION

In 1957, a team under the leadership of Professor Liu Dagang carried out a scientific investigation of the saline lakes of the Qaidam Basin. This represented the first systematic study of the geology and chemistry of this area. During the search for potassium and boron resources, pinnoite, ulexite and carnalite deposits were discovered (Liu et al., 1985), leading to a larger expedition in 1958. In the scientific report of this expedition (Zheng et al., 1958) the origin of the Qarhan potash lake was discussed and estimates were made of the reserves of potash in the lake brines, stressing the areas resource potential. The same report identified the boron salts in the lake bed of Da Qaidam as ulexite and pinnoite. The origin of the boron was thought to be "related to the activities of ancient and modern boron–bearing hydrothermal fluids in the South Qilian fault zone". The high concentration of lithium in the lakes brines was also noted. At approximately the same time another detailed survey in the Da Qaidam lake area resulted in the discovery of a salt bed rich in ulexite and other boron minerals on the eastern bank of the lake. Boron deposits were also studied in the Yashatua and Mahai areas.

During the period 1959–1963 major advances were made in the study of Quaternary saline mineral resources throughout the Qaidam Basin. For example, a rich lake–bed boron deposit of ulexite and pinnoite was found in the Xiao Qaidam lake which providing substantial reserves for the local production of borax. The Dalangtan K–Mg salt lake and the magnesium sulphate–type K–Mg–Li salt deposits of the East and West Taijnar lakes were also discovered at this time. The results of this comprehensive survey of the saline mineral deposits of the Qaidam basin were published in series of reports (Xie et al., 1964, 1965; Qu et al., 1964, 1965; Han, 1981; Gao and Li, 1982; Chen et al., 1986; Qian, 1985). At the same time a preliminary exploration was carried out of the borate deposits of the Da Qaidam and Xiao Qaidam lakes to calculate their boron reserves and saline resources.

Based on a systematic comparative study of the geological features of boron deposits in China and elsewhere in the world it was suggested that the boron deposits of the Da Qaidam lake represent a new type of borate deposit (Xie et al., 1965). Detailed studies of the boron deposits on the banks of the Da Qaidam lake and the halogen minerals from Yashtu led to the discovery of hungchaoite, macalisterite and caloborite (Qu et al., 1964, 1965; Xie et al., 1964). Experimental evaporation studies in the field and laboratory were carried out on brines from the Da Qaidam and Xiao Qaidam lakes (Gao and Li, 1982;Chen et al., 1986) and these showed that borates could be precipitated by diluting the concentrated brines from Da Qaidam lakes. The mechanism of formation of borates by dilution was used to interpret the genesis of the nest–like pinnoite from the Xiao Qaidam lake (Qian, 1985).

During 1958–1966, prospecting and exploration work on a scale of 1:100,000 was carried out in and around Qarhan playa in the search for solid and dissolved K–Mg salt deposits. Substantial reserves were proven in this comprehensive assessment of saline resources in this area. In addition, a large amount of valuable geological data were accumulated which provided the scientific basis for the development and utilization of the K−Mg salt deposits in that playa (Zheng et al., 1989).

From the end of the 1950's and through the 1960's a number of groups carried out thematic studies of the geochemistry and mineralogy of the Qarhan playa, together with studies of the formation of carnallite, hydrochemistry, water−salt equilibrium, saline karst, salt−crust landform, and engineering geology with the playa. This research was continued in the late 1970's and the early 1980's, during which further research was carried out on the formation of carnalite, ^{14}C and U−series dating of the lake deposits, geochemistry, mineralogy and hydrochemistry of the playa. As a result of these studies systematic ^{14}C age dates were obtained on the saline and clastic deposits, which allowed new hypotheses to generated concerning the deposition and evolution of the Qarhan playa. These suggested that the playa entered the salt lake stage 25 ka and that a new lake developed on the surface of the playa since 9 ka (Sun, 1974; Huan et al., 1980; Chen et al., 1981; Yu, 1984).

A new "high−mountain, deep lake" salt forming model was proposed (Yuan, 1983) and, together with the application of comparative sedimentological techiniques, this allowed the construction of models describing the formation of Qarhan lake and the conditions of deposition of potash salts within the lake in ancient times (Sun, 1974; Yang, 1982; Yuan et al., 1985). These studies represented important progress in the theoretical study of mineral deposits in the area and had a major impact in guiding salt productive practices in the area.

Since the 1980s a number of cores have been drilled to aid in comprehensive studies of the saline lakes, with particular reference to sedimentation, hydrochemistry, micropalaeontology, trace element and isotope geochemistry, remote−sensing geology, and dynamic regimes within the brine. As a result important new understandings were gained into these areas (Wu et al., 1986).

K−bearing brines have also been the subject of prospecting expeditions in the Dalangtan, Qagan Silatu and Kunteyi lakes. This led to an expansion in the proven subsurface K−bearing brine reserves of the Dalangtan playa and produced new findings concerning the geology of these lakes. For example, Neogene−Quaternary paleomagnetic data from the salt−bearing formations of Dalangtan playa allowed for a quantitative study of the evolution of saline lakes in the basin.

Specialised investigations of the saline lakes of the Qinghai−Tibet Plateau began in 1956 with a reconnaissance for borax in the Bangkog Co area. During this expedition the shallow part of the lake was investigated and a 1:50,000−scale topographic and geologic map was produced. During 1956−1957 geological mapping of the northern part of Tibet was conducted on a 1,000,000 scale and a geological traverse was carried out in the western Ngari lake area. This resulted in the first geological data concerning the borax deposits at Chalaka and Langmari Lake, Ngari.

In 1958 investigations were centred on Bangkog Co, and a reconnaissance was carried out on the Dujiali and Pongcê Co lakes. These studies resulted in the correct identification of the "white gypsum", which is widespread on the terraces of the

Bangkog Co, as hydrous magnesium carbonate (hydromagnesite). The first data concerning the Li contents of saline lakes in Tibet were also collected. The information collected during the investigation allowed for a discussion of the lacustrine origin of Bangkog Co and identified the North Tibet lake district as a promising area for economic boron deposits.

In the period 1959–1961 further investigations were carried out in the north and west of the plateau (Fig. 1.1). These included studies of an additional 8 lakes (the largest of which were Siling Co and Zigetang Co) as well as further studies on the geochemistry, mineralogy and genesis of Bangkog Co. At the same time there were reconnaissance surveys of 9 other lakes, including Dung Co. These expeditions resulted in a report on the boron mineral resources of the area, which made a preliminary assessment of the boron potential of the various lakes and, in particular, noted the relatively high boron content of Langmari Lake (Liu et al., 1959).

The geological features of these lakes were discussed in a separate report (Zheng, 1959), which also included the first data concerning the chemical composition of the lakes and geothermal springs in the area. This study noted the high boron and lithium concerntrations of the springs and suggested that the deposits on saline lake terraces represented a potential target for new boron resources. Other surveys at this time included searches for borax in the Shuguo Lake area and a more general survey of 30 lakes of various types in the area to the north of Dujiali Lake. This latter survey included studied of Kongkong Caka in the north, Sêrnug Co in the south, Yangpeng Lake in the west and Jiarebu Lake in the east. A more detailed survey focused on Pongyin Co (originally named the Pangyu Caka), which let to a preliminary report on the extent and nature of the mineral resources in the saline lakes to the north of Bangkog Co.

In 1960 an investigation was carried out on 13 lakes, including Pongyin Co, to the north of the Bangkog Co. Later this was extended to include an investigation of the boron–bearing potential of thermal springs and the contact zones of granites in the Lunpola Basin and the southern part of the Bangkog Co area. The expedition then moved to the Dinggyê–Kambu area, where it conducted a preliminary survey of 3 lakes including the Ganggang Lake. Finally, it carried out a reconnaissance along the Hei–A and Xinjiang–Tibet highways and collected water samples from 11 lakes along the way. The report resulting from these expeditions (Zheng et al., 1960) noted in particular the relatively high Li contents of some of the 28 lakes. They also discovered ulexite and borax deposits in Southern Tibet. High boron concerntrations were also reported in dacite within the Tertiary red beds at Lunpola to the north of the Bangkog Co.

Prospecting teams made extensive traverse reconnaissance surveys in the Nagqu and Ngari areas during 1961 leading to the discovery of a saline lakes. They also made more detailed studies of important individual saline lakes, such as Chalaka Caka and Zabuye Caka (Chabyêr Caka), during which exploratory shallow drilling was conducted. Reports on the general survey and the more detailed study of Chalaka Caka and Zabuye Caka emphasised the great potential of the latter as a borax resource.

Also in 1961, extensive investigations were carried out of over 40 lakes of various types along the Hei–A highway from Dujiali Lake westward to Gar. This resulted in the discovery that Chagcam Caka represented a new type of saline lake that was rich in Li–Mg borates. On the basis of an estimate of the mineral reserves of the saline lakes of Tibet, and the geology and economic geography of the area a comprehensive analysis was made of the saline lake and associated resources of Tibet. A preliminary scheme for the two–stage development of these abundant saline–lake resources was then made (Zheng et al., 1963).

In a summary report of the search for borax in the saline lakes of Tibet (Fan et al., 1965) data from 90 lakes were used to give a comprehensive review of the Quaternary geology, geomorphology, hydrogeology, evolution and minerogenesis of lake basins of the North Tibet lake district. This study concluded that the boron mineralization is related to thermal and cold springs activity in the area. Unfortunately, the enormous amount of data collected during these expeditions could not be fully analysed and studied in depth because to time constraints.

In another report a relatively detailed study was made on the occurrence, distribution and hydrogeological setting of the boron deposits and hydromagnesite orebodies in the areas of Dujiali Lake and Bangkog Co. In these areas there were substantial drilling and pitting operations, as well as long–term hydrodynamic observations. These proved the existence of large quantities of commercial mineral reserves, thus making important contributions to the local production of borax and the future development of the boron mineral resources.

The genesis and distribution of saline lake mineral deposits in Tibet was discussed in depth in a series of reports (Xie and Zheng, 1964; Zheng and Jin, 1964; Xie et al., 1965). These led to a mineralogical–hydrochemical classification scheme of the saline lakes and identified 10 new salt minerals that had not previously been recognised in China.

During the period 1958–1961 the various groups had travelled ~ 10,000 km, made detailed studies of 54 lakes (including 20 saline lakes) and collected preliminary data from a further 80 lakes. As a result of this painstaking work and the reports that followed, the nature of the saline lake deposits of Tibet received there first detailed scientific study, and the enormous potential of the saline resources of the area for boron, lithium and other elements, was brought to light.

Two expeditions were mounted to northern Tibet in 1976 and 1978. These groups studies over 50 saline lakes, including a small number that had not been previously surveyed, such as the Margog Caka and the Neier Co and discovered two new occurrences of boron mineralization. They also obtained new ^{14}C age data which suggested that most of the saline lakes in Tibet formed in the Holocene. The scientists within these teams analysed a large number of water samples, which yielded the first series data concerning the levels of U–series radioisotopes and trace elements, such as Cu, Zn, Pb and As, in northern Tibet. They also made the first discoveries of hydroglauberite and chlorite in the saline lakes of Tibet (Huang et al., 1980; Zheng, 1981, 1982; Yu and Tang, 1981).

A detailed search was also carried out for halite in the South Zabuye Caka, and as a result the rock salt reserves of the lake were established. Relatively high K, B and Li contents were also discovered and an estimate was made of the resources of these elements, together with Mg, within the lake.

In the 1980's, there were number of studies of the geothermal activity, Quaternary geology, geomorphology and limnology (with an emphasis on fresh water and brackish lakes) of Tibet, which increased the understanding of the origin and development of lakes (Chen, 1981; Tong et al., 1981; Li et al., 1983; Yang et al., 1983; Guan et al., 1984; Xu, 1985). At the same time glacialogical and cryopedological studies were conducted along the Qinghai–Tibet highway, which produced the first paleomagnetic data of Tertiary–Quaternary lacustrine formations (Qian et al., 1982).

A 10–year petroleum prospecting campaign in northern Tibet, centered on Lunpola, resulted in the compilation of a 1:100,000 scale geological map of the Bangkog Co area and provided a large amount of valuable basic geological data for the subsequent saline lake studies of the area. This was supplemented by studies of the regional geology, particularly Cenozoic stratigraphy, tectonics and sedimentology, that also provided valuable basic geologic data for subsequent saline lake studies of the Qinghai–Tibet Plateau.

Since the 1980s studies have concentrated on the geologic structure and distribution of mineral resources on the plateau, together with studies of the sedimentary environment of the saline lakes of the area and their K, B and Li resources. In summer of the 1980 further investigations were carried out in the area of the Bangkog Co and Fenghuo Mountain, including the first survey of the deep water of Siling Co by boat. The first reconnaissance survey of the Shuanghu Lake area was made in 1982 during more detailed studies of Zabuye Cake and Zhacang Caka in the Ngari area. This expedition also included the first study of the deep water area of Taro Co. An additional deep water survey was made in 1984 on North Zabuye Caka. Supplementary survey were also made of several saline lakes in the Qaidam Basin and the first investigations (including water chemistry studies) were made of the lakes in the unpopulated Mount Fenghuo and Hoh Xil areas.

During this round of expeditions the various teams travelled more than 5,000 km, studied 53 lakes of various types, drilled 15 shallow boreholes and excavated 169 shallow pits. Analytical work was carried out in subsequent laboratory studies, that included experiments on the culture of halophilic algae. These studies resulted in the utilisation of a new type of cesium mineral deposit and the discovery of a halophilic algae resource.

This book is largely based on the results of the author's own field and laboratory studies of the saline lakes on the Qinghai–Tibet Plateau over the past 30 years.

In particular, it concentrates on the field traverse and concentrated surveys carried out in the periods of 1956–1961 and 1980–1986 (Fig. 1.1). During these years, the author made field investigations of 102 lakes of various types and collected additional valuable data on a further 352 lakes, including relatively complete water geochemical data of 114 lakes (Table 1.1 and Appendix 1). A brief summary of some the advances in scientific and applied research is presented here.

1. Basic research:
    a) A stratigraphic has been established for the Late Cenozoic in the Tibet lake region.
    b) A "block upwarping" model for the plateau has been established on the basis of temporal and spatial distribution of lake basins on the plateau.
    c) A tectono–geochemical model has been proposed for the origin of B, Li and Cs in saline lakes via geothermal activity associated with neotectonic activity on the plateau.
    d) The first map (1:2,500,000 scale) of hydrogeochemical zoning of saline lakes on the plateau has been produced. This has formed the basis for a division of metallogenic belts of saline lakes and agriculture and animal husbandry activities.
    e) A model of "Supply source" has been proposed for controlling the formation of water types in the lake basins.
    f) A multi–stage model of lake sub–basins has been proposed for controlling salt formation. This indicates that soluble salts tend to be concentrated in low order saline lakes due to topography, chemical differentiation and other processes.

2. Applied research:
    a) The author and his colleagues have played a part in the discovery of the Qarhan continental potash salt lake and the Da Qaidam lake–bed boron deposit.
    b) A continental origin was suggested for the first time for the K and Mg salts from the Qarhan playa and an estimate made of these potash salt lake reserves Zheng et al., 1994.
    c) The economic significance of the Qarhan deposits has been stressed, together with the observation that the generation of this deposit appears to be related to a structural depression.
    d) Da Qaidam lake has been identified as containing a new type of boron deposit containing pinnoite and ulexite.
    e) It has been suggested that the Zhacang Caka boron deposit is also a novel Li–rich type containing kurnakovite and pinnoite (Zheng and Jin, 1964; Xie et al., 1965).
    f) As the result of a breakthrough in the study of microgranular Li–bearing sediments in saline lakes new minerals, such as zabuyelite (natural lithium carbonate) and Li–bearing magnesite and dolomite have been discovered. These discoveries have opened up a large new potential Li mineral resource in saline

INTRODUCTION

TABLE 1.1. List of salt lakes on the Qinghai–Tibet Plateau

Lake No. ①	Name	ASL ② (m) / Area (km²)	Water type	Hydrochem. type	Salinity (g/l)	pH	KCl (mg/l)	B₂O₃ (mg/l)
017	Cocholung L.	4300 / 1	Lake water	s.s. subt.	154.1	–	–	–
202	Langmari L.	4240 / 0.1	Surf. brine	m.s. subt.	322.261	7.53	12462.2	3664
203	Rabang Co	4324 / 32	Lake water	s.s. subt.	67.594	9.2	4843.02	1634.00
207	A'ong Co	4427 / 56	Lake water	m.c. subt.	58.809		2180.01	2078.8
208	Xia Caka	4426 / 3	int. brine / lake water	w.c. subt.	385.238 / 201.465	/ 8.45	49157.78 / 17209.9	4958 / 2984
209	Dubuke II L.	4585 / 1	Lake water	w.c. subt.	302.93	9.3	37990	5938.7
210	Nyêr Co	4399 / 24	Surf. brine	s.s. subt.	233.994	7.6	31536.82	4482
201–1	Chalaka I L.	4790 / 100	Lake water	s.s. subt.	40.02	–	–	2807.5
211	Chalaka II L.	4786 / 3	surf. brine / int. brine	m.c. subt.	107.222 / 332.006	9.54 / –	3032 / –	5689.5 / 7074.6
212	Chali Co	4796 / 48	lake water	m.c. subt.	46.584	9.38	1396.1	3712
214	Zabuye Caka (North L.)	4421 / 98	Surf. brine	m.c. subt.	377.21	9.1	47190.83	7800
214–1	Zabuye Caka (South L.)	4421 / 145	int. brine / lake water	m.c. subt.	435.657 / 398.949	9.3 / 9.2	74475.70 / 56228.58	12240 / 7628
214–3	Leen L.	4444 / 0.2	Lake water	m.c. subt.	58.298	–	5720.1	5685.23
215	Cam Co	4500 / 86.7	Surf. water	s.s. subt.	153.698	9.0	14895.14	1795
217	Jibucaka Co	4500 / 8.2	Lake water	s.s. subt.	82.719		9070.17	2422
218	Laggor Co	4500 / 92	Lake Water	w.c. subt.	59.794	8.8	4991.7	2355

① For the "numbers" of the lakes, the reader is referred to appended map I.
② The areas and elevations above sea level are mostly based on the 1:100,000–scale topographic map and in a few cases are measured.

227	Daqzê Co	$\frac{4459}{240}$	Lake water	s.c. subt.	40.691	9.63	1899.07	715.2
234	Pongyin Co	$\frac{4720}{48}$	Surf. brine	w.c. subt.	340.538	8.36 | 9.05	44797.92	3054.7
235	Pêli Co	$\frac{4800}{17}$	Surf. brine	w.c. subt.	215.374	9.02	9533.5	1869.53
236	Norma Co	$\frac{4695}{66}$	Lake water	s.c. subt.	35.418	9.74 9.84	1186.92	1591.9
240	Dujiali L.	$\frac{4515}{65}$	int. brine	m.c. subt.	220.499	8.8	21528.2	6052.8
244	Yagedong Co	$\frac{4530}{35}$	int. brine	w.c. subt.	67.97	9.8	4396.47	6v9.8
250	Ngang Co	$\frac{4510}{5}$	Surf. brine int. brine	m.c. subt.	93.498 98.90		806 761.5	792.1 748.5
251	Bangkog III L.	$\frac{4520}{80}$	Surf. brine int. brine	m.c. subt.	258.166 234.33	8.7 10.3	40573.45 39352.31	3126.13 4667.4
251-1	Bangkog II L.	$\frac{4522}{50}$	Surf. brine int. brine	m.c. subt.	187.07 402.525	8.6	21393.17 55184.66	2555.85 7650
253	Ben Co	$\frac{4650}{20}$	Lake water	m.c. subt.	82.8	–	–	1399.7
254	Angdar L.	$\frac{4850}{34}$	Surf. brine	w.c. subt.	180.868	9–10.5	5910.27	706.4
255	Zainzong Caka	$\frac{4550}{9}$	Surf. brine	s.s. subt.	412.484	–	32919.14	3069.78
261	Kyêbxang Co	$\frac{4610}{142}$	Lake water	Carbonate	63.35	10.2	1538.71	724.90
262	Xogor Co	$\frac{4603}{23}$	Lake water	m.c. subt.	35.58	–	–	136.4
263	Dungqag Co	$\frac{4616}{46}$	Lake water	Carbonate	36.383	–	–	442.3
302	Kayi Co	$\frac{4280}{3}$	Surf. brine	m.s. subt.	167.020	7.8	7717.58	1584.42
303	Changmu Co	$\frac{4288}{6}$	Lake water	c. subt.	63.144	8.2	1464.54	590
308	Chagcam Caka III L.	$\frac{4338}{32}$	Surf. water	m.s. subt.	57.251	7.5	4654.3	1814.12
309	Chagcam Caka II L.	$\frac{4326}{57.6}$	int. brine surf. brine	s.s. subt.	384.780 305.123	8.0	41642 26369.66	2143 2088

314	Chagnag Co	$\frac{4529}{6}$	Surf. brine	m.s. subt.	349.874	7.6	65176.73	1935
317	Gangmar Co	$\frac{4710}{14}$	Surf. brine	w.c. subt.	369.067	9.5	22346.52	3877
318	Buerga Co	$\frac{4607}{12}$	Surf. brine	s.s. subt.	135.477	7.9	8986.28	571.00
319	Domar Co	$\frac{4500}{18}$	Surf. brine	m.s. subt.	117.122	8.49	9533.5	1706.83
320	Riwan Caka	$\frac{4730}{1.26}$	Surf. brine	m.s. subt.	342.835	–	14477.57	612.5
322	Cêmar Co	$\frac{4580}{36}$	Surf. brine	m.c. subt.	214.552	8.8	40994.1	2450.1
323	Dong Co	$\frac{4394}{88}$	Surf. brine	s.s. subt.	114.284	8.68	17210	1958
325	Youbu Co	$\frac{4638}{62}$	Surf. water	s.s. subt.	35.836	–	–	846.3
326	Dawa Co	$\frac{4636}{80}$	Surf. water	s.s. subt.	35.677	9.3	1339.65	878.00
328	Garing Co	$\frac{4650}{64}$	Surf. brine	s.s. subt.	277.66	7.4	1722.51	102.10
334	Ning Co	$\frac{5037}{8}$	Surf. brine	s.c. subt.	317.9	–	–	4080.2
335	Gomo Co	$\frac{4668}{66}$	Surf. brine	s.s. subt.	313.36	–	–	829.22
337	Wuming L.	$\frac{4870}{9}$	Surf. brine	s.s. subt.	51.961	9.01	953.35	–
339	Gangtang Co	$\frac{4866}{11}$	Surf. water	Carbonate	72.429	9.3 9.5	7568.51	1549.81
340	Yibug Caka	$\frac{4557}{88}$	Surf. brine	s.s. subt.	105.102	8.2 8.7	2219.4	219.00
341	Kangru Caka	$\frac{4766}{9.5}$	Surf. brine	m.s. subt.	322.548	7.0	5201.48	286.00
342	Margog Caka	$\frac{4830}{76}$	Surf. brine	s.s. subt.	323.018	7.7	10734.72	1009.00
343	Xo Caka	$\frac{4800}{11}$	Surf. brine	m.s. subt.	320.202	7.3	3910.64	381.40
345	Kungkung Caka	$\frac{4771}{38}$	Surf. brine	m.s. subt.	350.614	7.4	10677.52	948.71
347	Qagam Co	$\frac{4750}{20}$	Surf. brine	m.s. subt.	184.277	–	5159.5	528.8
349	Darngo Congoin L.	$\frac{4976}{40}$	Surf. brine	m.s. subt.	136.156	–	4452.140	251.9

350	Amu Co	4970/32	Surf. brine	s.s. subt.	218.843	8.1	7150.33	462.39
352	Ngoyêr Co	4817/52	Surf. brine	m.s. subt.	111.181	—	8902.4	459.2
353	Bilong Co	4820/22	Surf. brine	s.s. subt.	331.072	8.3 (crys.)	9247.5	382.21
354	Cêdo Caka	4822/36	Surf. brine	s.s. subt.	168.388	8.3	3527.4	296.88
355	Têrang Pun Co	4840/20	Lake water	s.s. subt.	70.415	8.9	2126.75	406.24
357	Cgcom Co	4910/4.5	Lake water	s.s. subt.	85.967	8.3	—	897.2
358	Yaggai Co	4865/92	Surf. brine int. brine	m.s. subt.	196.825 324.757	8.11	4356.8 24501.1	1266.1 2759.9
401	Lungmu Co	5002/96	Surf. brine	m.s. subt.	172.505	7.6 7.8	8112.82	45.55
402	Yanghu L.	4778/76	Surf. brine	c.t.	115.649	8.4	1688.78	—
406	Yurba Co	4880/54	Surf. brine	m.s. subt.	314.029	—	10258.05	—
407	Margai Caka	4785/76	Surf. brine	m.s. subt.	78.724	—	1154.87	1309.10
408	Co Nyi L.	4902/60	Lake water	m.s. subt.	62.730	8.5	1599.66	445.93
410	Tuoba L.	4928/30	Surf. brine	c.t.	271.333	6.85	8732.69	—
413	Taiku L.	4892/8	Lake water	c.t.	45.076	8.25	39.47	—
414	Rala Co	4808/56	Surf. brine	m.s. subt.	215.341	7.52	6215.84	—
415	Changjing L.	4974/11	Surf. brine	c.t.	93.095	7.06	2200.33	203.8
416	Yupan L.	4892/9.5	Surf. brine	m.s. subt.	143.715	7.26	3554.09	4269
417	Dogai Coring	4814/350	Surf. brine	c.t.	226.354	6.5	—	—
418	Cheyue L.	4814/28	Lake water	c.t.	60.285	6.7	—	—
421	84Yan I L.	4200/0.02	Surf. brine	m.s. subt.	97.183	—	3889.67	320.28
	84Yan II L.	4200/0.01	Lake water	m.s. subt.	23.055		406.13	67.92

INTRODUCTION 15

515	Hajiang L.	$\frac{4250}{9}$	Surf. brine	m.s. subt.	381.993	7.9		
517	Jianbaizi L.	$\frac{4400}{0.001}$	Hot lake water	Carbonate	98.349		8325.6	1174.49
519	Caka L.	$\frac{3059}{104}$	int. brine	m.s. subt.	323.109	6.8	8527.95	150.55
601	Aqqikkal L.	$\frac{4286}{320\pm}$	Lake water	m.s. subt.	78.193	–	2515.89	852.8
602	Jingyu L.	$\frac{4708}{165}$	Lake water	m.s. subt.	61.957	–	1601.63	–
603	Gas Hure L.	$\frac{2853.7}{103}$	Surf. brine	m.s. subt.	359.848	7..56	12736.76	547.5
604	Dalangtan P.	500	int. brine	m.s. subt.	382.433	6.81	8770.82	254.3
606	Kunteyi P.	$\frac{2723.9}{1667}$	int. brine	m.s. subt.	333.266	6.20	15768.41	301.7
607	Niulang Zhinu L.	1	int. brine	c. t.	554.996	5.25	1296.56	863
610	Yiliping P.	$\frac{2683}{360}$	int. brine	m.s. subt.	333.967	7.32	22365.59	1040
611	Xi Taijnar L.	$\frac{2678}{82}$	Surf. brine	m.s. subt.	338.260	7.70	12984.63	996.3
612	Dong Taijnar L.	$\frac{2681}{116}$	Surf. brine	m.s. subt.	331.535	7.75	6787.85	652.97
613	Dezong Mahai L.	$\frac{2740.1}{11}$	Lake water	m.s. subt.	39.48	7.42	4165	88.9
614	Baluan Mahai L.	$\frac{2743.7}{4.5}$	Surf. brine	c. t.	257.849	7.4	4385.41	250
615	Suli L.	$\frac{2675.57}{69-85}$	Surf. brine	m.s. subt.	333.387	6.9	13861.71	804.4
616	Da Biele L.	$\frac{2676.57}{7.38}$	Surf. brine	m.s. subt.	259.88	7.00	5345	175.2
617	Xiao Biele L.	$\frac{2676.37}{6.25}$	Surf. brine	m.s. subt.	205.169	6.20	130.76	80.01
618	Dabsan L.	$\frac{2675}{184.334}$	Surf. brine	m.s. subt.	318.562	7.35	14929.46	220
618–1	Xi Dabsan L.	$\frac{2675}{30}$	Surf. brine	m.s. subt.	262.10	7.35	4035	73.3 385
619	Da Qaidam L.	$\frac{3148.1}{36}$	int. brine	m.s. subt.	340.573	7.99	9743.24	2580
620	Xiao Qaidam L.	$\frac{3172}{35.91}$	int. brine	m.s. subt.	239.530	7.80	2097.37	1273.7

621	Tuanjie L.	$\frac{2675}{6}$	int. brine	m.s. subt.	426.277	5.40	13766.37	756.5
622	Xiezuo L.	$\frac{2691}{17}$	int. brine	c. t.	359.501	5.50	14719.72	467.16
623	Nan Hulsan L.	$\frac{2675.57}{33.41}$	Surf. brine	c. t.	312.688	7.85	695.6	24.18
624	Bei Hulsan L.	$\frac{2675}{90.44}$	Surf. brine	c. t.	243.067	7.50	1506.29	69.6
625	Qarhan P.	$\frac{2677}{5856}$	int. brine	m.s. subt.	323.104	7.00	23261.74	223.5
628	Gahai L.	$\frac{2851.2}{37.4}$	Lake water	s.s.. subt.	90.594	8.28	476.68	—
629	Hoh Yanhu L.	$\frac{3010}{90}$	Surf. brine	m.s. subt.	335.656	6.75	8406.55	167.6
630	Xiligou L.	$\frac{2938}{19.5}$	Surf. brine	s.s.. subt.	261.320	7.62	3126.99	215
632	Dajiuba L.	$\frac{4078}{5}$	Surf. brine	m.s. subt.	217.876		1923.86	1223
635	Chaikai L.	$\frac{3010}{18}$	Surf. brine	m.s. subt.	324.135	6.85	2995	95
636	Ayakkum L.	$\frac{3876}{535.5}$	Surf. brine	m.s. subt.	145.900	—	1785.62	800.6
637	Chahanshilatu P.	$\frac{2710}{2000\pm}$	int. brine	m.s. subt.	351.308	—	12164.75	177.6
638	Potash L.	$\frac{2726}{1.5}$	int. brine	c. t.	321.20	—	—	57.8
639	Dongling L.	$\frac{2690.3}{7.2}$	Surf. brine	c. t.	356.90	6.90	36	334.2
640	Gansenquan L.	$\frac{2735}{16}$	Surf. brine	s.s.. subt.	51.35	7.90		6.5

Notes: 1. Surf. brine—surface brine; int. brine—intercrystal brine; 2. s.s. subt.—sodium sulfate subtype; m.s. subt.—magnesium sulfate subtype; w.c. subt.—weak carbonate subtype; m.c. subt.—middle carbonate subtype; s.c. subt.—strong carbonate subtype; c. t.—Chlorite type.

lakes.

g) The plateau has been divided into metallogenic belts to aid in the search for new saline lake mineral resources. This has been coupled with a comprehensive survey and statistical analysis of the commercial resources of 94 saline lakes over the whole plateau region. As a result it has been possible to calculate the total mineral resources of the various kinds of saline lakes of Tibet, Qinghai and southern Xinjiang. This has provided a base for long-term mineral exploration and utilisation in the area.

h) The extensive geological survey of the key saline lakes of the area, such as Zabuye Caka, Zhacang Caka and Bangkog Co, and preliminary experiments concerning technological processing of salt deposits have also provided important information necessary for the utilisation of these economic resources.

3. Miscellaneous:

a) A large new caesium deposit, composed of Cs—bearing geyserite, has been discovered. Preliminary investigations suggest that the geothermal zone along the Yarlung Zangbo River is a promising area for further discoveries of this type of mineral deposit.

b) An extensive discovery of natural halophilic algae has been discovered in Zabuye Caka in Tibet. This rare species of algae (*Dunaliella salina*) is of particular interest because it is rich in kerotene and protein and has greater cryophilic tolerance than other reported species of varieties of this species. It is also possible that biomineralisation may play a significant role in the development of some salt lake deposits and research is now being undertaken in the area.

CHAPTER 2 SALINE LAKES AND LAKE DISTRICTS

Qinghai–Tibet Plateau in southern Asia has an average elevation of over 4500 m and fully deserves its description as the roof of the world. Numerous bright and colourful lakes of various sizes are scattered over the vast plateau and constitute an important feature of the natural landscape of the region. There are over 1600 lakes with an area greater than 1 km^2, and there are 5 lakes that exceed 1000 km^2 in water surface area; namely, Qinghai Lake, Nam Co, Siling Co, Tangra Yumco and

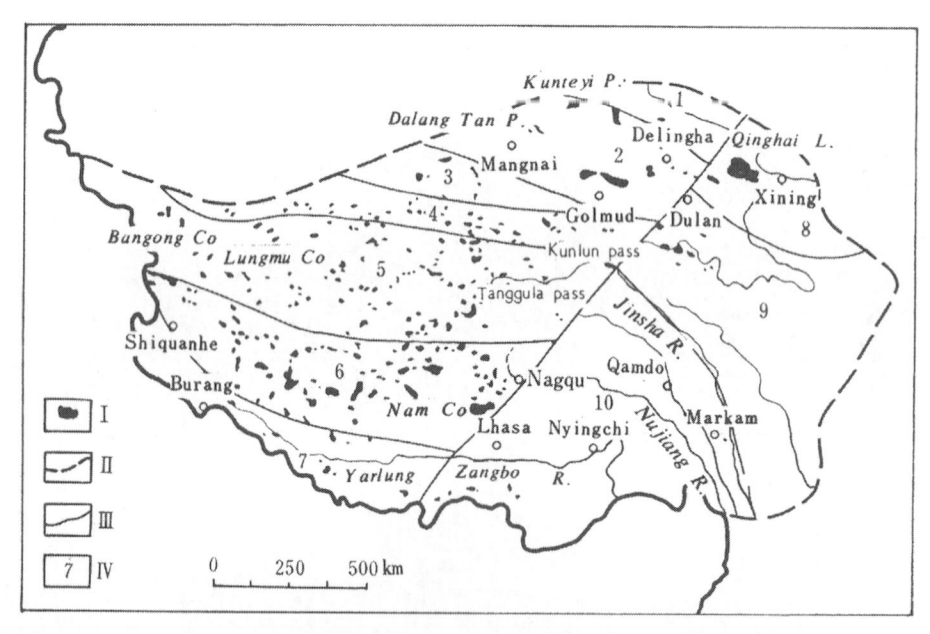

Figure 1.2. Lake districts on the Qinghai–Tibet Plateau. I—Modern lake (including playa); II—Plateau boundary; III—Boundary of lake district; IV—Lake district No.; 1—Qilian inflow district; 2—Qaidam inflow district; 3—Kumkol inflow district; 4—Hoh Xil inflow district; 5—Qiangtang inflow district; 6—North Tibet inflow district; 7—South Tibet inflow–outflow district; 8—Qinghai Lake–Gonghe outflow–inflow district; 9—Bayan Har outflow district; 10—Southeast Tibet inflow–outflow district.

TABLE 1.2. Lake districts on the Qinghai–Tibet plateau

Lake district	Area of lake districts (km²)	No. of lakes with different areas (km²) ≥1000	<1000 -500	<500 -100	<100 -10	<10 -0.5	Sum	Total area of lakes (km²)	Lake ratio (%)	Saline lake and playa No.	Area (km²)	Brackish lake No.	Area (km²)	Fresh-water lake No.	Area (km²)	Saline lake ratio (%)	Brackish lake ratio (%)	Fresh-water lake ratio (%)
Qilian inflow	62631	0	1	0	0	9	10	631	1.01	0	0	0	0	10	631	0	0	100
Qaidam inflow	208161	3	1	6	15	10	35	11813	5.67	26	11422	6	137	3	254	96.69	1.16	2.15
Kumkol inflow	74619	0	1	1	5	10	17	973.5	1.30	3	864.5	11	97	3	12	88.80	9.96	1.23
Hoh Xil inflow	151956	0	0	7	42	208	257	3087.7	2.03	65	992.9	126	1762.3	66	332.5	32.16	57.07	10.77
Qiangtang inflow	432264	0	0	17	134	308	459	9196.5	2.13	177	5146.3	137	3277.6	145	772.6	55.96	35.64	8.40
N. Tibet inflow	365715	4	1	23	99	191	318	15524.3	4.24	72	2898.5	96	8886	150	3739.8	18.67	57.24	24.09
S. Tibet outflow–inflow	141588	0	0	4	11	43	58	1502.1	1.06	2	14.0	9	880	47	608.1	0.93	58.58	40.48
Qinghai Lake–Gonghe outflow–inflow	120078	1	0	1	1	3	6	4796	3.99	3	113	1	4635	2	48	2.36	96.64	1.00
Bayan Har outflow	419517	0	2	1	12	303	318	1885.2	0.45	4	14	2	46	312	1825.2	0.74	2.44	96.82
SE Tibet inflow–outflow	469197	0	1	1	11	121	134	1502.5	0.32	0	0	3	129	131	1373.5	0	8.59	91.41
Total	2445726	8	7	61	330	1206	1612	50911.8	2.08	352	21465.2	391	19849.9	869	9596.7	42.16	38.99	18.85

Notes: 1. Base maps used for the statistics: 1:1,000,000–scale Topographic Map compiled by Institute of Geography for Tibet; the same map for the Qinghai Lake–Gonghe lake district in Qinghai; 1:3,000,000–scale Map of the Qinghai–Tibet Plateau for the Qilian lake district. 2. The division of lake types is based on salinity data collected in field investigations, with reference to satellite image data. 3. Lake ratio = $\frac{\text{Total area of lakes}}{\text{area of lake district}} \times 100\%$; saline lake ratio = $\frac{\text{Total area of saline lakes}}{\text{Total area of all lakes}} \times 100\%$; brackish lake ratio = $\frac{\text{Total area of brackish lakes}}{\text{Total area of all lakes}} \times 100\%$; fresh–water lake ratio = $\frac{\text{Total area of fresh - water lakes}}{\text{Total area of all lakes}} \times 100\%$.

Zhari Namco. The Qarhan Playa has an area of 5856 km^2, and is one of the largest modern inland playas in the world. The total area of the lakes (including playas) on the whole plateau amounts to 50,900 km^2, which is approximately 2% of the total land area. There are 352 saline lakes of various types, which have an aggregate area of 21,465 km^2, making up 42% of the total area of all lakes in the region (Table 1.2). The lakes in the region show a wide range of hydrochemical types and are complicated an varied in the composition of their lake water chemistry. Apart from the relatively common lakes, in which there are large accumulations of halite, mirabilite, magnesium salts and torona, there are also more unusual saline lakes in which there is a great abundance of elements such as potassium, boron, lithium and caesium. Hence, these saline lakes are of great interest from both a scientific and economic point of view.

The lakes on the plateau can be divided into 10 districts on the basis of their geographical location, geological features, drainage systems and physiochemical characteristics (Fig. 1.2). From north to south and west to east these districts are; the Qilian, Qaidam, Kumkol, Hoh Xil, Qiangtang, and North Tibet inflow lake districts, the Qinghai Lake—Gonghe, South Tibet, and Southeast Tibet outflow—inflow lake districts, and the Bayan Har outflow lake district (Table 1.2) (Zheng et al., 1983).

The Qaidam and Qinghai Lake—Gonghe lake districts are located on the first step of the plateau at a relatively low altitude (2700—3200 m). The Qaidam inflow lake district lies in a large closed "high mountain—deep basin". The lakes in this district have a long evolutionary history and are now largely at an advanced stage of development. The district is dominated by playas, with the lakes covered by surface water making up slightly less than 10% of the aggregate area of all the lakes (including playas) in the district. This district also contains the greatest area of saline lakes on the Qinghai—Tibet Plateau, comprising approximately 23% of the total. The Qinghai Lake—Gonghe outflow—inflow district, at the eastern end of the plateau, is also a "high—mountain deep basin", but it is smaller in size. During the Pleistocene this district contained an outflow lake system. At present, all the lakes in the Gonghe Basin in the southern part of the lake district have been drained, except for Caka Salt Lake in the western part of the district. Where brine still remained. The northern part of the lake district was closed as a result of the uplift of Riyue Mountain, which has resulted in the formation of Qinghai Lake, which is the largest lake on the plateau. At the moment, this lake contains brackish water, but it is becoming more saline as a result of continued evaporation. Indeed, Gahai Lake, an interbank lake on the northeast edge of Qinghai Lake, has already become saline lake in recent times.

The North Tibet, Qiangtang and Hoh Xil inflow lake districts are located on the thrid step of the plateau at an elevation of 4000 m. They are moderate— to low—mountain or moderate— to high—mountain shallow basins and comprise a series of basin systems controlled by graben systems of various dimensions and small lakes formed by the silting of glaciers and rivers. These lake districts contain a mixture of saline, brackish and fresh—water lakes, most of which are in the middle or late stages of their development. The North Tibet district contains the largest area of lakes on the plateau (15,500 km^2), which the Qiangtang and Hoh Xil lake districts are relatively smaller. Lakes with areas of greater than 1000 km^2 comprise 72%, 41% and 47% of the total area of lakes in the North Tibet, Qiangtang and Hoh Xil lake districts, respectively. The number of all the different types of lakes in the three districts comprise 20%, 28.% and 16%, respectively, of the total number of lakes on the plateau.

The Qilian (17 lakes) and Kumkol (10 lakes) inflow lake districts lie to the northeast and southwest of the Qaidam lake district. They are moderate—shallow basins surrounded by moderate to high mountains and lie on the front margins of the second step of the plateau at an altitude 1000 m greater than that of the Qaidam lake district. The lakes in these two districts receive abundant recharge in the form of snow—melt from the Qilian and Kunlun mountains, so that they contain a high proportion of initial stage (fresh—water) lakes and a smaller proportion of middle (brackish) and late stage (saline) lakes (Table 1.2).

The South Tibet outflow—inflow lake district is located between the Gangdise and Himalaya mountains and is also situated on the third step of the plateau. The lakes in this district are mostly located on the northern slope of the Himalaya mountains at elevations of 4000—5000 m. The lakes are relatively small in size, with an aggregate area of 1500 km^2, that accounts for slightly less than 3% of the total area of lakes on the plateau. The district contains predominantly inflow lakes, which have a accumulative area of slightly less than 1400 km^2 and account for 93% of the total area of lakes in the district. Analysis of the evolution of the lake basins indicates that there were a greater number of lakes in the district during the Pleistocene than at present, and that most of the present—day large lakes are remnant lakes that have been transformed from inflow to outflow lakes.

The area to the east of the Yangbajain—Dulan line and to the south of the Qinghai Lake—Gonghe lake district is divided into the Bayan Har outflow lake district and the Southeast Tibet inflow—outflow lake district. These two districts are separated by the Jinsha River. The Bayan Har district is part of the exoric drainage systems of the Yellow River and the Yalong River, while the Southeast Tibet district is contained within the Sanjiang (the Nujiang, Jinsha and Lancang rivers) and the Yarlung Zangbo River drainage systems. Within these two districts there are a series of mountains that define separate E—W and N—S trends. Between the mountain systems the plateau surface is fragmented by headward incised rivers and deep river—cut valleys. Hence, in the central and eastern parts of the two lake districts, where the high—mountain and canyon landforms prevail, most of the ancient Pleistocene lakes (which originally developed on the plateau surface at an elevation of 4200—4600 m) have been drained. The remaining lakes are generally small in size formed either by recent glacial activity or from the silting and contraction of rivers.

However, the Southeast Tibet district does contain some large lakes on the remnant plateau surface in its southwestern part.

The Bayan Har district has a total drainage area of nearly 420,000 km^2. At present there are three lakes with areas greater than 100 km^2 each that make up 73% of the total area of lakes in the district. They are Ngoring (611 km^2) and Gyaring (526 km^2) Lakes (both of which are alternately endoric—exoric lakes within the Yellow River drainage system) and the fresh-water Toson Lake (223 km^2). The other lakes in the district are relatively small, which are no more than several square kilometres in size. Most of the lakes contain fresh water, with the exception of few sandbank contracted lakes, which are newly emerged saline lakes formed in a locally closed environment, e.g. Hajiang Salt Lake (about 7.5 km^2 in area) lying by Gyaring Lake.

The Southeast Tibet lake district is the largest on the Qinghai—Tibet Plateau, with a drainage area of 469,000 km^2. There are two lakes with areas greater than 100 km^2; namely the famous Yamzho Yumco (649 km^2) and Puma Yumco (295 km^2), which are located on the southwestern remnant of plateau surface and account for 63% of the total area of lakes in the district. Both of the lakes are early stage outflow lakes formed by local blocking on the plateau surface. The district is dominated by fresh water lakes, except for the southwestern region where there are a few brackish lakes (e.g., Degên Co, Nari Yumco, and Chen Co) and saline lakes (e.g., Cocholung Lake and Mazhongshan Lake) In addition, Jianhaizi Lake in the Chongsar geothermal field (Baxoi County, Qamdo) is an alkaline lake formed from pooling of geothermal water and has a salinity of 93.33 g/l (Zhu, 1982), but it is very small in area.

CHAPTER 3 EVOLUTION OF CENOZOIC LAKE BASINS AND THE FORMATION OF SALINE LAKES

3.1. Cenozoic Salt-bearing Strata and Salt-forming Periods

3.1.1. CENOZOIC SALT-BEARING STRATA

Cenozoic lacustrine strata are well developed on the Qinghai–Tibet Plateau and salt beds can be traced back as far as the Eocene. At this time salt deposits were widespread across the plateau, with gypsum salt basins developed from Gonjo (Eocene) in the southeastern part of the region northwestwards to Fenghuo Mountain (Oligocene). Eocene–Oligocene salt deposits are also found in the western part of the Qaidam Basin and in the area of the Dalangtan Playa①. The salt forming basins retreated northwestwards throughout the Miocene and Pliocene, but the Pliocene witnessed an increase in the intensity of salt forming processes. For example, relatively thick (Ca. 1–5 m) halite beds have been found in the western Qaidam Basin and in the Kumkol Basin. The discussion that follows will be focused on the Quaternary salt forming periods and the nature of their cyclicity.

3.1.2. QUATERNARY SALT FORMING PERIODS AND PERIODICITY

The Quaternary salt forming processes on the Qinghai–Tibet Plateau were initiated in the Early Pleistocene. In 1957 numerous salt beds were found in Middle and Lower Pleistocene strata in petroleum prospecting boreholes. Since the 1980's exploratory drilling and geomagnetic studies in the Dalangtan Playa and

---

① Salt beds were encountered in boreholes drilled in 1984 by the Tibet Bureau of Geology in the Gonjo Basin, Qamdo. There were different opinions on the age of the red beds in the area of Fenghuo Mountain due to the lack of fossils, but Sun Mengrong's analyses of samples collected on the peripheries of the salt domes in the area during profile mapping in 1980 indicated that the mudstone, sandstone and gypsum rock yielded sporopollen assemblages which Sun assigned to the Oligocene. Oligocene salt beds were also encountered in boreholes drilled by the Qinghai Bureau of Petroleum Industry in the western Qaidam Basin.

TABLE 1.3. Salineness of the Quaternary strata on the Qinghai–Tibet plateau

Lake district	Era	Palaeo-magnetic age (Ma)	^{14}C age of salt-bearing member (a.B.P.)	Thk. of strata (m)	Salt sediment	No. of beds	Aggregate thk. (m)	Av. thk. (m)	Salt bed ratio $\left(\dfrac{\text{Total thk. of salt beds}}{\text{Total thk. of strata}}\right)$	Remarks
Dalangtan Playa (Qaidam lake district)	Q$_1$	247–100	–	230	Mainly halite, occasionally intercalated with mirabilite	15	90	6	0.39	Based on borehole ZK336 data from Qinghai 1st Geol. Party
	Q$_2$	100–30	–	206	As above	15	112	7.47	0.54	
	Q$_3$	30–2.9	31000~10000(?)	>98	Mainly halite, containing K–Mg sulphates	4	>42	10.5	>0.43	Top part of salt bed eroded
	Q$_4$	–	–	–	–	–	–	–	–	Wanting in most cases
Kunteyi Playa	Q$_1$	–	–	116	Mainly halite, with occasional mirabilite intercalation	9	46.4	5.2	0.4	
	Q$_2$	–	–	90	As above	7	62.4	8.91	0.69	Statistics based on data from Qinghai 1st Geol. Party
	Q$_3$	–	–	38	Mainly halite, with K–Mg salts	3	27.2	9.07	0.72	
	Q$_4$	–	–	–	–	–	–	–	–	
Qarhan Playa	Q$_1$	–	–	–	–	–	–	–	–	
	Q$_2$	–	–	–	–	–	–	–	–	
	Q$_3$	30–1.1	–	360±	Mainly halite, with K–Mg salts in upper part	3	14.25	4.75	0.04	After Qinghai 1st Geol. Party and Chen Kezao et al.
	Q$_4$	–	10000–6000±	20	Mainly halite, with K–Mg salts	1	17	17	0.85	

EVOLUTION OF CENOZOIC LAKE BASINS AND THE FORMATION

Qaidam L dist	Da Qaidam Salt L	Q_1	—	—	—	—	—	—	—	After Qinghai 1st Geol. Party
		Q_2	—	—	—	—	—	—	—	
		Q_3	19500–16800 ±	50 ±	Mainly halite, with thin–bedded mirabilite	2	15.3	7.65	0.31 ±	^{14}C age data after Gao Shiyang, 1982
		Q_4	10000–present	11.05	Mainly halite, with mirabilite and boron salt	2	9.2	4.6	0.83	
North Tibet lake district	Bangkog Co	Q_1^1	—	10–65	—	—	—	—	—	Calcareous, sand–bearing conglomerate in lower part and clay in upper part
		Q_1^2	—	20–120	—	—	—	—	—	Sand and gravel beds intercalated with sandy clay
		Q_2^2								
		Q_3	18900–11000	30 ± (salt–bearing mem. 8.51)	Mirabilite and silt–bearing mirabilite	3	4.28	1.43	0.14	Sand and gravel bearing clay and carbonate clay in lower and middle parts
		Q_4	10000–present	16.22	Mirabilite and borax compound salt	10	12.84	1.28	0.79	Statistics based on borehole CK606
	Chacam Caka	Q_1	—	about 40–50	—	—	—	—	—	Party wanting (after the authors)
		Q_2	—	>30	—	—	—	—	—	
		Q_3	—	about 30	—	—	—	—	—	After the authors
		Q_4	10000–present	5.5	Mainly mirabilite and partly halite and boron salt	2	2.2	1.1	0.4	After the authors and borehole 78CK 2 data from Saline Lake Institute

—continued

25

other saline lake areas have indicated that the deposition of evaporites began in the Early Pleistocene and the precipitation of soluble salts occurred in each of the Early, Middle and Late Pleistocene stages. Through this time there was a general increase in the thickness of the salt beds, the salt rock ratio and the amount of soluble salts precipitated. An unpenetrated salt bed in the Dalangtan Playa has been dated at 4.0 Ma. To the south and southeast of the Dalangtan Playa the salt forming periods tends to be younger and the salt beds become thinner. The age of the oldest salt bed is 2.86 Ma at the Yiliping Playa, compared to 0.83 Ma at Kunteyi Lake and 0.53 Ma at the Mahai Playa. In the area from Mahai southeast to Taijnar Lake there is an increase in the thickness of the Early–Middle Pleistocene formation, but no salt beds have been observed within these strata. In this area the earliest salt forming episode was about 31 Ka[①]. Further to the southeast, at the Qarhan Playa, the earliest salt forming period was 24.4 ± 0.5 Ka (Chen, 1984).

From the Qaidam Basin south to the vast lake region on the plateau surface the Quaternary salt forming periods tend to be shorter and younger. At Bangkog Co in the North Tibet lake district the earliest Quaternary salt forming episode was during the terminal Late Pleistocene, with a ^{14}C age of about 18.9 Ka. However, more recent borehole data obtained from the Zabuye Caka area in 1990 shows that the precipitation of ulexite started at this location at 23 Ka. Further to the south, close to the North Tibet lake district, minor borax was not deposited (in the Yinghu Lake) until the Late Holocene (Zheng et al., 1983) (Table 1.3).

Periodic changes in climatic, hydrogeological and structural conditions in the saline lakes on the Qinghai–Tibet Plateau led to corresponding periodicity in the deposition of saline and non–saline beds. Rhythmites of various orders are also present. The scale of these rhythmites range from seasonal (annual), perennial (10–100 a) and long term (several hundred to tens of thousands of years), and are referred to as 3rd–order, 2nd–order and 1st–order rhythmites, respectively. These rhythmites are seen by changes in the colour, elemental composition, texture and structure of the argillaceous beds (gypsum–bearing and saliniferous silty clay, silt or carbonate clay) and salt beds (halite, mirabilite or K–Mg salts and borates), or by regular alterations in composition and thickness of the argillaceous and salt beds. These variations are though to be related to seasonal climatic changes and longer term changes in sunspot activity (Shen Zhen Shu et al., 1993). The 1st–order rhythms are manifest by periodic alternations in the relatively thick beds of salt sediments and argillaceous beds, with a single rhythm ranging from 20 cm to 30 m thick. These 1st order variations are probably related to advances and retreats of glaciers caused by long period, Milankovic style astronomical changes. Although

---

[①] According to data collected from the Qinghai First Geological Party, the intercalation of salt beds in the lower Shangyan Formation at Dalangtan is dated at 27.9 ± 0.7 Ka.

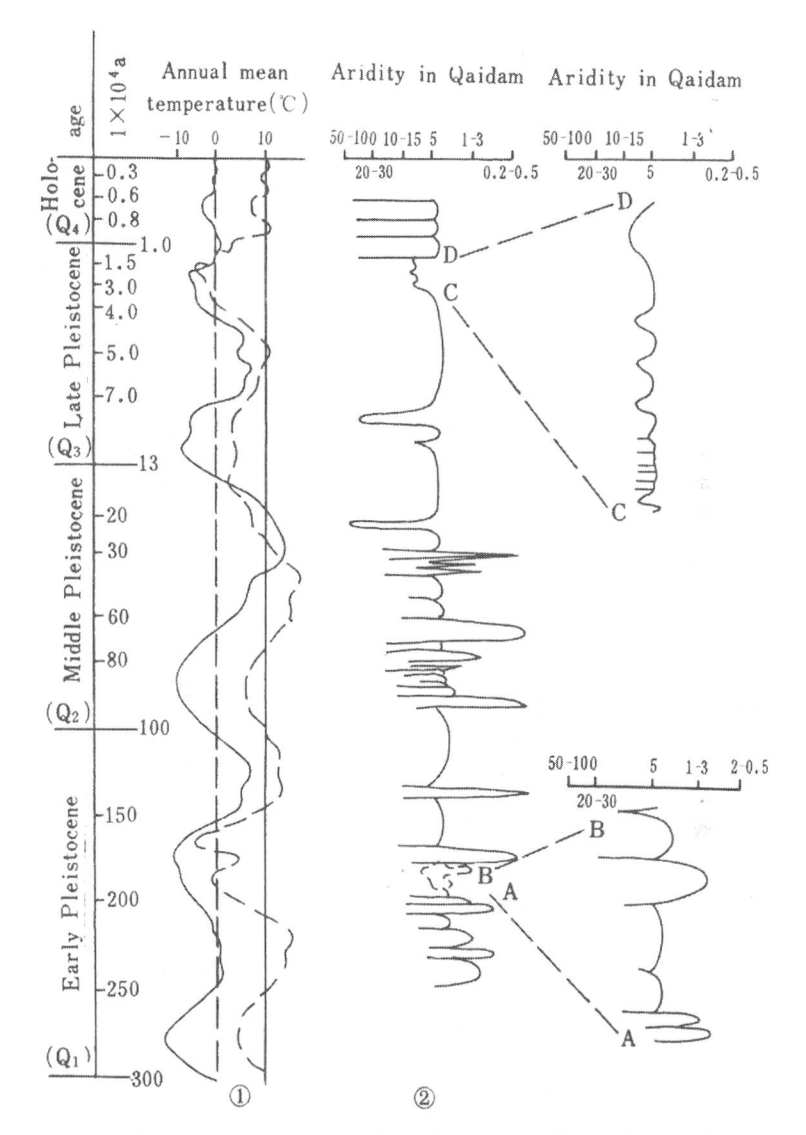

EVOLUTION OF CENOZOIC LAKE BASINS AND THE FORMATION 27

Figure. 1.3. Correlation of wet–dry periods in the Qaidam Basin with the changes of average temperature in other regions of China during the Quaternary. ① After Guo Xudong (1984); ② Based on the general relationship between the composition of the modern lacustrine sediments and the changes of relative aridity in China, the presumed ranges of the general palaeoaridity are inferred by the authors for the sedimentary mineral associations in this section as follows: 0.2–1 for argillaceous matter; >3–5 for gypsum; 5–15 for halite; 15–30 for epsomite and astrakhamite–halite; >30–100 for potash salt.

structural controls also play a role in controlling the formation of salt deposits.

As a result of palaeomagnetic and ^{14}C data from a few borehole, Huang et al. (1980) and Wang et al. (1985) were able to distinguish 8–10 Early Pleistocene 1st–order rhythmites, in which a single cycle commonly spanned several tens of thousands of years, with a maximum of 270 Ka. The duration of the deposition of the argillaceous beds, representing wet periods, was far longer than that of the salt beds (representing dry periods), with the ratios of the wet to dry periods being about 20:1 to 70:1. There are about 5–7 Middle Pleistocene 1st–order rhythmites, with a single cycle generally spanning tens of thousands of years, up to a maximum of 130 Ka. In this case the ratio of wet to dry periods ranges from 6:1 to 65:1. Between 2 and 7 Late Pleistocene 1st–order rhythmites have been identified with time spans ranging from tens of thousands of years at the base of the sequence to several hundred years in the middle and upper parts. Similarly the ratios of wet to dry periods range from 112:1 in the lower part to 0.65:1 in the upper part of the sequence. In the Holocene there are 2–3 1st–order rhythmites, with the cycles lasting a few thousand years and the ratio of wet to dry periods being 44:1 in the lower part and 0.63:1 in the upper part of the sequence (Tables 1.4 and 1.5).

This interpretation of the cyclicity of wet and dry periods is based on a relatively small amount of age data (largely palaeomagnetic data) available in the area and its accuracy requires further checking. Nevertheless, certain trends can be discerned. From the Early Pleistocene to the Holocene there was repeated deposition of as many as 40 or more 1st–order rhythmites, consisting of salt and argillaceous beds. These show a general trend of increases in the thickness of single salt beds and in the ratio of the thickness of salt beds to argillaceous beds towards more recent times. This has led to a corresponding increase in the abundance of elements such as K, Mg, Li and B, and substantial deposition of K–Mg salts.

During salt formation in the Quaternary there were five periods during which the ratio of salt to argillaceous beds was relatively high. These occurred during the middle part of the Early Pleistocene (ca. 2.01–1.62 Ma), the early part of the Middle Pleistocene (ca. 0.90–0.76 Ma), the middle part of the Late Pleistocene to the Holocene (ca. 80–42 Ka, 31 Ka, and 24 Ka to the present), with the highest ratio (as high as 5) during the Holocene. The final period of salt deposition was marked by substantial precipitation of carnalite and sylvite, and the contents of the economically important elements, K, Mg, B and Li, showed a large increase in the brine relative to previous periods. This is mainly related to long–continued chemical differentiation of the Late Cenozoic evaporite basins led to later concentration of more soluble salts. Besides, concentration of B and Li was also very likely to be related to Late Cenozoic hydrothermal–tectonic events in the region. Each of the five periods of increased salt deposition was followed by an interval of decreased deposition of salts. It is interesting to note that the periods marked by decreased salt deposition coincide with high–temperatures on the Tibet–Qinghai Plateau during interglacial stages (2.47–2.00 Ma, 1.40–1.00 Ma, 0.70–0.25 Ma, 56–35 Ka and

EVOLUTION OF CENOZOIC LAKE BASINS AND THE FORMATION   29

TABLE 1.4. Cyclothemic analysis of saline sediments from borehole ZK336 at the Dalangtan Playa

Era (Ma)	Rhythm No.	Palaeo-magnetic age (10 ka)	Age value of salt bed (a) / Age of argillac. bed (a)	Age of 1-order rhythmite (a)	Thk. of strata (m)	Thk. of salt bed (m)	Thk. of argillac. bed (m)	Thk. of salt bed / Thk. of argillac. bed	Deposition rate (cm/a) Salt bed	Deposition rate (cm/a) Argillac. bed
L. Pleistocene (Q₃)	35	1±	2571± / 61856±	64427±	46	10	12	0.83	0.389±	0.019±
	34		5143 / 20619	25762		20	4	5.00	0.389±	0.019±
	33	10	885 / 99115	100000	33	5	28	0.18	0.565	0.028
	32	20	1124 / 89837	90961		4	16	0.25	0.356	0.018
0.3	31		2247 / 16844	19091	52	8	3	2.67	0.356	0.018
	30		1685 / 22459	24144		6	4	1.50	0.356	0.018
	29		843 / 44919	45762		3	8	0.38	0.356	0.018
M. Pleistocene (Q₂)	28	38	15173 / 121457	136630	42	15	6	2.50	0.099	0.005
	27		6069 / 101215	107284		6	5	1.20	0.099	0.005
	26		5058 / 101215	106273		5	5	1.00	0.099	0.005
	25	73	1381 / 34519	35900	90	8	10	0.80	0.579	0.029
	24		518 / 20711	21229		3	6	0.50	0.579	0.029
	23		2417 / 13807	16224		14	4	3.50	0.579	0.029
	22		690 / 20711	21401		4	6	0.67	0.579	0.029
	21		1208 / 17259	18467		7	5	1.40	0.579	0.029
	20		1554 / 24163	25717		9	7	1.29	0.579	0.029
	19	88	690 / 10356	11046		4	3	1.33	0.579	0.029

	18	88	1596/44686	46282	34	10	14	0.71	0.627	0.031
	17		957/12767	13724		6	4	1.50	0.627	0.031
	16	94	2991/359281	362272		3	18	0.17	0.100	0.005
1.0	15		9970/59880	69850	60	10	3	3.33	0.100	0.005
	14		3988/279441	283429		4	14	0.29	0.100	0.005
	13	172	4985/59880	64865		5	3	1.67	0.100	0.005
	12		2128/31915	34043		8	6	1.33	0.376	0.019
E. Plei- sto- cene	11		798/31915	32713	51	3	6	0.50	0.376	0.019
	10		798/47872	48670		3	9	0.33	0.376	0.019
	9		1596/31915	33511		6	6	1.00	0.376	0.019
(Q₁)	8	188	532/10638	11170		2	2	1.00	0.376	0.019
	7		1399/55944	57343	30	4	8	0.50	0.286	0.014
	6	201	2797/69930	72727		8	10	0.80	0.286	0.014
	5		2688/69124	71812	50	14	18	0.78	0.521	0.026
	4	214	576/57604	58180		3	15	0.20	0.521	0.026
	3		5607/93493	99100		6	5	1.20	0.107	0.005
	2		2804/37398	40202	30	3	2	1.50	0.107	0.005
2.47	1	247	3738/186986	190724		4	10	0.40	0.107	0.005

Calculations based on borehole data from the Qinghai 1st Geological and Hydrogeological Party.

TABLE 1.5. Cyclothemic analysis of saline sediments from borehole ZK2022 at the Qarhan Playa

Era (Ma)	Rhythm No.	^{14}C age (a. B.P.)	Age value of salt bed / Age of argillac. bed	Age value of rhythmite (a)	Thk. of strata (m)	Thk. of salt bed (m)	Thk. of argillac. bed (m)	Thk. of salt bed / Thk. of argillac. bed	Deposition rate (cm/a) Salt bed	Deposition rate (cm/a) Argillac. bed
Holocene (Q₄)	14	6000 ±	523/836	1359	13.13	5.7	0.57	10.0	1.091	0.0682
	13		485/1261	1746		5.0	0.86	5.8	1.091	0.0682
	12	9170 ± 100	92/4540	4632	7.72	1.0	0.79	1.27	1.091	0.0174
	11	15700 ± 340	1991/1566	3557	13.57	6.93	0.93	7.45	0.348	0.0594
	10		224/1195	1419		2.66	0.71	3.75	1.188	0.0594
	9		391/1195	1586		4.64	0.71	6.54	1.188	0.0594
	8	20600 ± 410	330/1132	1462		3.92	2.36	1.66	1.188	0.2084
	7		77/341	418		3.21	0.71	4.52	4.1687	0.2084
	6		135/413	548		5.64	0.86	6.56	4.1687	0.2084
L. Pleistocene (Q₃)	5		69/341	410	27.38	2.86	0.71	4.03	4.1687	0.2084
	4		43/288	331		1.79	0.60	2.98	4.1687	0.2084
	3		45/341	386		1.86	0.71	2.62	4.1687	0.2084
	2		43/341	384		1.79	0.71	2.52	4.1687	0.2084
	1	24780 ± 510	60/513	573		2.50	1.07	2.34	4.1687	0.2084
	0									

Calculations are based on borehole data from Chen Kezao et al. of the Qinghai Saline Lake Institute; the single-bed deposition rates for rhythmites Nos. 1–12 are inferred from the ratio of argillaceous bed / salt bed equal to 1/20 and those for rhythmites Nos. 13–14— from the ratio of argillaceous bed / salt bed equal to 1/10.

9.8–8.5 Ka). Likewise, the five periods of increased salt deposition are coincident with low–temperature glacial intervals (Fig. 1.3).

3.2. Genesis and Classification of Lake Basins

There are a variety of different types of lakes in Qinghai and Tibet that are the products of the prolonged actions of internal and external processes. Genetically they can be grouped into 8 different types: 1) glacial lakes, 2) silted–river valley lakes (including barrier and oxbow lakes derived from rivers and debris flows), 3) interbank lakes, 4) dissolved–salt lakes, 5) meteorite crater lakes, 6) hot–water lakes, 7) volcanic lakes, 8) tectonic lakes. The first five types of lakes are largely formed by external processes, whereas the final three types are essentially generated by internal processes. The tectonic lakes include those within deep intermontane fault–block basins, intrazone downwarped–faulted lake basins, micro–rift lake basins and strike–slip lake basins (Table 1.6). Field investigations and interpretation of satellite images of the entire region indicate that the saline lakes on the Qinghai–Tibet Plateau are mainly of tectonic origin and are dominated by downfaulted basins and inherited depressional basins (Qian et al., 1979; Zheng et al., 1983). There are relatively few saline lake basins that formed as the result of exogenic factors, such as the actions of glaciers and rivers.

3.2.1. GLACIAL LAKES

Glacial lakes include glacier–barrier lakes (which are lakes formed by the damming of rivers by ancient and modern moraines), and glaciated, superglacial and subglacial lakes that formed by glacial excavation. The lakes falling in this category are all relatively young, being mostly of Holocene and Recent age, and they are all fresh water lakes. They are generally small in size, ranging from 1 km^2 to 20 km^2 in area, although they are amongst the most numerous of the different types of lakes. They are distributed extensively in the glaciated areas of the eastern part of the plateau, except for a few which occur in the high mountain areas. For example, Basang Co lies on a tributary of the Nyang Qu (river), is 26 km^2 in area and is situated in a glacially excavated trough that has been barred by a terminal moraine ridge (Qian et al., 1979).

3.2.2. SILTED–RIVER VALLEY LAKES

Silted–river valley lakes include drowned–river valley lakes in depressions within river valleys. They have been formed by the shrinkage of rivers, rivers dammed by debris flows, landslides and proluvial fans, and oxbow lakes formed by the meandering of rivers. The overwhelming majority of lakes of this type are recent fresh

EVOLUTION OF CENOZOIC LAKE BASINS AND THE FORMATION   33

water of relatively small sizes, ranging from less than 1 km² to scores of square kilometres in area. They are mostly distributed in marginal areas of the plateau and are particularly widespread in the Bayan Har and East Tibet lake districts on the eastern part of the plateau. The drowned-river valley lakes are generally characterised by a series of fluvial, waterlogged depressions distributed around alternate inflow-outflow lakes (e.g., Gyaring and Ngoring Lakes) on the upper reaches of the Yellow River drainage system (Fig. 1.4). An example of a lake formed by the damming of a river is provided by Dinggyê Co, which is 10 km² in area and is an exhoric lake formed 30 years ago by water accumulation on the west branch of the Yarlung Qu river valley during a flash flood. Oxbow lakes have a more limited distribution. Langmari Lake, in the western part of the plateau, is an example of an oxbowlake formed from the westward migration of the Gar Zangbo River.

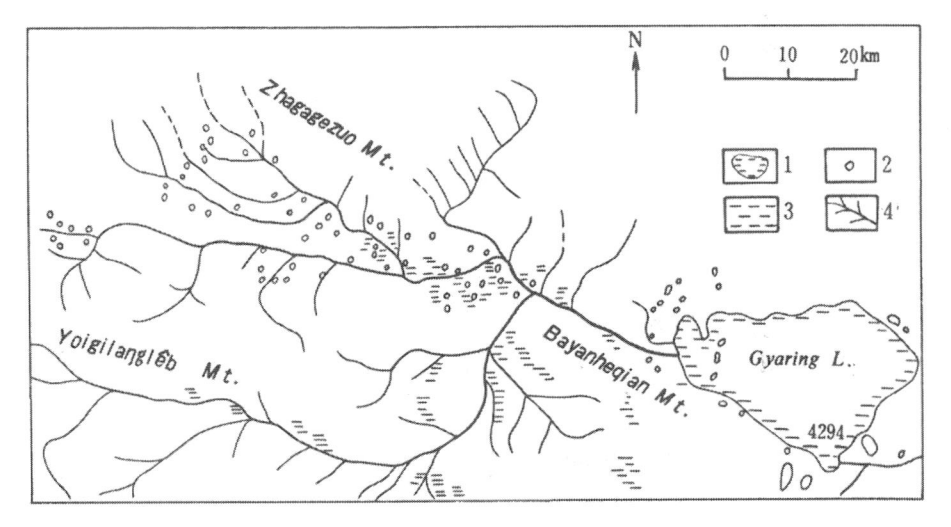

Figure 1.4. Distribution of silted-river valley lakes around the Gyaring Lake. 1-Alternately inflow-outflow lake; 2-Silted-river valley lake; 3-Swamp; 4-River.

3.2.3. INTERBANK LAKES

Interbank lakes are derived from the contraction or overspill of other lakes. Typically, they are "multi-step lakes" formed by the storage of water in depressions among sand banks created during the transgression and regression of lakes. Lakes of this type are widespread in lake districts where large lakes are developed, particularly in the North Tibet lake district. In the area from Bangong Co in the west to Siling Co and Nam Co, the large lakes are surrounded by thousands of interbank lakes, most of which are less than 1 km² in area. For example, approximately 100 interbank lakes of various sizes (such as Yagedong Co) have formed within the Siling Co area (Fig. 1.5).

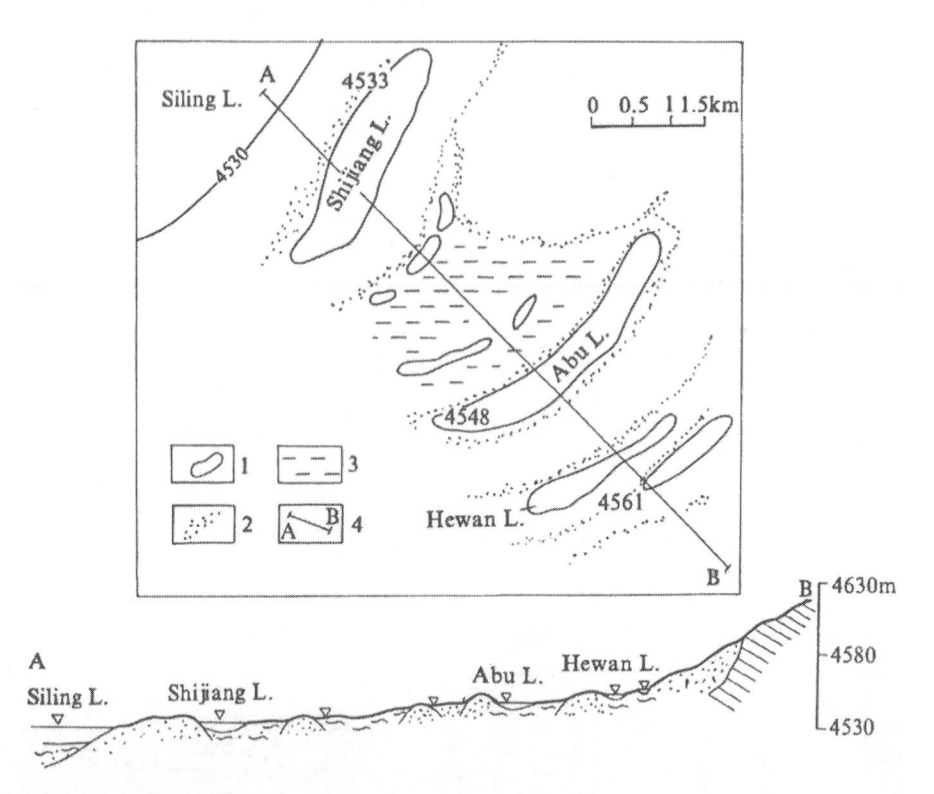

Figure. 1.5. Interbank lakes to the southeast of the Siling Co. 1—Interbank lake; 2—Sand bank; 3—Swamp; 4—Section line.

3.2.4. DISSOLVED SALT LAKES

Lakes of this type are restricted to the plateau salt–domes areas and the Qaidam Basin. They formed from the redissolution of playa salt deposits. An example of such a lake is provided by Rierlama Lake in the Kunlun lake district. This lake was formed by accumulation of water in a local depression resulting from the uplift of Eogene salt domes and the subsequent dissolution of the salt bodies. Rierlama Lake is saline and has an area that ranges from several hundred square metres to tens of thousands of square metres. In general, lakes of this type are sparsely distributed in salt dome areas.

3.2.5. HOT WATER LAKES

Hot water lakes are restricted to particular area within geothermal areas and are formed by the storage of hot-spring water in depressions resulting from hydrothermal explosions or localised depressions. They are always relatively small in area, for example, Yangbajain lake has an area of only 7350 m^2, with a water depth of 15.5 m and a water temperature of 49-57℃. Another example is provided by Jianhaizi Lake (Baxoi County, eastern Tibet), which has an area of 2300 m^2, a water depth of 0.3-0.5 m and a water temperature of 49-57℃ (Coulson, 1933).

3.2.6. VOLCANIC LAKES

There have been a large number of periods of Cenozoic volcanism on the Qinghai-Tibet Plateau, although Quaternary volcanic activity has been largely confined to the Kunlun lake district. Hence, well-preserved deposits of volcanic ash and lavas and volcanic craters are widely distributed across the plateau. In the Kunlun lake district there ara a number of lakes that have been formed by damming of river valleys by lava flows. As early as 1913, Von Zugmayer referred to two volcanic barrier lakes in the headwaters of the Keriya River in the West Kunlun Mountains. An example of this type of lake is provided by Wulukele Lake in the Kaerdaxi volcanic swarm. This lake has an area of 14 km^2, a maximum water depth of 40 m, and was formed by the coalescence of lavas from two volcanoes that blocked a river valley (Tong et al., 1981).

3.2.7. TECTONIC LAKES

Most of the large and medium-sized lakes on the Qinghai-Tibet Plateau fall into this category. These lakes were formed by compression of the continental crust during the third phase (initial to middle stages of the Early Pleistocene) or fourth phase of collision (middle to late stages of the Early Pleistocene) of the Indian and Eurasian plates. On the basis of the differences in the tectonic settings of the lakes and the style of associated faulting it is possible to distinguish four types of tectonic lakes (Fig. 1.6).

Deep intermontane fault block basins.
Basins of this type have been inherited from much tectonic events in the past. During the Mesozoic, the formation and evolution of the ocean in the south of the area and the resistance of the Paleozoic block in the north resulted in the fragmentation of the Paleozoic block (Zhu, 1983). As a result of the differential movement of the Mesozoic mountain chains the fragmented blocks on the northern and northwestern margins of the Palaeozoic block subsided to form local lake basins. During renewed collision of the Indian and Eurasian plates in the Middle-Late Holocene, the fragmented blocks were further separated from their parent body to the north (the Alxa block) and that to the northwest (the Tarim block). They

36 CHAPTER 3

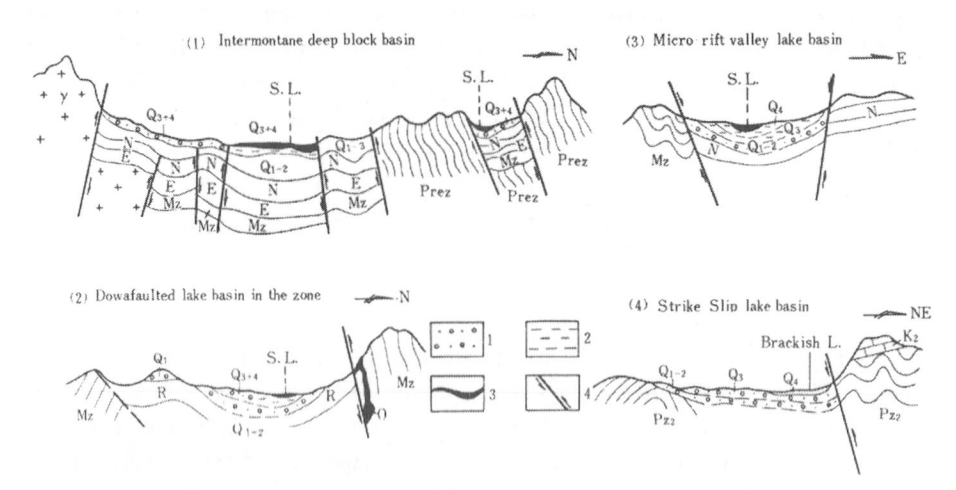

Figure. 1.6. Types of tectonic lake basins on the Qinghai–Tibet Plateau. Q_4–Holocene; Q_3–Upper Pleistocene; Q_{3+4}–upper part of the Quaternary; Q_{1+2}–lower part of the Quaternary; Q_1–Lower Pleistocene; N–Neogene; E–Eogene; R–Tertiary; K_2–Upper Cretaceous; M_2–Upper Paleozoic; Pre_2–Pre–Sinian; γ–granite; O–ophiolite. 1–Sand and gravel; 2–Argillaceous sediment; 3–Saline sediment; 4–Fault.

EVOLUTION OF CENOZOIC LAKE BASINS AND THE FORMATION 37

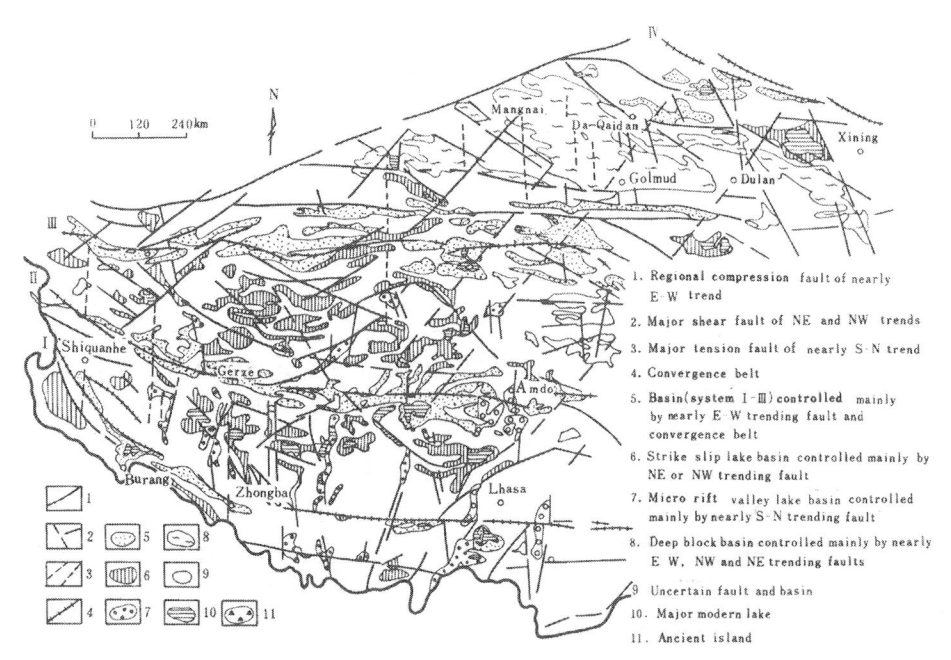

Figure 1.7. Sketch of Quaternary basins and linear structures on the Qinghai–Tibet Plateau. 1–Nearly E–W regional compressional fault; 2–NE and NW–trending principal shear fault; 3–Nearly N–S principal tensional fault; 4–Convergence zone; 5–Basin (series I–III) controlled mainly by nearly E–W fault and convergence zone; 6–Strike–slip basin controlled mainly by NE or NW–trending fault; 7–Micro–rift basins controlled mainly by nearly N–S fault; 8–Deep fault–block basin controlled mainly by nearly E–W to NW–trending and NE–trending fault; 9–Fault of unknown nature and basin; 10–Major modern lake; 11–Ancient island.

became part of the flexible body of the Qinghai–Tibet structural system. Under the strong compression from the south and north these fault blocks subsided at an anomalously high rate, leading to large vertical movements, horizontal compression and strike–slip shearing (Zhu, 1983). Example of this type of basin are seen in the Qaidam lake district. In this district the stage of geosynclinal development terminated in the Proterozoic. The district then entered the platform development stage, when most of the district (together with the Tarim and Altun blocks) belonged to the same ancient crystalline basement (Jiao, 1984; Pan et al., 1984). As a result of the opening of the Mesozoic ocean to the south during the Jurassic, the Paleozoic folded mountain system in the district underwent differential grabening, leading to the formation of intermontane basins. At the same time, an extensional downfaulted basin of Cretaceous–Jurassic age was formed along the Altun block (Pan, 1984). This extensional downfaulting resulted in the separation of the Qaidam block from the Tarim block. During the first phase of plate collision in the Triassic (Zheng and Xiang, 1985), following the development of the intracontinental convergence of the Qinghai–Tibet Plateau, the Tarim block was subducted southwards. The Altun fragmented block was an intermediate buffer zone and was sheared sinistrially and locally arched up①. Hence, the entire Qaidam block was involved in the Qinghai–Tibet system and subsided to give rise to a large intermontane fault block basin. The Altun shear–arched structure (referred to by Pan Guitang as the "arch–type structure") formed under compressions from the north and south, and controlled the migration of the depocentre of the Qaidam basin from the northwest to the southeast since the Eocene. In the meantime, the compressional force that originated from the Alxa block, and was transmitted by the Qilian mountain system, caused the formation of the WNW–trending compressional fault system of the Qilian mountain chain (Fig. 1.7). This caused the salt–bearing Qaidam basin to be increasingly elongated with the lapse of time. The late–stage saline lake was driven to the synclinal basin and down faulted basin that hosted the earliest deposits, and thus became an internally overlapping inherited saline basin (Fig. 1.6). The earlier deposited $Tr-Q_{1-2}$ unconsolidated sediments were folded and upwarped from the west and formed a series of depressions and rises along the WNW and ENE palaeo–faults. In the middle and late stages of the Quaternary, a saline lake formed and developed in the downfaulted depression. This depression had formed as the result of N–S compressional stress. In summary, the migration of the Qaidam Cenozoic saline lake basin from the northwest to the southeast and from the southwest to the northeast, the narrowing of the salt basin with time, the deposition of the very thick salt beds and the change in the source regions from the

---

① According to regional survey data, Early Quaternary lacustrine deposits have been found in the Altun mountains, between the Qaidam and Tarim basins, suggesting that the Altun mountains were not totally uplifted in the Tertiary.

northeast to those derived from the southwest can all be explained by the structural setting and evolution described above. Similar large basins that are controlled mainly by nearly E–W to NW and NE trending faults include the Kumkol and Gonghe basins (Fig. 1.7).

TABLE 1.6 Genetic classification of lake basins on the Qinghai–Tibet Plateau

	Lake type	Principal features	Examples
	Glacial lake	Lakes formed by exaration, barrage or corrosion in the course of development of glaciers	Yi'ong Co.
	Silted–river valley lake	Lakes formed by shrinkage and course change of rivers or barrage of rivers by debris flows, landslides and proluvial fans	Langmari L.
	Interbank Lake	Water–logged depressions on sand banks formed by contraction and migration of lakes	Xiaochu'anmu L. and Hajiang L.
	Dissolved–salt lake	Local water–logged depressions formed by dissolution of salt beds or domes in ancient lake areas	Rierlama L.
	Hot–water lake	Barred depressions formed by hydrothermal explosion or local topographic rises, and logged with hot water, which has been concentrated to precipitate salt	Xianhaizi L.
	Volcanic lake	Lakes enclosed by Cenozoic volcanic craters or volcanic lavas	Aqqikkol L. and Wulukele L.
Tectonic lake	Intermontane fault–block deep basin	"High–mountain deep basins" (back–arc basins) formed by overall uplifting caused by post–suture collision of the Indian and Eurasian plates	Salt lakes in the central and southern–central Qaidam Basin
	Intrazone downwarped–faulted lake basin	Tectonically inherited downwarped (or downfaulted) depressions within Mesozoic and Cenozoic suture zones, which are mostly E–W trending lake basins.	Bangkog Co, Chagcam Caka and Margai Caka
	Micro–rift lake basin	N–S tensional (tenso–shear) basins, often accompanied by magma–type geothermal activity	Daggyai Co, Dung Co and Chalaka Co
	Strike–slip lake basin	NW or NE compresso–shear (tenso–shear) basins formed by strike–slip shear movement	Gyaring Co and Yibug Caka
	Meteorite–crater lake (?)	Water–logged depressions formed by meteorite impact, characterized by marginal ring–shaped rises	Four lakes including the Yuanhu lake (to the north of Baingoin)

Intra-zone downwarped faulted basins
The basins are essentially formed by tectonically inherited downwarped (or downfaulted) basins, formed by block resulting from reactivation of the Mesozoic–Cenozoic suture zones (the Lungmu Co–Jinsha River, Bangong Co–Nujiang River, and Yarlung Zangbo zones) and the Kunzhong arcuate fracture zone (central Kunlun Mountains) under the influence of the collision between India and Asia since the Eocene. Hence, these basins mainly occur as superimposed basins developed on Tertiary sedimentary basins extending in a nearly east–west direction. These basins are largely controlled by E–W compressional structures, with thrust faults commonly observed in the north and overlaps or downwarps often occurring in the south (Fig. 1.6). They are also occasionally governed by NW to NNW trending and NE trending structures (Zheng et al., 1983). Three lake basin series can generally be distinguished from north to south, namely; the Hoh Xil lake basin (I), Bangong Co–Siling Co lake basin series (II), and the Yarlung Zangbo lake basin series (III) (Figs. 1.6 and 1.7). Basins of this type are usually quite large, except for the Yarlung Zangbo lake series whose ancient lake basins were largely destroyed as a result of their capture by the newly formed Yarlung Zangbo River.

Micro-rift valley basins
These are nearly N–S trending tensional downfaulted basins formed by the E–W extensional processes caused by the N–S compression since the second phase of collision in the Neogene. These micro–graben basins are structurally similar to intracontinental rift valleys and they also have associated geothermal activity driven by epigenic magma bodies at depth (Huang, 1984) (Fig. 1.6). These basins are mainly distributed in the southern part of the plateau, probably as a result of the soft basement in the area. To the south of the Gangdisê and Bangong Co–Nujiang River there are 8 basins series, namely; the Luerdapa Co (Co Nag Co) (1), the Rinqin Xubco (2), the Dawa Co (3), the Tangra Yumco (4), the Yueqia Co (5), the Gomang Co (6), Yangbajain (7), and Qungdo'gyang (8) (Fig. 1.7). From this area east to western Yunnan these N–S trending basins are still well developed and arranged with a spacing of 60–180 km. Examples of this type of basin can also be found in the central and northern plateau, although they become much less abundant to the north. For example, the Yueqia Co lake basin series (5) can be traced north to Linggo Co at the southern foot of the Tanggula Mountains and Aqqikkol Lake at the southern foot of the Kunlun Mountains. The Qungdo'gyang lake basin series (8) lies on a N–S trending fault that extends northwards to Mangnai Lake in Qaidam Basin.

In addition to the 8 lake basins noted above, several nearly latitudinal faults with nearly equal spacings can be seen to the east of the Qungdo'gyang lake basin series. Lake basins are developed in the northern part of this area. For example, the Mêdog–Bomi latitudinal fault can be traced northwards to Qaidam Basin and may become the boundary delimiting the eastern part of Dabsan Lake and the western part of Da Qaidam Lake. It may also play a role in controlling the alignment of Nan (South) Hulsan and Bei (North) Hulsan Lakes.

Overall the patterns of lake and fault distributions suggest that the plateau basement is more rigid at shallower levels in its northern part. They also suggest that the northern part of the Qinghai–Tibet Plateau has been affected by E–W extension since the Quaternary.

Strike–slip basins

These are NW or NE–trending tenso–shear downfaulted lake basins formed by strike–slip shearing, largely in the third phase of collision. They were formed at the same time as the micro–rift valley basins, and they both result from the same stress field. These lake basins are more clearly developed to the north of the Bangong Co–Nujiang River deep fault. In this area, the northwest sides of the NE–extending downfaulted lake basins are largely controlled by tenso–shear faults, which are often compounded by faults in other directions. Their north and west sides are marked by sedimentary overlaps. These lake basins are well developed to the east of 85° E, and are characterized by sinistral movements on the controlling faults. For example, the northeast sides of the NW–extending downfaulted lake basins, such as the Yangbajain palaeo–lake and Yibug Caka, are largely controlled by tenso–shear faults, which are again often compounded by faults in other directions. The southwest sides show sedimentary overlaps, as observed in the lake swarm of the Gyaring Co (Fig. 1.6). The NW–trending strike–slip lake basins are relatively well developed to the west of 85° E and are mostly marked by a dextral shear movement. Hence, the two directions of the strike–slip movement and the N–S tensional lake basins probably represent a conjugate structure, formed under a nearly N–S opposing compression. These complicated tectonic patterns result from heterogeneities in the physical properties of different parts of the plateau (Fig. 1.7). Conjugate basins controlled by X–shaped faults may be present in the same lake basin. For example, the Nam Co and Xiangdao Lake lake basins have borders controlled by NE and NW–trending faults.

3.2.8. METEORITE CRATER LAKES

There is increasing recognition that meteorite craters are more widely distributed on the Earth's surface than has previously been thought. On the basis of satellite images it has recently been suggested that some of the circular lakes with ring–shaped marginal rises in Tibet are possibly of meteorite–impact origin (K. Burke, pers. commun.). These lakes are all in remote areas, so that these suggestions have not been verified by field data. Hence, the possible existence of meteorite crater lakes requires further investigation.

3.3. Evolution of Lake Basins and Formation of Saline Lakes

The Qinghai–Tibet Plateau is a highly unusual tectonic unit. It is a recently activated tectonic system formed by intracontinental convergence and N–S compression during the Eocene, after repeated suturing of the continental crust. The N–S shortening, E–W extension and block subsidence led to the formation of a series of structural basins. As a result of uplift of the plateau, atmospheric circulation over the area underwent repeated vertical and horizontal differential changes that caused regional climatic changes. This caused variations in the salinity of waters in the lake basins, the nature of sediments deposited in these basins, the formation of numerous glacial lakes and the general evolution of lake basins. As a result the lake basins on the plateau have undergone an evolutionary process that has led to N–S zoning and E–W division of the lake districts. There have also been frequent vertical and lateral changes in the nature of the lakes, remarkable migrations in depocentres and salt–forming processes, warm–cold alternation in the water bodies, changes in the salinity of lake waters from fresh to saline, and gradual transformation of large areas of lake waters to partially or completely dry lake basins. In particular, evolution of the Qaidam Basin, in the northern part of the plateau, can be traced from the Quaternary back to the Oligocene. The duration of the salt–forming processes is so long that the Qaidam Basin is unique among the lake basins in the world. The study of the history of this, and other ancient lake basins, can provide information concerning the controls over modern salt lakes. Hence, in a discussion of the development of saline lakes on the Qinghai–Tibet Plateau it is first necessary to examine the evolutionary history of the lake basins. With particular attention to the evolution of the Late Cenozoic lake basins, the authors have recognized four stages of evolution of the lake basins on the plateau: 1) Eocene–Miocene (E–N_1) or Eocene–Early Pliocene (E–N_2^1); 2) Pliocene–early or middle Early Pleistocene (N_2–Q_1^1 or Q_1^2); 3) middle to late Early Pleistocene–early Late Pleistocene (Q_1^{2-3}–Q_3^1); 4. late Late Pleistocene–Holocene (Q_3^2–Q_4) (Yuan, 1946).

3.3.1. EOCENE–MIOCENE OR EARLY PLIOCENE

This stage corresponds to the initial stages of intra–continental convergence relating to the collision of the Indian and Eurasian plates, and is termed the first phase of post–suture collision (E_2–N_1). The lake basins of this age on the plateau can be grouped into 3 districts on the basis of their dimensions and structural features.

The south district was located on the southern margin of the Gangdise Mountains. During the Palaeocene–Eocene there was still an E–W elongated relic sea basin in the Gamba area of southern Tibet. In the middle and terminal Eocene, the suturing and collision of the Indian and Qinghai–Tibet plates at the Yarlung Zangbo River resulted in the formation of the Gangdise epicontinental mountain chain. In the E–W trough on the southern edge of the mountain chain there were a series of discontinuous, limited, fluviolacustrine, coal–bearing clastic formations and molasse formations (Qian, 1982). This is indicative of a humid and hot tropical–subtropical sedimentary environment in this area, e.g., in the Qiuwu and Miocêr basins. This area remained relatively warm and humid until the Miocene (the Lower Member of the Wulong Formation at Namling) (Li and Song, 1980).

The central lake district was located in the central part of the plateau, where the lake basins were mainly spread in an E–W tensional (shear) downwarped (down–faulted) zone and some broad depressions. Large and extensive lake basins occurred particularly in the already sutured Bangong Co–Nujiang River and Lungmu Co–Jinsha River zones. At that time, the lake basins were located in extensional and downwarped basins controlled by the E–W fault (compression below and tension above) which had been formed by the action of the regional principal stress produced at the initial collision stage of the two plates upon the lower part of the Mesozoic sedimentary terrain or folded mountain system. Although all these basins were located in the same subtropical–tropical zone, palaeo–climatic differences resulted in basins in the southern part of the area being occupied by fresh water lakes in a warm and humid climate, while those in the northern part of the area from Hoh Xil to Gonjo were occupied by saline gypsum–salt lake basins in a dry and hot climate.

The northern district consisted of the Qaidam and Kumkol basins. The Qaidam basin was an intermontane basin in the Jurassic–Cretaceous. It became an inherited downwarped depression at the beginning of the Tertiary when the WNW–trending fault system was initiated. This caused strong and relatively extensive subsidence in the district. During the Eocene the lake basin was located at the west end of the Qaidam Basin, covering an area of several thousand square kilometres, where red beds were deposited under a subtropical climate. In the Oligocene the lake basin migrated northwest and became a gypsum salt lake basin under an arid and hot climate. From the Miocene to the early Pliocene the lake basin expanded southeast to cover an area of 10–20,000 km^2. At that time, a semi–arid temperate–zone climate probably prevailed with a flourishing coniferous forest–steppe vegetation, and gypsum–salt sediments were still being deposited in the western part of the basin. As noted above, the Tertiary Qaidam Basin is probably an inherited block–faulted basin, and the same is probably true of the Kumkol Basin to its southwest[1]. Both the basins contain WNW and NE–trending fault blocks. These were formed within the original intermontane basins by opposing

---

[1] According to Li Guodong of the Xi'an Institute of Geology, Cretaceous marine formations may be present in this area.

wedging of the Sino–Korean and Tarim blocks caused by the northward transmission of the NNE principal compressional stress of the two plates. Hence, the Qaidam and Kumkol Basins can be considered to be the products of collisional tectonism.

3.3.2. PLIOCENE–EARLY–MIDDLE EARLY PLEISTOCENE

At the start of the Pliocene, the lake basins on the plateau was subject to large scale tectonic forces. This resulted in the migration of the depocentres andthe folding and uplift of the basins. In addition, there were changes in the distribution pattern of the lake basins leading to the formation of a series of N–S Cenozoic lakes basins in the region. This stage was equivalent to the second collisional phase of N–S intracontinental convergence, which became more intense at this time (Tapponnier et al., 1982). Because of the resistance of the Sino–Korean block to the north of the plateau, the soft basement material on the Qinghai–Tibet, particularly those in the southern part of the plateau, migrated eastward and a group of N–S– or NNW– and NNE–trending new lakes were formed in different geological settings. Examples of such basins to the south of the Gangdise Mountains include (from east to west); the Zanda, Moincêr, Burang, Gyirong, Nieniexiongla, Pagri, Qingtü, Chigu, Quigdo' gyang, and Subanxili (extending south into India) basins. Regional Pliocene, nearly N–S tensional (shear) downfaulted basins were formed in further east in western and central Yunnan. In northern Tibet the E–W extensional stress was largely offset by the presence of the Precambrian basement and the more resistant material in the Yanshanian area. Hence, the motion of the various geological features was not entirely to the east, as has been suggested by some (Tapponnier et al., 1982). For example, in the Lunpola Basin the Eocene–Oligocene (or possibly to Miocene①) deposits were folded and uplifted, and the Pliocene lake basin migrated south to the area of Qiling Lake–Bangkog Co, with the lake basin extending along a nearly E–W fault. A large number of N–S trending basins have been located on the eastern and western sides of the southern and central parts of the plateau. To the east of the Amdo (To'gyü)–Sangxung–Kunlun Pass NE–trending fault line, nearly N–S extending Pliocene–Lower Pleistocene lake basins formed, such as (from east to west), the Xagquka, Amdo, Tuotuohe, Kunlun Pass, Sêrxong, Dêngqên and Markam–Baxoi basins, together with a few small lake basins in the northern part of western Yunnan (Fig. 1.8). To the north of the Gangdise Mountains, on the edge of the western part of the plateau, Pliocene subtropical basins were formed. The common features of these basins is that they contain a continuous record of Pliocene–Early Pleistocene lacustrine deposits, con-

---

① According to the studies of sporopollen and ostracods, Sun Mengrong and Xia Jinbao suggest that Pliocene and even Miocene deposits are absent in this basin.

sisting of stable fine—grained clastic sandy and argillaceous sediments, with the base lying unconformably upon the pre—Tertiary or Eocene formations (Zheng, 1980; Ji et al., 1980; Wu, 1983). No new lake basins were formed in the Qaidam Basin, to the north, but the lake basin migrated continually southeast and the lake area greatly expanded to the southeast①.

The second stage of the evolution of the Cenozoic lake basins on the plateau reflect the clockwise rotating (up to 40°) and the eastern expansion of the Altun—Gansu—Weihe River fault (Xie et al., 1965). This middle stage of the intracontinental convergence of the Qinghai—Tibet Plateau is approximately equivalent to the second collisional stage of 10—20 Ma proposed by Tapponnier et al. (1982). All of these collisional—compressional tectonic effects are reflected in the evolution of lake basins on the plateau.

The lake deposits on the plateau are usually relatively fine—grained (mainly clay and silt). The Pliocene sediments contain *Hipparion* (Zheng, 1980; Ji et al., 1980) and hyprophilic and thermophilic faunal colonies, such as rhenoceros, suggesting a relatively low level (500—1000 m above sea level) of the ancient lake surface and a relatively humid climate to the south of the Tanggula Mountains. An analysis of the palynological, botanical and sedimentological data from the plateau (Zhou, 1984; Comprehensive Expedition of Academia Sinica, 1984) show that at this stage the climate was tropical—subtropical, with abundant rainfall. As a result the lakes were dominantly fresh water in both southern and northern Tibet.

To the north of the Tanggula Mountains there are stable lacustrine deposits that were laid down under a subtropical climate, indicating alternating arid—humid palaeoclimatic conditions. This led to the development of brackish and saline lakes, with the salinity increasing from south to north. Widespread gypsiferous deposits are found from the Chainjoin Co in the Qiangtang area in the southern part of the plateau, eastward to the Tuotuo River and Wudaoliang along the Qinghai—Tibet Highway and the area south of the Yellow River (Li et al., 1983; Jin et al., 1984; Xu, 1985). Relatively extensive gypsum—salt deposits (the Upper Youshashan Formation) are observed northward in the Qaidam and Kumkol basins (Zhu, 1985).

3.3.3. MIDDLE—LATE EARLY PLEISTOCENE TO EARLY LATE PLEISTOCENE

In the middle or late Early Pleistocene, the lake basins on the plateau underwent another period of intense and large—scale changes, which were marked by the following characteristics.

---

① Based on the Neogene lithofacies map compiled by Yang Zhilin of the Qinghai Bureau of Petroleum Industry (1983).

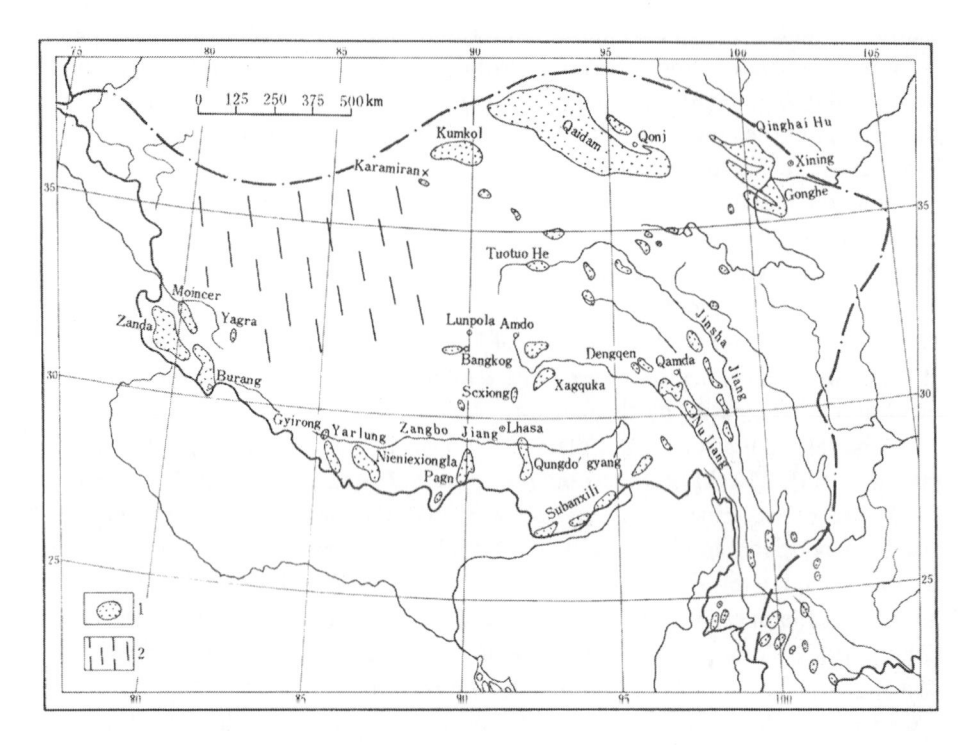

Figure 1.8. Distribution of Pliocene–initial Early Pleistocene lake basins on the Qinghai–Tibet Plateau. 1–Pliocene–initial Early Pleistocene lake basin; 2–Tertiary undifferentiated.

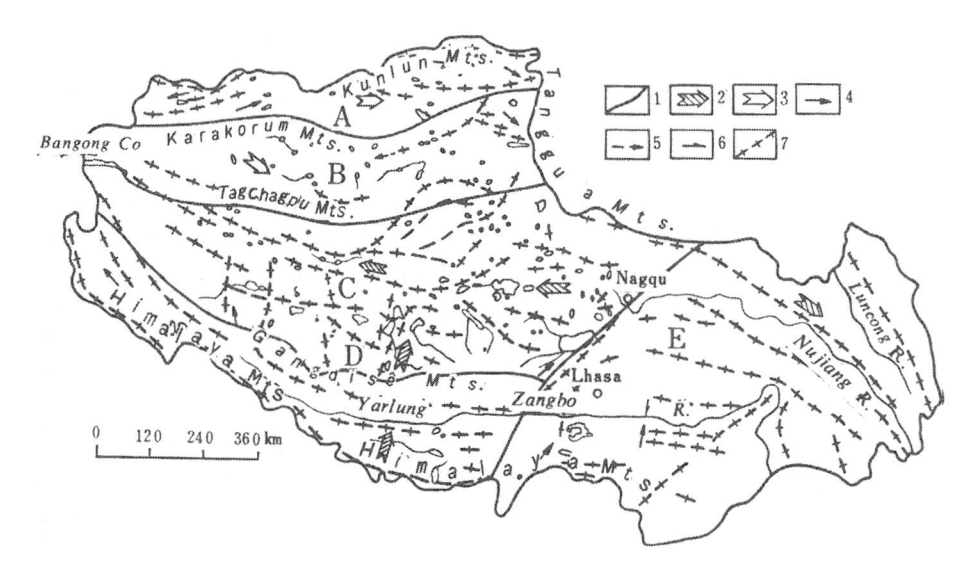

Figure 1.9. Quaternary lake districts and lake migration in Tibet. A—Hoh Xil inflow lake district; B—Qiangtang (outflow) inflow lake district; C—North Tibet inflow (outflow) lake district; D—South Tibet inflow–outflow lake district; E—East Tibet outflow lake district. 1—Boundary of lake district; 2—Major direction of lake migration or drainage of lake water; 3—Major direction of lake migration (inferred); 4—Minor direction of lake migration; 5—Minor direction of lake migration (inferred); 6—Minor direction of drainage of lake water; 7—Mountain range.

(1) The drastic differential tilting of the Himalaya, Gangdise–Nyainqêntanglha, Kunlun and Altun–South Qilian mountains and their variable uplift along the Yang (Yangbajain)–Du (Dulan) NE–trending fault zone resulted in the successive loss of a part of the lake basins on the edge of the plateau that had formed during the second stage. The sedimentary layers were nearly horizontally uplifted, but were generally upfolded. The tilted surface on the northern slope of the Himalayas in the southern part of the plateau was upwarped from the south and southwest to the north and northeast (Chen et al., 1981; Huang et al., 1980). This caused the water in the ancient lakes to drain to the north and northeast. As a result, the Zanda, Gyirong, Pagri and Qungdo'gyang ancient all lost their water during this stage. The waters of the ancient lakes in eastern Tibet, western Yunnan and eastern Qaidam began to drain away to the east–southeast along the NE–trending Yang–Du fault and gradually dried out completely.

(2) The lakes in some large fault–block basins, such as the Qaidam and Gonghe basins, reflected the trend of migration to the southeast or east. For example, the depocentre of the Qaidam ancient saline lake basins migrated from the

Chahansilatu Playa in the northwestern part towards Yiliping Playa and Suli Lake.

(3) New lake systems generally formed along the suture zone, and along several nearly E–W downfaulted zones and depressions. These formed the basis of the pattern of lake swarms scattered over the modern Qinghai–Tibet Plateau. The Tanggula–Amdo ancient lake uplift, where the Quguoquan Formation in the Pliocene–Lowest Pleistocene consists of fine–grained lacustrine deposits, has a peneplained top at an altitude of 5010 m above sea level. There, the ancient lake water has drained away to the southeast, and the southwestern denuded area has subsided to give rise to the Nam Co river–lake system.

The drastic change of the lake basins on the plateau in the middle and late parts of the Early Pleistocene corresponds to the very thick molasse deposits (the Xiyu conglomerate, Upper Xiwalike Formation, etc.) in the peripheral pasts of the plateau. Sand and gravel beds (the Gongba conglomerate in southern Tibet and Zhutoushan conglomerate in northern Tibet) formed in the interior part of the plateau. This suggests that this period of tectonism (which has been assigned to the third phase of collision, known as the Qinghai–Tibet movement) caused drastic uplift of the Qinghai–Tibet region. During this period the intracontinental opposing convergence changed from a horizontal movement into a vertical differential movement within the plateau.

The remarkable uplift–subsidence that occurred between the main part of the plateau and the Qaidam fault–block basin led to the deposition of the very thick Early–Middle Pleistocene lacustrine (mainly silt and clay) sediments in the Yiliping–Qarhan area of the Qaidam basin. There were deep and moderately deep, fresh water and brackish lakes in the Qaidam basin. Saline lakes were still more or less extensively developed in the area of the Dalangtan Playa, Kunteyi Playa and Mahai Playa in the northwest of the basin. Here, cold facies deposits of astrakhanite, and even mirabilite, have been observed in addition to the abundant gypsum–salt beds. This suggests an arid and cool climate. From this area northward, there was an alternating cold–warm climate in the vast plateau region as a result of the dramatic uplifting and the three glacial–interglacial cycles. This led to the formation of the glaciofluvial–lacustrine deposits that are widespread on the ancient plateau surface. However, at that time the Himalayan mountains to the south had not been uplifted sufficiently to block the warm and humid air currents from the south. Hence, precipitation was abundant, resulting in fresh water arenaceous and argillaceous deposits in most of the lake basins in this part of the region, except in the area of the Tuotuo River where saline gypsiferous sediments were deposited locally.

3.3.4. MIDDLE–LATE PLEISTOCENE TO HOLOCENE

At the start of the Late Pleistocene, the lake basins on the plateau were subject to great change once again. As a result, the depocentres of the lake basins in the inte-

EVOLUTION OF CENOZOIC LAKE BASINS AND THE FORMATION   49

rior of the plateau underwent their third migrational stage. This lead to an unconformable contact between the Late Pleistocene strata and the underlying formations that can be observed in the marginal parts of some lake basins. As the lakes on the plateau are chiefly of tectonic origin, the lakes of various sizes in different fault–block areas show similar migrational features in the different lake districts (Zheng et al., 1983). Hence, the approximate migration direction of the ancient lake waters in the various lake basins on the plateau can be traced by the well preserved lake basin geomorphology and sedimentary features of the different lake districts (Figs. 1.9 and 1.10).

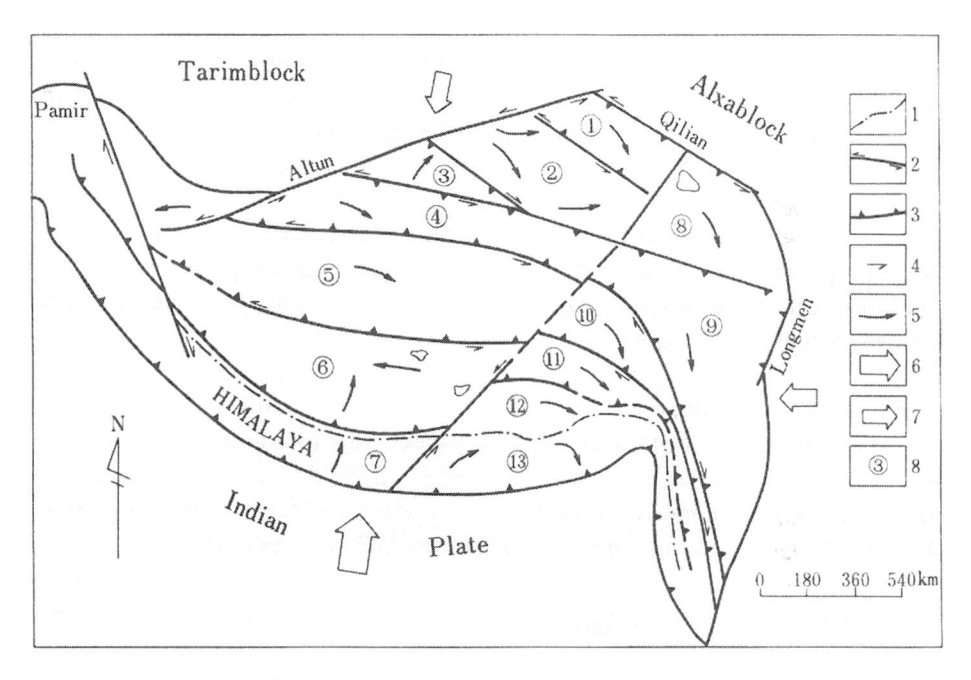

Figure 1.10. Migration of lake districts since the Late Pleistocene on the Qinghai–Tibet Plateau. 1–Yarlung Zangbo suture line; 2–Strike–slip fault; 3–Thrust fault; 4–Migration direction of fault; 5–Direction of lake water migration; 6–Direction of principal compressive stress; 7–Direction of minor compressive stress; 8–Lake district No.: (1) Qilian inflow lake district; (2) Qaidam inflow lake district; (3) Kumkol inflow lake area; (4) Hoh Xil inflow lake district; (5) Qiangtang inflow lake district; (6) North Tibet inflow lake district; (7) South Tibet outflow–inflow lake district; (8) Qinghai Lake–Gonghe outflow–inflow lake district; (9)–(13) East Tibet (inflow) outflow lake district.

Qinghai Lake–Gonghe outflow–inflow lake district
In the latter part of the Late Pleistocene, the Gonghe basin rose in the north and the ancient lake water was drained southeast. The Caka sub–basin became enclosed by a local rise in the western part of the district, giving rise to the Caka Salt Lake (104 km^2 in area) in the eastern part. There were only local depressions in the sand dune

area to the south of Sazhubxi River, which resulted in the formation of a small saline lake (Dalian Lake), which is the eastern most saline lake on the Qinghai–Tibet Plateau. The Qinghai Lake basin was then transformed from an exoric lake into and endoric one. This evidence indicates that this lake district migrated from the northwest toward the southeast.

Qaidam inflow lake basin
The evolution of the depocentre of Late Pleistocene–Holocene saline sediments is characterised by the succession of Dalangtan to Yiliping to Dong (East) and Xi (West) Taijnar Lakes to Qarhan. Hence, it can be seen that in the Late Quaternary the Qaidam lake basin migrated from the northwestern part of the basin to the eastern–southeastern parts and from the southern part to the northeastern part. There are traces to indicate that it continued to migrate from the west to the east, or from the southwest to the northeast, as recently as the Holocene.

Kumkol inflow lake district
The northward advancement and development of the Ayakkumkol Lake and the subsequent formation of the Quaternary lake terraces, which are high in the south and low in the north (Zhao, 1984), suggests that the ancient lake water migrated northward in the Quaternary.

Hoh Xil inflow lake district
In this area, the recent lake basins mostly advanced from west to east (Fig. 1.9).

Qiangtang (outflow) inflow lake district
Here, the recent lake basins mostly migrated from northwest to southeast.

North Tibet inflow (outflow) lake district
This lake district is characterized by the obvious migration of lake basins from south to north and from east to west. On the northern edge of the Gangdise Mountains, a series of lakes lie at the north end of a nearly N–S ancient lake basin. In the central part, the traces of westward migration of the ancient lake water are recognisable in the lake basin along the Hei–A Highway. Multi–step broad and gentle terraces and large areas of lacustrine plains are developed in the eastern and southern parts of the lake areas of Siling Co–Nam Co and Dong Co, whereas in the western and northern areas the terraces are steeper and fewer in number. The base terraces stand like walls by the lakeside. The lake areas were uplifted in the east and south and downthrown in the west and north as a result of the neotectonic movements. There was an apparent rhythmic succession of stage–by–stage contraction and separation from the marginal to the interior part of the lake system.

Taking the Siling Co–Nam Co area as an example; in the early and middle parts of the Late Pleistocene the two lakes remained connected as one exoric lake with a total area of 28,900 km^2. In the Late Pleistocene (ca 30 Ka) the lake contracted, leading to the latest massive regression, which in turn resulted in the separation of Dung Co from the Nag Co outflow lake swarm. This is referred to as "the regression stage of Dung Co". During this stage a lacustrine sand–gravel terrace, or base terrace (order III) formed to a height of ~ 200 m. This regression ended the Nam Co–Siling Co area as an outflow lake system. As the Siling Co–Bangkog Co area became closed, it was the first lake area to develop into a brackish lake. The second massive regression occurred at ca 16 Ka, and is referred to as "the regression stage of Siling Co", and is marked by the separation of Bangkog Co from Siling Co and the formation of the 126 m high saddle–shaped terrace of lacustrine sand and gravel containing calcareous material (order II –10). This regression caused a drastic change in hydrogeological conditions at Bangkog Co and a sharp decrease in its fresh water replenishment, hence the lake became to develop towards saline conditions and was the prelude to the generation of saline lakes in Tibet. The third massive regression took place at ca 5.2–3.6 Ka, and resulted in the universal formation of inter–lake I–1–3 subterraces in all the lake systems in Tibet, the disintegration of sub–basins into progressively smaller lake basins, and the further salinization of the lake basins. For example, Bangkog Co contracted and disintegrated into Bangkog–I Co, Bangkog–II Co and Bangkog–III Co. This led to the formation of the inter–lake, 14–27 m high lake–shore I–3–step terraces of hydromagnesite and clay between the lakes and caused the substantial reduction of water sources for the Bangkog I and II lakes and their subsequent development into playas. This stage is referred to as "the regression stage of Bangkog Co".

South Tibet outflow–inflow lake district.
This lake was sandwiched between the Himalaya and Gangise Mountains. The upwarping of both mountains systems from the south (Yang and Liu, 1974) caused a general trend of northward migration of the lake systems there. The Yarlung Zangbo River formed the middle boundary of this migration as the river had intensely subsided and incised downward along a fault.

East Tibet outflow (inflow) lake district
The district may have been controlled by the Sanjiang (Nujiang–Jinsha–Lancang rivers) arcuate structure. The lakes were totally drained to the southeast as a result of the headward incision of the three rivers (Fig. 1.9).

These changes in the different lake districts of Tibet since the Late Pleistocene indicate that the Qinghai–Tibet Plateau is characterised by an overall uplifting of the plateau and differential movement of fault blocks (Fig. 1.11).

The middle–late Late Pleistocene to Holocene tectonic movement is characterised by an upward vertical motion that inherits the intracontinental opposing convergence, and is referred to as the fourth phase of collision. It is within this stage that the most important thermal event on the plateau since the

Quaternary occurred. Because of the thrusting, accompanied by stacking and thickening of crustal fault slabs, and the intensification of thermal expansion (associated with intense hydrothermal activity), the uplifting of the plateau showed a tendency of marked acceleration. Following the intensified uplifting, the Himalaya and Gangdise mountains became higher and consequently barred off the warm and moist air from the south, thus allowing the Central Asian arid belt to expand to the central and southern parts of the plateau. The western saline lake area of the Qaidam Basin, that had long been in existence, expanded southeastward to the Taijnar Lake at ca. 31 Ka; Qarhan Playa was transformed from a fresh water–brackish lake into a saline lake at ca. 25 Ka (Chen, 1984); Da Qaidam Lake, on the northern margin of the Qaidam basin, became a self–precipitating saline lake at ca. 20 Ka (Huang, 1980); later Bangkog Co on northern Tibet changed to a saline lake at ca. 19 Ka; Chagcam Caka became a self–precipitating saline lake at ca. 11 Ka (Zheng et al., 1983) (Table 1.2). However, in southern Tibet saline lakes were not formed until the last several hundred years.

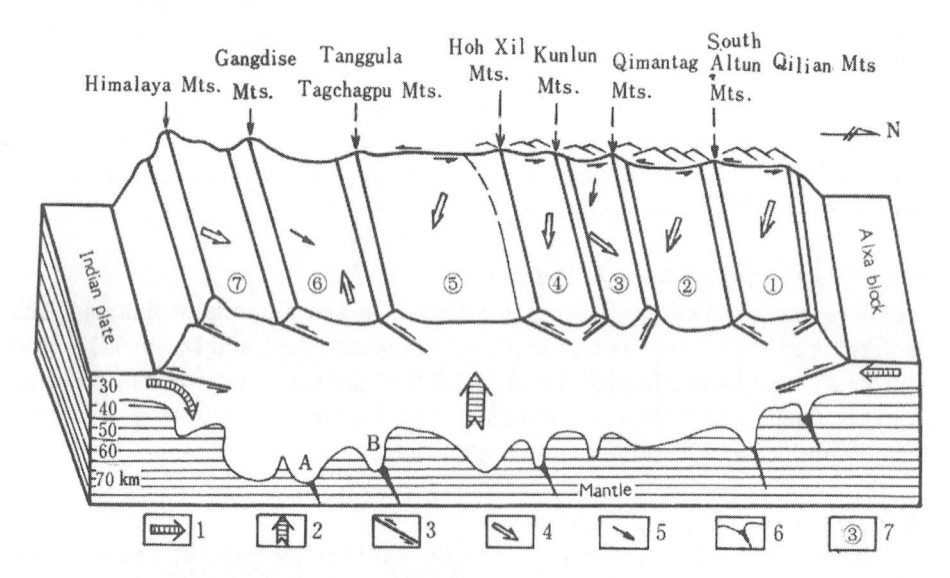

Figure 1.11. Ideal model of the Plateau's uplifting during the Quaternary. 1—Movement direction of plate or block; 2—Direction of upper mantle adjustment and plateau's uplifting; 3—Movement direction of fault; 4—Main direction of lake system migration; 5—Minor direction of lake system migration; 6—Late Jurassic (A) and Early Tertiary (B) suture zones; 7—Block No.: ①—Qilian block; ②—Qaidam block; ③—Kumkol block; ④—Hoh Xil block; ⑤—Qiangtang block; ⑥—Northern Gangdise block; ⑦—Southern Gangdise block.

In conclusion, the formation of saline lakes on the Qinghai–Tibet Plateau can be traced back to the first development stage of the lake basins in the region and it has continued discontinuously until the present. With the lapse of time the salt—forming basins showed an "accordion—type process of reciprocal expansion— contraction and migration in terms of their spatial distribution. Overall this resulted in:

—widespread distribution of lakes from the northwest (Qaidam Basin—Qilian Mountains) to the southeast (Qamdo) during the Eocene—Oligocene

—northwestward migration and contraction of lakes in the Miocene

—expansion of lakes from the northwest (Qaidam Basin) to the south (Qiangtang—south of the Yellow River) in the Pliocene

—another contraction of lakes in the Qaidam Basin in the Quaternary (Q_1–Q_3^1)

—an advance of lakes from Qaidam Basin to the southern part of the plateau at ca. 31—25 Ka.

It should be emphasised that at present the principal salt—forming processes in the saline lakes of the Qaidam Basin have terminated and most of the lakes there have evolved to the stage of shrivelled or extinct playas, or sand—covered lakes. In contrast, salt—forming processes in the North Tibet, Qiangtang and Hoh Xil lake districts, that lie on the plateau surface, are still active. Since the Holocene a considerable number of lakes in these districts have successively reached the stage of strongly self—precipitating lakes. According to statistics (Table 1.1), the saline lakes in the North Tibet, Qiangtang and Hoh Xil lake districts contribute 19%, 56% and 32% respectively, of the total number of salt lakes on the plateau.

The characteristic features of the evolution of lake basins and the generation and development of saline lakes on the plateau indicate that, in comparison with the regional and areal climatic factors, global climatic evolution and the migration of its climatic zones have played a leading role in the formation of saline lakes. In addition, the uplift of the Qinghai—Tibet Plateau and the resultant local climatic changes have been important during certain stages. In particular, the late stage massive uplift of the Qinghai—Tibet Plateau was extremely important in the formation of saline lakes. The "accordion type" space—time distribution of the Cenozoic salt lakes in the plateau region is closely related to the changes in the Cenozoic palaeoclimatic zones in China and East Asia. For example, the southward extension of the ancient saline lakes in Qinghai and Tibet during the Early Tertiary was concordant with the expansion of the Chinese ancient saline lake zone, and the northward regression of the Miocene saline lakes is also coincident with the extensive northwestward migration of the ancient saline lakes in China (Zheng et al., 1975). The southward expansion of the saline lake districts in Qinghai and Tibet during the Pliocene is the result of the southward expansion of the Central Asian desert climate. The alternation of thick argillaceous beds and salt beds in ancient lake sediments also provides envidence for the controlling effect of global climatic changes. However, during the period from the late part of the Late Pliocene to the Holocene, the surface of the Qinghai—Tibet Plateau attained an average altitude exceeding 4000 m above sea level. This gave rise to an arid climatic region, completely different from other regions at the same latitude. At this stage the development of saline lakes on the plateau was characterised by a sudden change, although there were also dry—wet rhythmic changes. For example, when the world—wide Atlantic humid and hot period arrived (Li et al., 1983). The saline lakes on the plateau continued to become more saline (in northern Tibet) and developed into playas (in the Qarhan lake chain). These observations all suggest that the local factors arising from the uplift of the Qinghai—Tibet Plateau played a leading part in the formation of the recent plateau saline lakes.

Although it is beyond the scope of this monograph it is worth noting that the problem of whether high lake levels and low lake levels correspond, respectively, to interglacial or glacial epochs has been a subject of controversy in Quaternary geology and palaeoclimatology. The evidence above indicates that the last Quaternary high lake level roughly corresponds to the "Dali" interglacial epoch (otherwise called the terminal interglacial epoch) of ca. 36—25 Ka. The initiation stage of modern saline lakes corresponds with the arrival of the "Late Dali" (otherwise known as the "Rongbushi") glacial epoch of ca. 25—10 Ka. This suggests that glaciation led to a fall in lake water levels and the formation of arid air masses. Following melting of the glaciers in the interglacial epoch the lake waters levels rose, leading to high moisture contents the air masses.

CHAPTER 4 HYDROCHEMISTRY AND MINERAL ASSOCIATIONS OF SALINE LAKES

4.1. Salinity and pH Values

The salinity of lakes on the Qinghai–Tibet Plateau is closely related to the natural environment and in particular the climatic conditions for their evolution. The vast amount of data from 259 lakes and interpretation from satellite images show that there is a general decrease in the salinity of the lakes from the northern and northwestern parts to the southern and southeastern parts of the plateau. This is a coincident with the annual aridity (annual aridity = annual evaporation loss / annual precipitation) of the plateau (Fig. 1.12).

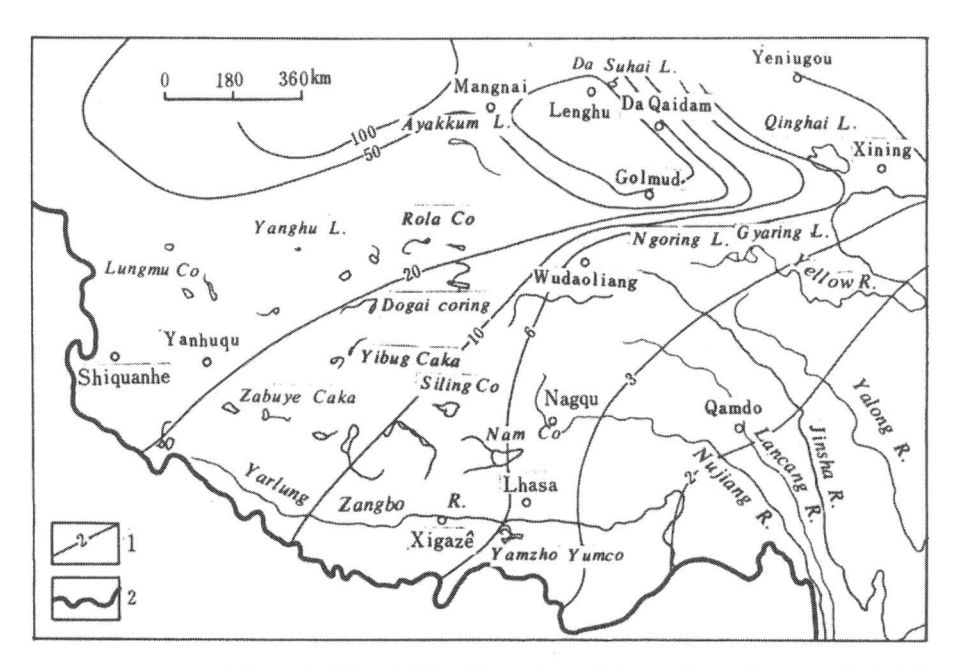

Figure 1.12. Annual aridity of the Qinghai–Tibet Plateau (compiled according to the meteorological data of 1980–1985 of the State Meteorological Bureau). 1. Isoline of annual aridity; 2. National boundary.

CHAPTER 4

As shown in Fig. 1.12, the annual aridity of the Qaidam Basin in the northern part of the Qinghai–Tibet Plateau may reach a maximum of 100. This basin has a long history of evolution and accumulation of saline lakes and ranks first on the plateau in terms of aridity. Hydrochemically, the basin belongs to subzones IV$_3$ and IV$_4$ (see Appended Map I; for hydrochemical zoning, see the next section). The salinity is highest in the ancient saline lake districts in the northwestern

TABLE 1.7. Averages of pH values and salinities of natural water in the I–V hydrochemical zones of lakes on the Qinghai–Tibet Plateau

Water type	Item	I	II	III	IV$_1$	IV$_2$	IV$_3$	IV$_4$	V
Saline lake	pH	9.06 (19)	8.21 (26)	7.78 (12)		7.62 (1)	7.35 (16)	6.64 (10)	7.90 (1)
	Salinity	156.66(33)	192.14(33)	145.85(13)	125.98 (4)	261.32 (1)	287.14(18)	352.68(10)	240.14(2)
Brackish lake	pH	9.33 (33)	9.02 (17)	8.35 (7)		8.10 (3)	8.95 (1)		8.25 (4)
	Salinity	9.71 (58)	13.76(23)	9.56 (10)	2.68 (1)	18.36 (3)	18.28 (3)		5.04 (4)
Freshwater lake	pH	8.06 (10)	8.61 (2)			7.95 (1)			7.46 (8)
	Salinity	0.47 (18)	0.70 (4)		0.73 (1)	0.89 (1)			0.27 (12)
Various kinds of lake	pH	8.82	8.61	8.07		7.89	8.15	6.64	7.87
	Salinity	55.61	68.87	77.71	43.13	93.52	152.71	352.68	27.88(18)
Geothermal water	pH	7.61(88)	6.44(13)	6.0 (1)		7.65 (2)		7.27 (172)	
	Salinity	1.64(100)	1.13(14)	28.52 (2)		1.86 (6)		2.03 (177)	
River water	pH	7.82 (19)	7.88 (13)	8.19 (5)		7.83 (3)			7.74 (8)
	Salinity	0.81 (36)	0.46 (15)	0.91 (6)		1.235 (28)			0.21 (11)
Groundwater (cold spring)	pH	7.83 (12)	7.52 (25)	7.86 (6)		7.24 (8)			
	Salinity	0.57 (12)	1.11 (29)	1.08 (6)		6.70 (55)			

Note: Numbers in parentheses refer to number of samples that take part in the statistics.

and central parts, e.g., the salinity of lake waters is as high as 555 g/l in Niulangzhiniu Lake, 408 g/l (surface brine, measured in 1968) in Dabsan Lake, and 382 g/l (intercrystalline brine) in Dalangtan Playa. On the eastern and northern margins of the basin the salinity tends to decline, e.g., the salinities in Da Suhai Lake and Dezong Mahai Lake are 32 g/l and 39.5 g/l, respectively. The salinity of the lakes in the basin is the highest on the Qinghai–Tibet Plateau. The average salinity of subzones IV_3 and IV_4 is 153–353 g / l (Table 1.7). The annual aridity decreases abruptly to 6–3 from the Qaidam Basin northward to the Qilian Mountains lake district (IV_2), where most of the lakes contain freshwater.

To the southwest of the Qaidam Basin, there is an ENE–trending area with an annual aridity of 20–50, whose extent is largely coincident with that of the Qiangtang lake district, the western sector of the Hoh Xil lake district, and the Kumkol lake district. Hydrochemically, it falls largely in the western sectors of hydrochemical zones IV, III and II (see Appended Map 1). The western sector of zone IV contains the Hoh Xil magnesium sulphate subtype subzone (IV_1), with an average salinity of 84.5 g / l. The lakes of zone III have an average salinity of about 77.7 g / l, but in its western sector the average salinity is more than 101 g / l. Although there has been limited investigation of the lakes in this area, playas have been found in many places, e.g., Kushuitan Playa where the salt bed is over 3 m thick (Regional Geological Party of Xinjiang, 1985), so the salinity in the western sector of zone III ranks second on the plateau. The average salinity of the lakes in zone III is 68.9 g / l, but the average salinity in its western sector is about 125 g / l. There has been more detailed investigation of this area than of zone III. Lakes that are known to have relatively high salinities in this area at present include Chagcam Caka II Lake (117–305 g / l for surface brine and 310–385 g / l for intercrystalline brine) and Langmari Lake (322–365 g / l for lake water).

To the east of the above mentioned area lies an area with an annual aridity of 20–6. This area largely covers the inflow lake district of northern Tibet and the inflow–outflow lake district of southern Tibet (Fig. 1.2). Hydrochemically, most parts of this area belong to the carbonate zone (I), including the high salinity lake subzone north of Gangdise (subzone I_2) and the low salinity subzone south of Gangdise (subzone I_1). The average salinity of the whole zone is about 55.6 g / l. As it took longer for lake closure to take place in northern Tibet (subzone I_2) the salt forming process is at an earlier stage than in southern Tibet, so the salinity in northern Tibet is commonly higher, averaging 157 g / l. The lake with the highest salinity in the area is south Zabuye Lake (the salinity of the intercrystalline brine is 512 g / l and that of the surface brine is 320–425 g / l). High salinities are also found in Bangkog Co II Lake (salinity of intercrystalline brine = 402 g / l), Zainzong Caka (salinity of intercrystalline brine = 416 g / l), Bilong Co (salinity of surface brine = 331 g / l) and Yaggai Co (salinity of surface brine = 196 g / l, intercrystalline brine = 324 g / l). Zabuye Lake, Zainzong Caka and Bangkog Co all have a relatively long history of salt formation, which is a major reason for their high salinity relative to their neighbouring areas. The lakes of the southern Tibet inflow–outflow lake districts (subzone I_1) have had a relatively long outflow time, and the aridity there is also low, so salts have not easily accumulated and the salinity of the lakes is generally low, averaging 17.3 g / l. The lakes with relatively high salinity are Malashan Lake (117 g / l), Paiku Co (18.3 g / l) and Yinhu III

Lake (6.9 g / l for the lake water).

East of the various above mentioned zones (largely with the western margins of the Qinghai Lake — Wudaoliang and Yamzho Yumco as the boundary), there is an area with an annual aridity lower than 6. It includes the Qinghai–Gonghe, Bayan Har and eastern Tibet lake districts, corresponding to the sulphate subtype discharge area (zone V). Most of the ancient lake basins in this area have been captured by outflow rivers and there are still outflow lakes and many glacial lakes. Hence, the salinity of the lake waters in this area is the lowest on the Qinghai–Tibet Plateau, averaging about 27.9 g / l (Table 1.6). The exceptions are for a few small saline lakes and brackish lakes with relatively high salinities (e.g., the levee lake on the edge of Gya Lake, Hajiang Salt Pond and China's largest salt lake—Qinghai Lake, with a salinity of 12.4 g / l). The glacial lakes have the lowest salinities, e.g., the salinity of Buchong Co in Q'amdo is only 0.057 g / l (Guan et al., 1984). In addition, the salinity of the discharge lakes is also low, e.g., the salinity of Ngoring Lake is 0.376 g / l.

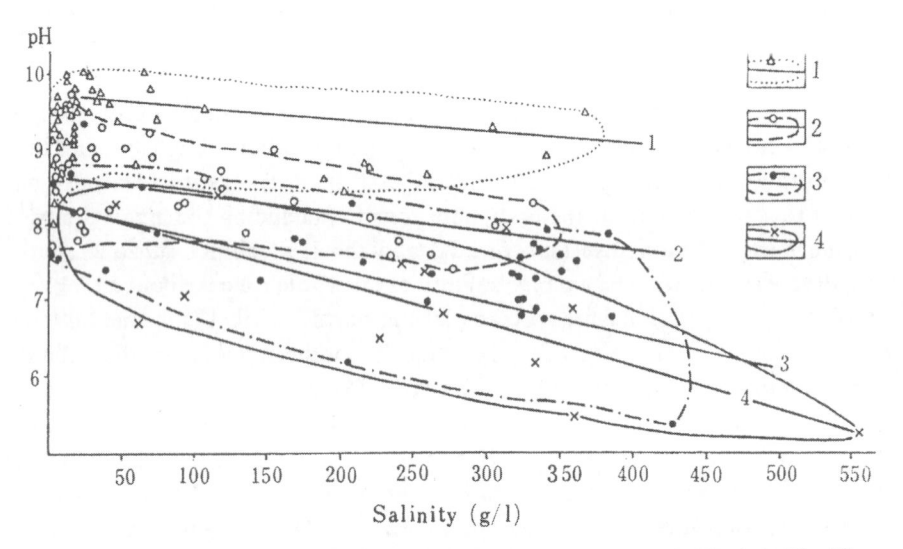

Figure 1.13. Relation of pH to salinity and hydrochemical types of lakes on the Qinghai–Tibet Plateau. 1. Carbonate type; 2. Sodium sulphate subtype; 3. Magnesium sulphate subtype; 4. Chloride type.

The pH value is also an important characteristic of lake waters and determines the chemical processes that occur in natural waters. The pH of a lake is also closely related to its hydrochemical type and thus largely determines the nature of the saline mineral association forming salt sediments. Studies indicate that the pH of lake waters (saline and brackish lakes in particular) on the Qinghai–Tibet Plateau decreases in response to the change in their hydrochemical state from carbonate type to sodium sulphate subtype to magnesium sulphate subtype to chloride type. For lakes with different salinities, the pH of brackish lakes is the highest, and the pH of the freshwater lakes is higher than that of saline lakes (Table 1.6). An inverse correlation exists between the pH and salinity of various types of saline lakes, and this inverse correlation exhibits regualr changes in response to different hydrochemical types, i.e., the slopes become higher in the order of carbonate type to sodium sulphate subtype to magnesium sulphate subtype to chloride type (Fig. 1.13). As shown in the plot, the slope of carbonate type in the region is relatively low and the pH of the lake waters only decreases by a small amount with increasing salinity. This is because the boron content of lakes of the carbonate type in the region is higher than that of saline lakes of other types and boron acts as a buffer preventing large decreases in the pH.

The characteristics discussed above lead to an increase in the pH of lake waters of the Qinghai–Tibet Plateau from north to south (Table 1.7). In the chloride–sulphate zone (zone IV), the pH of the lake waters is lowest, e.g., the lake pH of the Qaidam Basin averages 7.7, with most lake waters being neutral, with a few weakly alkaline or acidic (Table 1.6)①. In the chlorde type subzone (IV$_4$), the pH of the saline lakes is relatively low, e.g., 6.2 for Kunteyi, 7.0 for Qarhan and 5.5 for Xiezhu Lake. From Qaidam southward to the chloride–sulphate subtype zone (zone III), the pH averages 8.0, and the lake waters are mainly weakly alkaline and in a few cases alkaline, e.g., the pH is 9.5 in Ma'ergai Caka and 9.3 in lake No. 419. Further south in the sodium sulphate subtype zone (zone II), the pH averages 8.6 and the lakes are dominantly slightly alkaline and in a few cases strongly alkaline. There are no weakly acidic or acid lakes in this zone. Even further to the south are the carbonate type zones north of Gangdise (zone I$_1$) and south of Gangdise (I$_2$). Here the average pH is 8.8 and the lakes are mainly alkaline, while a few are strongly alkaline with pH > 10, e.g., the pH is 10.2 for Kyêbxang Co and Sênco and 10.1 for Damxung Co. No weakly acidic or acid lakes are found in these zones. Finally, in the sodium sulphate subtype zone (zone V) located in the east the averages pH is 7.9, and the lakes are mainly neutral and in a few cases weakly alkaline (e.g. the pH of Qinghai Lake is 8.8) and weakly acidic (e.g. the pH of the glacial Yi ong Lake is 5.6).

4.2. Hydrochemical Types and Their Zoning

---

① The pH ranks referred to in this book are as follows: acid, pH = 3–5; weakly acid, pH = 5–6.5; neutral, pH = 6.5–7.5; weakly alkaline, pH = 7.5–8.5; alkaline, ph = 8.5–9.5; strongly alkaline, pH > 9.5.

It is well recognised that the common ions in lake waters are Ca^{2+}, Mg^{2+}, Na^+, K^+, CO_3^{2-}, HCO_3^-, SO_4^{2-} and Cl^-. However, the proportions of the various ions show marked variations resulting from differences in salinity and geological setting. Various classifications of hydrochemical types based on compositional variations have been proposed, but in this discussion we will use the Kurnakov–Valyashko classification. The carbonate type is further divided into three subtypes (Ellis et al., 1977), according to the different total alkalinities;

$$KC = \frac{Na_2CO_3 + NaHCO_3}{\Sigma_{salts}} \times 100\%$$

and saline mineral assemblages. The large amount of data available from 259 lakes on the Qinghai–Tibet Plateau (see Appended Table 1. Hydrochemical compositions of lakes of the Qinghai–Tibet Plateau) give the following percentages of lakes in the different hydrochemical classes: 46% of the lakes are of the carbonate type (of which 12.5% are of strong carbonate subtype, 16.9% moderate carbonate subtype, 23.4% weak carbonate type, and 8.9% undivided subtype), about 25% of the lakes are of sodium sulphate subtype and only 6.5% are of the chloride type. If the percentages are calculated on the basis of lake area, those of magnesium sulphate subtype are the most abundant and those of the chloride type are also account for a higher proportion of the total.

The lakes in the saline, brackish and freshwater hydrochemical types show regular distributions. From north to south four zones and one area can be distinguished. Eash zone or area may be further subdivided into several subzones, however, there is insufficient data to divide all the zones in this manner. From north to south the hydrochemical zones are successively; the sulphate–chloride type zone (zone IV), the chloride–bearing magnesium sulphate subtype zone (zone III), the high salinity sodium sulphate subtype zone (zone II), and the carbonate type zone (zone I) (which includes the high salinity lake subzone (I–2) north of Gangdise and the low salinity lake subzone (I–1) south of Gangdise). To the east of these four zones is the sodium sulphate subtype (low salinity) discharge area (area V). They are described here from north to south and from west to east as follows (Appended Map 1).

Sulphate–chloride type zone (zone IV)

This zone is nearly rhombic in shape (575 km long and 125–400 km wide) and is distributed in an E–W direction. It includes the Kumkol magnesium sulphate subtype subzone (IV_1), the sodium sulphate subtype subzone (IV_2) on the peripheries of Qaidam, the Qaidam magnesium sulphate subzone (IV_3) and the chloride subzone (IV_4) distributed in Kunteyi and Huobuxun. Dat on water chemical compositions were first obtained during the investigation of subzone IV_1 in 1984. Among six lakes investigated in this subzone, five of are the magnesium sulphate subtype, of which the larger lakes include Ayakkam Lake (No. 636 on Appended Map 1 and Appended Table 1), Aqqikkol Lake (No. 601) and Jingyu Lake (No. 602), which are all saline lakes. Only the brackish Dashaziquan Lake (No. 633) — a small lake on the margins of Ayakkum Lake — is of the sodium sulphate subtype. There are also brackish lakes adjacent to Beileke Lake and Xianhu Lake. The surrounding areas of the Qaidam basin belong to subzone IV_2; to the south many glacial lakes are scattered in the vast Kunlun Mountains, of which only a few are saline or brackish lakes. To the north there are freshwater lakes, such as the Hala Lake situated in the remote lofty Qilian Mountains, but there are no water chemistry data available for these lakes. Only Da Suhai Lake (No. 608) and Xiao Suhai Lake (Nos. 609), Xiligou Lake (No. 630), Mangnai Lake (No. 605) and Hurleg Lake (No. 626) are known to be of the sodium sulphate subtype, belonging to saline, brackish or freshwater lakes. Most of subzone IV_3 are distributed in the Qaidam Basin. There are a total of 21 lakes in this subzone; of which 16 are of the magnesium sulphate subtype (all of which are saline), including the famous East and West Taijnar Lakes (Nos. 612 and 611), Da Qaidam (No. 619) and Xiao Qaidam (Nos. 620), and Dalangtan Playa (No. 604); 4 lakes are of the sodium sulphate subtype or sulphate subtype, all of which are distributed along the margins of the salt precipitation area, including Gansenquan Lake (No. 640), Nabei Lake (No. 641) and Tart Lake (No. 642) (all three are saline lakes) as well as Solon Lake (No. 627) (which is a brackish lake); Dongling Lake (No. 639) is of the chloride type. Subzone IV_4 is distributed in two opposite corners northwest and southeast) of the rhombic Qaidam Basin (Appended Map 1). Kunteyi Lake (No. 606) and Balun Mahai Lake (No. 614) are located in the northwestern corner and South and North Huobuxun Lakes (Nos. 623 and 624) and Xiezuo Lake (No. 622) are located in the southeastern corner. The two chloride saline lakes districts are not the result of normal concentration chemical differentiation, nor are they encircled by the above—mentioned sulphate subzone (IV_3), but they are situated on the eastern side of two magnesium sulphate saline lake series, i.e., Dalangtan Playa—Qagan Silatu Lake (No. 637) and Yiliping Playa (No. 610; its northern part contains Qagan Silatu) to the East and West Taijnar Lakes (Nos. 612 and 611) —Dabsan Lake (No. 618). This asymmetric hydrochemical differentiation pattern is the hydrochemical manifestation of the effect of tectonic collision on modern saline lakes. It also vividly reflects another aspect; namely the migration of saline lakes from west to east and from west and northwest to southeast in the Qaidam Basin. In addition, the lake waters in the two chloride type subdistricts may be mixed with deep—seated calcium chloride—bearing water that ascends along fractures. Such directional hydrochemical differentiation and mixing are still proceeding now.

Chloride—bearing magnesium sulphate subtype zone (zone III)
This zone extends in a near E—W direction and is located south of the Kunlun Mountains. It is up to 1800 km long and 75—200 km wide. Its main part is repre-

sented by the Hoh Xil lake district extending from Lungmu Co (No. 401) and Kushui Lake eastward through Margai Caka (No. 407) and Yupan Lake (No. 416) to the 84 Salt Lake (No. 421) on the Qinghai–Tibet Highway. In this zone lakes of the magnesium sulphate subtype are dominant, but there are some lakes of the chloride type concentrated in the area between Taiku Lake (No. 413) and Dogai Coring Lake (No. 417) in the central part. Their origin will be discussed in Section 5 of this chapter.

Sodium sulphate subtype zone (zone II)
Lying largely south of the line between Lungmu Co and the Tuotuo River, this zone has a "T–shape" that results from the southward protrusion from the central part of the east–west zone. It is up to 1500 km long and generally 75–200 km wide, and the protruding part in the center is as wide as 440 km. The zone extends from Bangong Co (No. 301–1) eastward through Chagcan Caka (No. 308–310), Gomo Co (No. 335) and Margog Caka (No. 342) up to Laorite Lake (No. 360) and Qoima Co. In this zone lakes of the sodium sulphate type are dominant, but there are many saline lakes of the carbonate and magnesium sulphate subtypes (Chagcam Caka), showing that this district is a hydrochemical transition area. At the northwestern end of this zone there is a group of lakes of the carbonate type distributed in a nearly N–S direction. From south to north the lakes are Aru Co (No. 307), Mêmar Co. (No. 306) and Luotuo Lake (No. 305). To the north of these lakes there is a saline lake (Shuogu Lake) of the carbonate type that lies to the north of the Ka'erdaxi volcanic area at the southern bending portion of the Kunlun Mountains. Hence, a small local carbonate type area, distributed in a N–S direction, likely exists in this area. However, the data are scarce so no detailed division will be made at this time.

Carbonate type zone (zone I)
The western sector of this zone is located south of the Hei–A Highway. Its central sector is shifted south and its eastern sector is shifted north, roughly falling south of the foot of the Tanggula Mountains. The southern boundary of the carbonate type zone is the Himalayas. Hence, it resembles an asymmetric saddle, 1350 km long and approximately 200–250 km wide, but its central part is widest at up to 650 km. This zone is mainly of the carbonate type, and may be further divided into the northern and southern subzones, with the Gangdise Mountains as the boundary (Zheng, 1983). The northern subzone (entitled the north of Gangdise high salinity lake subzone I_2) extends from Mandong Co (No. 201) eastward through such boron lakes as Chalaka Lake (No. 211), Zabuye Caka (No. 214), Bangkog Co (No. 251) and Dujiali Lake (No. 240) up to Co Nag (No. 275)–Co Ngoin (No. 277). The salinity of this subzone is far higher than that of the southern subzone. The lakes are mainly of the carbonate type (including three subtypes). In addition, there are a few scattered lakes of the sodium sulphate subtype, and occasionally lakes of the magnesium sulphate subtype may be found, e.g., Langmari Lake (No. 202) on the

western boundary. The southern subzone is termed the south of Gangdise low salinity lake subzone (I_1). Its salinity is far lower than that of other zones (subzones). It extends from the famous Mapam Yumco (God Lake, No. 002) and La'nga (Ghost Lake, No. 001) eastward through Kunggyü Co (No. 003) and Paikü Co (No. 004) up to Yinhu Lake (Nos. 010−011). Lakes of the carbonate type are dominant, although they are occasionally intercalated with a few lakes of the sulphate type. Among the 19 lakes investigated in this subzone, 12 are brackish lakes, 5 are freshwater lakes and 2 are saline lakes. The brackish lakes are relatively large, with Ghost Lake and Paikü Co covering areas of 265 and and 250 km^2, respectively. The freshwater lakes are relatively small and generally of glacial origin, but fewer investigations have been conducted on them. Of these, only God Lake is of any significant size, covering an area of 394 km^2 and is the largest lake in this subzone. A few lakes in this subzone have developed into saline lakes (contrary to the view of some people that there is no saline lakes in southern Tibet). However, the saline lakes are small, only 1 km^2 or so, e.g., Malashan Lake (No. 012) has a salinity of 117 g / l and Cocholung (No. 017) has a salinity of 154 g / l. In addition, Yinhu III Lake is an ephemeral lake and the lake waters may completely evaporate, precipitating borax from the lake waters during the dry seasons.

Sodium sulphate subtype discharge area (V)
The area lies east of the above mentioned four zones, with the NE−trending Yangbajain−Dulan fault as the boundary. This area comprises Sanjiang (Three River area) and the middle and lower reaches of the Yarlung Zangbo River in the eastern part of the Qinghai−Tibet Plateau. The area extends from Qinghai Lake (China's largest inland lake) southwestward through Gyaring Lake (No. 513) and Ngoring Lake (No. 514) up to Yamzho Yumco (No. 504)−Puma Yumco (No. 503). Most parts of this area belong to an outflow area and the ancient lakes have mostly been completely emptied and have become dry valleys and locally transformed into inflow lakes. They are large and some of them have become saline and brackish lakes. Qinghai Lake has an area of up to 4635 km^2 and a salinity of about 12.8 g / l; Caka Lake (No. 519), a residual of the Qinghai Lake−Gonghe basin, has an area up to 104 km^2 and a salinity of 323 g / l. There are still a few outflow freshwater lakes in the area, e.g., Gyaring and Ngoring lakes in the upper reaches of the Yellow River with salinities of 0.48 and 0.38 g / l, respectively. The hydrochemical types in this area are dominated by the sodium sulphate subtype, with a few lakes of the magnesium sulphate and carbonate type.

4.3. Chemical Compositions

Typical lakes on the Qinghai−Tibet Plateau have been studied. The compositions of the brines have been analysed by several different methods. In addition to the major elements, trace elements have been analysed by flame spectrophotometer (Li, Rb, Cs etc.), ion spectrometry (REE, non ferrous metals, etc.; some samples were pretreated for salts) and atomic absorption spectrometry (Sr, Ba etc.). Some samples were checked using two different analytical methods. The results are discussed here as follows.

4.3.1. COMPOSITION OF LAKE WATERS

At present 56 elements have been analysed in lake waters of the Qinghai–Tibet Plateau, of which 15 of these were recently analysed for the first time. They are arranged according to their position in the periodic table as follows.

I A: H, Li, Na, K, Rb, Cs
II A: Be, Mg, Ca, Sr, Ba
III A: B, Al, Ga
IV A: C, Si, Ge, Sn, Pb
V A: N, P, As, Sb
VI A: O, S
VII A: F, Cl, Br, I
I B: Cu, Ag
II B: Zn, Cd
III B: Sc, Y, La, Ce, Nd, Sm, Gd, Dy, Yb, Th, U
IV B: Ti, Zr
V B: V, Nb, Ta
VI B: Cr, Mo, W
VII B: Mn
VIII B: Fe, Co, Ni

Of these elements, the concentrations of Na, K, Mg, Ca and Cl and SO_4^{2-}, CO_3^{2-} and HCO_3^- are the highest. Their total concentrations generally account for 97%–98% of the ion equivalents of lake waters, except for a few of the saline lakes. The high concentrations of B and Li (and complex anions) in some of the saline lakes may account for more than 2% of the total ion concentrations in some brines. These brines constitute a special case and are contained within borate– and lithium–rich saline lakes.

Relative to seawater the brackish and saline lakes on the plateau are enriched in a number of elements (after normalisation to a seawater salinity of 35 g/l) (Zheng, 1985). For example, the rare earth elements, As, Cs, and Nb may be more than 1000 times more enriched in the brines relative to seawater, Li, W, CO_3^{2-}, Be, Cr, Ga, U, Th, Bi, Y, Ta and Zr may be 100 times more enriched in lakes and B and Rb may be a few tens of times more enrched in the lakes. In some saline lakes, saline sediments (chloride, sulphate, and borate) or brine contain economically important concentrations of Li, B, Cs, K, Br, Rb, Sr, Mg and Na.

Table 1.8. Average chemical composition in lake waters from hydrochemical zones of the Qinghai–Tibet plateau (in mg/l)

Aver. Composition	Zone	I₁	I₂	II	III	IV₁	IV₂	IV₃	IV₄	V
Cations (mg/l)	Na	1679.23(13)	21295.469(72)	38973.409(51)	28448.597(21)	28734.883(6)	43520.001(2)	58611.683(23)	52659.84(5)	9323.772(9)
	K	800.27(13)	3551.953(72)	7185.418(51)	1249.337(21)	686.927(6)	1030.001(2)	4993.139(23)	3324.28(5)	516.832(9)
	Ca	59.05(15)	25.729(88)	158.557(56)	793.582(21)	115.098(6)	276.000(2)	6117.543(23)	5827.28(5)	35.842(14)
	Mg	828.65(15)	734.327(89)	3497.493(57)	3146.333(21)	5758.59(6)	6510.001(2)	27472.596(23)	21706.68(5)	1992.644(14)
	Li	7.92(9)	184.483(58)	106.611(45)	37.106(17)	27.252(5)	8.5(1)	87.725(21)	14.283(3)	4.425(2)
	Rb	0.03(6)	7.771(20)	14.15(6)	0.34(3)	1.336(5)	3.6(1)	5.066(16)	0.63(1)	0.23(1)
	Cs	<0.1(2)	15.463(8)	7.144(5)	0.12(3)	2.204(5)	0(1)	0.035(12)	0(1)	0(1)
	Sr	0.03(6)	0.738(25)	4.084(15)	11.333(3)	6.425(4)		12.933(6)	399.0(1)	
	As	3.78(3)	17.865(11)	2.033(4)		1.16(2)		0.09225(2)		
	U		1.085(7)	0.047(3)						
	Si	3.74(4)	9.333(9)	7.258(6)				2.8(5)	0.40(1)	0.9(1)
Anions (mg/l)	CO₃	261.58(15)	3914.673(89)	999.531(57)	32.942(21)	224.752(6)	225.001(2)	126.862(21)	125.303(4)	47.692(14)
	HCO₃	384.89(15)	1543.430(87)	769.184(57)	260.62(21)	451.317(6)	697.13(2)	305.481(21)	658.375(4)	155.963(14)
	SO₄	5683.59(15)	10075.127(89)	10510.78(57)	3407.511(21)	5318.405(6)	28990.0(2)	18906.535(23)	2181.5(5)	2170.866(14)
	Cl	749.88(15)	23184.152(89)	62520.736(57)	43446.925(21)	59167.083(6)	65460.0(2)	177003.409(23)	155942.58(5)	13811.147(14)
	B₂O₃	446.85(10)	1282.728(85)	801.638(51)	612.214(11)	597.9(5)	122.2(2)	523.483(21)	179.968(5)	1.60(1)
	Br	1.60(7)	39.932(27)	22.990(24)	12.717(15)		7.0(1)	23.327(17)	41.35(2)	52.0(1)

CHAPTER 4

continued

	I	0.03(5)	0.340(14)	0.216(20)	0.313(4)		0.08(1)	1.054(17)	1.02(1)	0.15(1)
Anions (mg/l)	F	2.45(7)	6.590(28)	3.534(24)	1.622(17)			6.783(6)	22.2(1)	2.57(1)
	HPO$_4$	7.13(4)	35.349(15)	10.083(4)			0.232	0.232(2)		
	NO$_2$	0(4)	0.558(9)	1.335(11)	0.103(13)				1.45	
	NO$_3$	0.7(4)	2.062(9)	155.091(15)	0.452(13)				88.4	
Total ion concentration (g/l)		10.92	65.93	125.75	81.47	34.539(6)	74.020(4)	268.522(27)	242.518(5)	28.12
Results of plasma spectral analysis (ppm)	Al	0.431(7)	0.136(25)	0.179(7)	1.228(4)			0.152		
	Fe		0.070(8)	0.198(25)	0.077(6)			0.231		
	Ti	<0.1(3)	0.097(4)	0.017(18)	0.052(5)	0.057(3)		0.0078		
	Mn	<0.1(3)	0.576(13)	0.343(25)	0.024(6)	0.468(4)		0.1114		
	Ba	<0.01(3)	0.030(14)	0.031(11)	0.030(6)	0.525(4)		<0.05(2)		
	Be	<0.1(3)	0.003(15)	0.0008(15)	0(5)	0.033(4)		0		
	Co	<0.1(3)	0.012(13)	0.003(13)	0(5)	0.388(4)		0.0005		
	Cr	<0.4(3)	0.064(18)	0.030(26)	0.018(7)	0.317(3)		0.0268		
	Cu	<0.1(3)	0.03(8)	0.026(17)	0.006(2)	0.093(4)		0.0213		
	Ga	0.01(1)	0.041(15)	0.023(6)	0.01(2)	0.583(3)		0		
	Ni	0.01(1)	0.152(19)	0.050(26)	0.011(6)	1.273(4)		0.0119		
	Pb	<1.3(3)	0.017(12)	0.101(23)	0.081(7)	1.13(3)		0.0244		
	Th	0.01(1)	0.034(15)	0.019(8)	0.03(4)	0.555(4)				
	V	0.01(1)	0.099(19)	0.009(20)	0(4)	0.618(4)		0.0042		
	Zn	<0.1(3)	0.165(6)	0.232(13)	0.195(2)			1.86		
	La	<0.2(3)	0.04(2)	0.035(2)		0.335(4)		0.00235(2)		
	Ce	0.02(1)	0.148(18)	0.075(11)	0.01(3)	1.215(4)		<0.002(2)		
	Nd	0.14(1)	0.545(19)	0.249(15)	0.07(6)	0.985(4)		0.00295(1)		

HYDROCHEMISTRY AND MINERAL ASSEMBLAGES

Results of plasma spectral analysis (ppm)

						continued
Sm	0.01(1)	0.008 (5)	0.01 (1)	0.005 (2)	0.453 (4)	0.0011 (1)
Gd	0.01 (1)	0.261 (18)	0.026 (13)	0.01 (3)	0.588 (4)	<0.001
Dy	0.01 (1)	0.027 (19)	0.012 (15)	0.008 (6)	0.16 (4)	
Y	<0.03 (3)	0.026 (6)	0.033 (4)	0.003 (1)	0.038 (4)	0.00035 (1)
Yb	<0.03 (3)	0.059 (4)			0.043 (4)	
Sc	<0.01 (3)	0.036 (8)	0.018 (13)	0.008 (5)	0.07 (4)	
Bi	<1 (3)	0.016 (8)	0.026 (9)	0.033 (3)	0.407 (3)	0.00043
Cd	<0.1 (3)	0.140 (10)	0.002 (5)	0 (2)	0.303 (4)	
Mo	<0.4 (3)	0.247 (24)	0.099 (23)	0.019 (6)	0.828 (4)	0.116
Nb	0.08 (1)	0.663 (17)	0.854 (11)	0.178 (5)	1.13 (3)	
Ta	0.01 (1)	0.229 (14)	0.010 (6)	0.014 (5)	0.31 (3)	
W		0.377 (4)	0.021 (1)		2.993 (4)	0.03125 (2)
Sb		0.0055 (2)	<0.002 (2)			
Sn		0.011 (8)	0.007 (10)	0.003 (1)	0.38 (3)	0.0096
Ge						0.0295 (2)
Zr	<0.1 (3)	0.067 (17)	0.014 (8)	0 (4)	0.99 (3)	0.0027 (2)

Note: Numbers in parentheses refer to number of lakes taking part in the averaging.

TABLE 1.9. Dominant elements in lake waters from hydrochemical zones of the Qinghai–Tibet Plateau

Element group	I$_1$	I$_2$	II	III	IV$_1$	IV$_2$	IV$_3$	IV$_4$
I A		Li (1), Cs (1) Rb (2) K (3)	Li (2) Rb (1) K (1) Cs (2)			Na (1)	Na (1) K (2)	
II A		Be (2)	Sr (2) Ba (2)		Be (1), Ba (1)		Ca (1) Mg (1) Sr (2)	Sr (1) Mg (2)
III A		B (1), Ga (2)	b (2)	B (3) Al (1)	Ga (1)		Al (2)	Al (2)
IV A	Sn (2)	C (CO$_3$, HCO$_3$) (1) Si (1)	C (CO$_3$) HCO$_3$ (2) Si (2) Pb (2)		Sn (1) Pb (1)			
V A		P (HPO$_4$) (1)	P (HPO$_4$) (2)	Bi (2)	Bi (1)			
VI A	As (2)	As (1), Sb (1)	As (3)			S (SO$_4$) (1)	S (SO$_4$) (2)	
VII A I B II B		Br (2) Cd (2)	Cu (2) Zn (2)		Cu (1) Cd (1)	F (2) I (1) Cl (1) Zn (1)	F (1) Cl (2) Br (1) I (2) Zn (1)	
III B		Sc (2), Y (1), Yb (1) La (2), Ce (2), Na (2), Gd (2), Dy (2), Th (2)	Sm (2)		Sc (1), La (1), Ce (1), Nd (1), Sm (1), Gd(1), Dy (1), Y (2), Yb (2) Zr (1)			
IV B		Ti (1), Zr (2)						
V B		V (2), Nb (2), Ta (2)			V(1), Nb (1) Ta (1)			
VI B		Cr (2), Mo (2), W (2)			Cr (1), Mo (1), W (1)			
VII B		Mn (1)			Mn (1)			
VIII		Co (2), Ni (2)	Fe (2)		Co (1), Ni (1)		Fe (1)	Fe (1)

Note: Dominant elements are determined according to the average values of brines of different concentrations. Numbers in parentheses refer to the places occupied by the concentrations of the elements in various zones.

In contrast, the concentrations of some of the alkali earth elements, such as Sr and Ba, are lower than in seawater. This is particularly true in the northern and southern parts of the plateau, where the concentrations of Sr and Ba in the brines are 1000–10,000 lower than in seawater of the same ionic strength. The K concentrations are a few times higher (in zone I$_2$) or lower (in zone IV) than in seawater. The variation in the composition of the saline lakes indicates that they are of non marine origin. Although, in the chloride–magnesium sulphate subtype zone in the central part of the plateau, the Sr abundance is similar to that of seawater, that may reflect some contribution from marine sources.

There is a significant difference between the concentrations of some elements in the northern and southern hydrochemical zones. For example, the Ca and Mg abundances in the Qaidam Salt Lake of subzone IV$_2$ are higher or close to those observed in seawater, whereas they are a few thousand to tens of thousand times lower than in seawater in lakes of zones I and II, especially in the saline–carbonate lakes. This is because the different water types have different Ca and Mg solubilities. In contrast, the P concentrations are higher in the northern zones than those in the south (Table 1.8).

4.3.2. DOMINANT ELEMENTS

As comparison of the element abundances of the lake waters from the various hydrochemical zones on the plateau is shown in Table 1.8 and yields the following associations of dominant elements in the various zones (Table 1.9): (1) Subzone I$_2$ and zone II have the highest concentrations of rare alkalis and boron, and also have high concentrations of C, Si, P and As. (2) Subzones I$_2$ and IV$_1$ have high concentrations of Be, Ga and elements from the II B and IV B–VII B groups of the periodic table, as well as some elements of the VIII group. Comparison of the two subzones shows that subzone I$_2$ is relatively enriched in the heavy REE (Y and Yb), while subzone VI$_1$ is more enriched in the light REE. Subzone IV$_1$ has higher concentrations of elements from the VB, VI B and VIII groups and most elements in the III B group. (3) Subzones IV$_2$, IV$_3$ and IV$_4$ are more enriched in the major salt–forming elements Na, Ca, Mg, Sr, S and Cl, F, I, Br as well as Zn. In addition, K is relatively enriched in K in subzones IV$_3$ and IV$_4$. Subzone I$_2$ is also relatively enriched in Br. (4) As and Sn concentrations are relatively high in zone I, Bi, Cu and B concentrations are relatively high in zone III, and Sn, Bi, Cu and Pb concentrations are highest in subzone IV$_1$. In many cases it was not possible to collect sufficient volumes to determine all the elements in all the samples, hence the distribution patterns of minor elements, such as the REE, are still uncertain.

4.3.3. FEATURES OF DISTRIBUTIONS OF B, LI AND K

In order to study the distribution of B, Li and K in the plateau saline lakes in more

70   CHAPTER 4

detail an analysis has been made of the trends in the concentrations of these elements in various types of lake waters. The study indicates that the center of high boron values in lake waters of the Qinghai–Tibet Plateau is located in the Zhari Namco–Shiquan River area at the northern foot of the Gangdise Mountains. From this area northeast to the Qaidam Basin there is a progressive decreases in B_2O_3 concentrations (Appended Map 1; Fig. 1.14). The principal centre of high Li

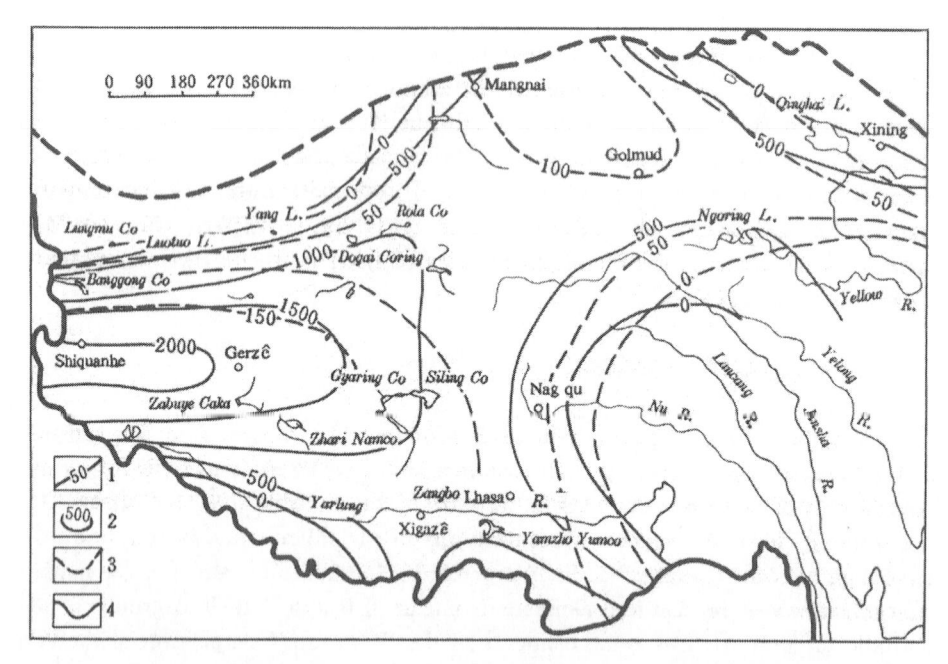

Figure 1.14. Trend surface analysis of B_2O_3 and LiCl of lake waters of the Qinghai–Tibet Plateau. 1. Isoline of B_2O_3 concentration; 2. Isoline of Li Cl concentration; 3. Plateau boundary; 4. National border.

concentrations (Li > 200 mg / l) also lies in the Zhari Namco–Shiquan River area, although their is a secondary center (Li > 100 mg / l) in the area of Taijnar–Da and Xiao Qaidam and Dalangtan Playa in the Qaidam Basin (Fig. 1.14). The principal center has high B, Li and K concentrations. Analysis of the surface trends in $B_2O_3 2 \times 10^3 / \sum_{salts}$ and $Li \times 10^3 / \sum_{salts}$ indicates that there is one high value centre ($B_2O_3 \times 10^3 / \sum_{salts} = 33$–$67$) on the plateau in the Yarlung Zangbo at the northern foot of the Gangdise Mountains. Qaidam belongs to a medium to low value trend area ($B_2O_3 \times 10^3 / \sum_{salts} = 0$–$33$) (Fig. 1.15), implying that the bulk of the Qaidam area belongs to an area with intermediate boron abundances on the Qinghai–Tibet Plateau.

Trend in the surface analyses of K and $KCl \times 10^3 / \sum_{salts}$ show that KCl > 12500 mg / l defines a high value area, while KCl > 10000 mg / l defines a secondary high value area (Fig. 1.15), and shows that the distribution of KCl is similar to that of Li and B.

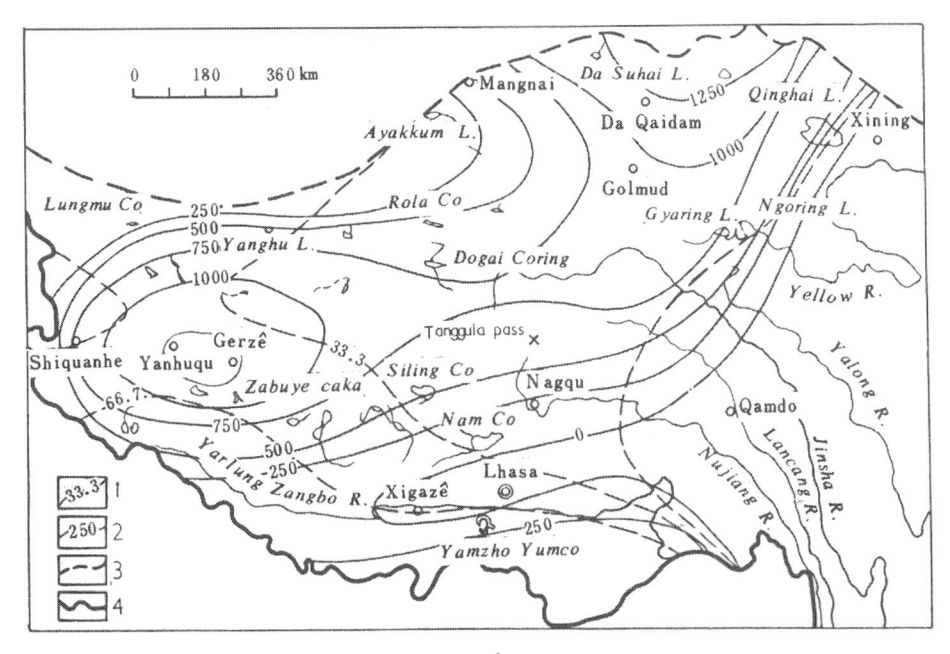

Figure 1.15. Trend surface analysis of $\frac{B_2O_3 \times 10^3}{\sum \text{salts}}$ and KCl of lake waters of the Qinghai–Tibet Plateau. 1. Isoline of $\frac{B_2O_3 \times 10^3}{\sum \text{salts}}$; 2. Isoline of KCl concentration; 3. Plateau boundary; 4. National border.

These analyses of the trends in B, Li and K concentrations, coupled with correlation of hydrochemical data of lakes on the Qinghai–Tibet Plateau, show that the plateau (and the adjacent Pamir Plateau) is a region of high values of B, Li, and K (and Cs and Rb) in East Asia and that the centre of this region is located in the area of Yarlung Zangbo at the northern foot of the Gangdise Mountains. This area is coincident with the younger plate convergence zone of the plateau. Away from this area there is a progressive decrease in the concentrations of B, Li, K, Cs and Rb. Hence, the traditional view that Qaidam is the centre of high levels of these elements in lakes on the Qinghai–Tibet Plateau should be revised. In some cases the absolute contents of B, Li, K and Cs in lakes of Qaidam are higher than in Kumkuli and Qiangtang to its south. This is largely due to evaporative concentration of these elements over a prolonged period of time. However, the ratio of these elements to the total salinity of the lake waters is lower in the Qaidam area.

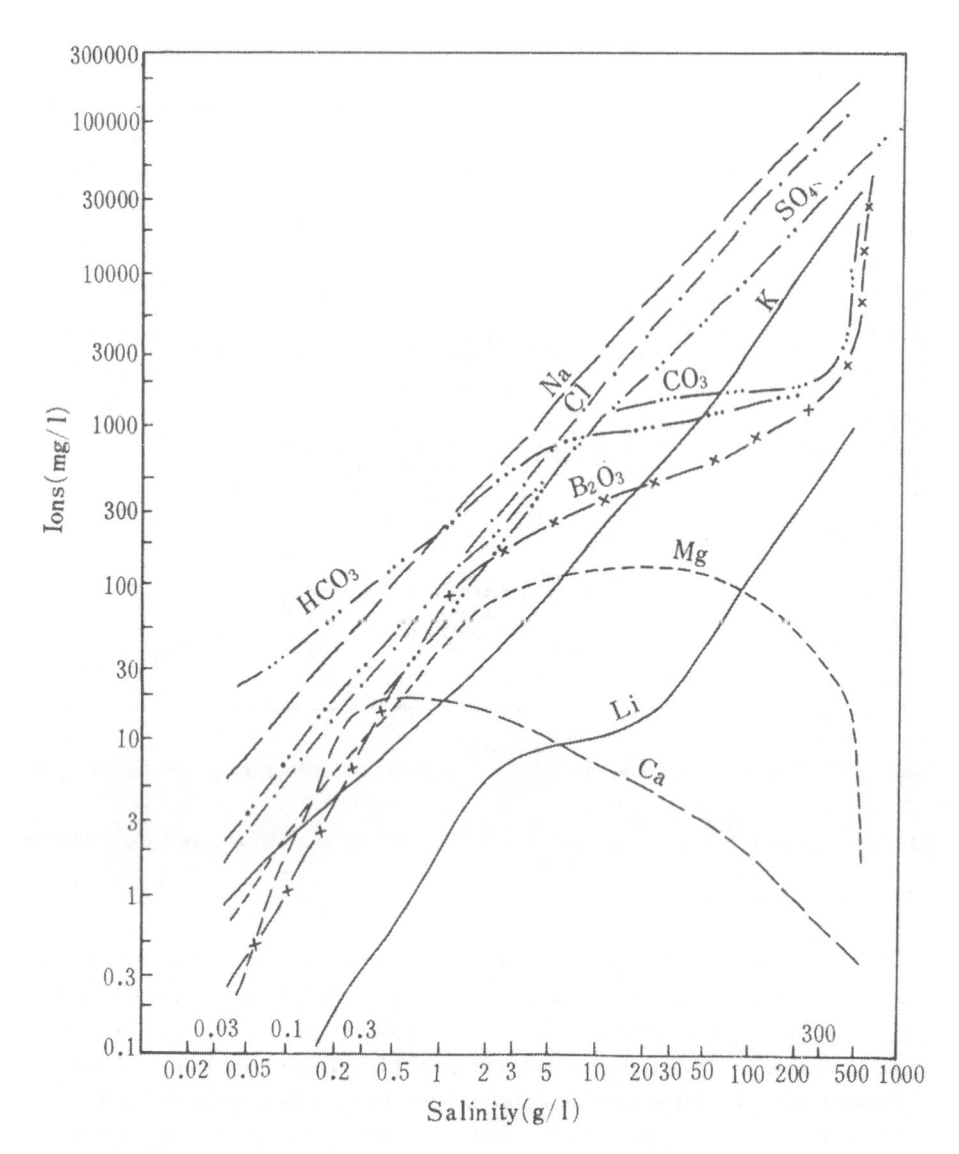

Figure 1.16. Relation between main ions and salinity of natural water of the Zabuye Salt Lake.

4.3.4. CORRELATION OF ELEMENTS

In plateau lake waters, B is closely associated with Li, Rb, Cs and K, and all these elements show a general positive correlation with the salinity of the lake waters.

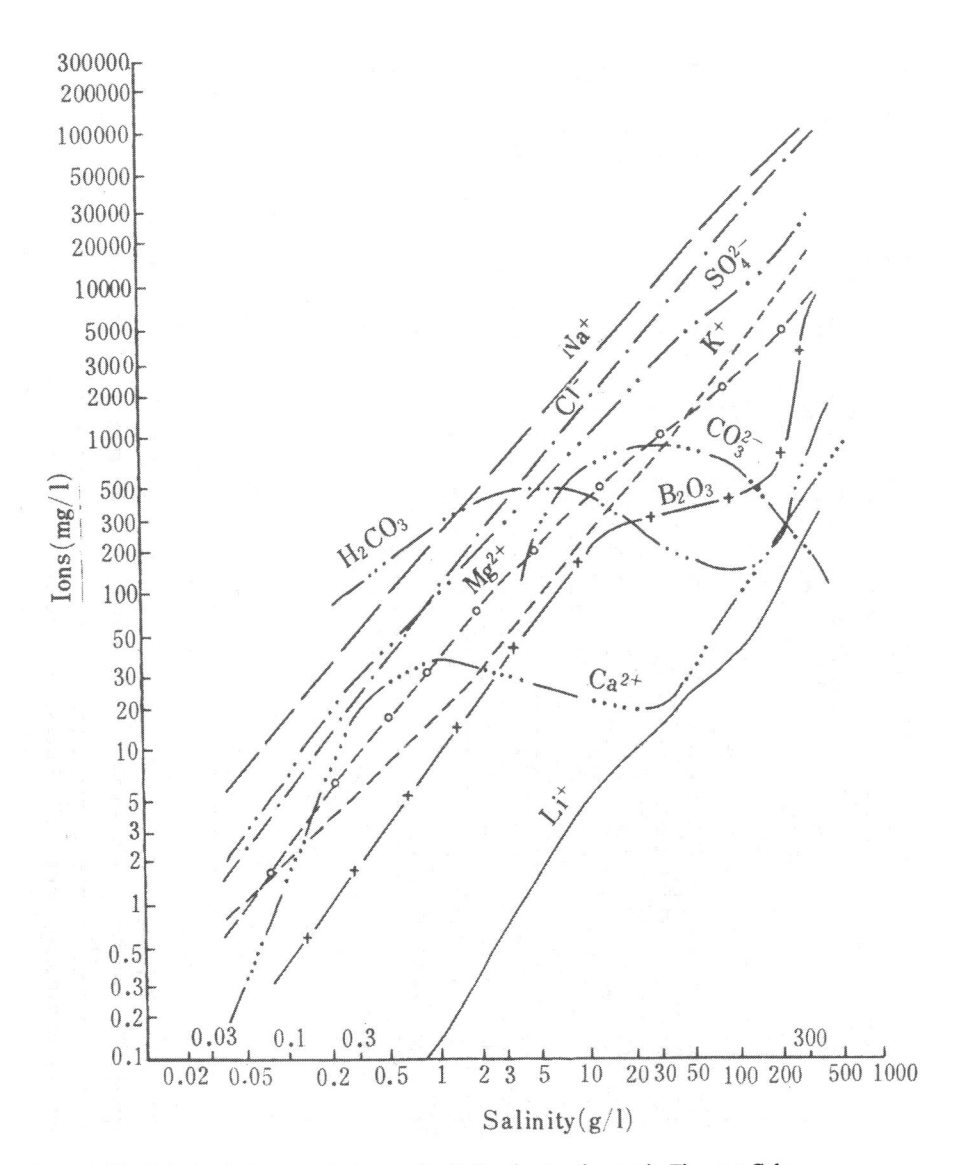

Figure 1.17. Relation between main ions and salinity of natural water in Zhacang Caka.

However, in saline lakes of different chemical types the trends of B and Li are somewhat different. In carbonate saline lakes, e.g., Zabuye Salt Lake, the ratios of B and Li to salinity are 3×10^{-3} and 1×10^{-3}, respecitvely. In addition, the covariations of Na, CO_3^{2-} and HCO_3^- are consistent with precipitation of borax, trona and lithium carbonate (Fig. 1.16). In saline lakes of the magnesium sulphate subtype the ratios of B and Li to salinity are 6×10^{-3} and 0.3×10^{-3}, respectively (Fig. 1.17). The coincident decreases in Ca, Mg, CO_3^{2-} and HCO_3^- concentrations

Table 1.10. Statistics of REE in lake waters of the Qinghai–Tibet plateau

Item Element	Crust (ppm)	Upper mantle (ppm)	Lower mantle (ppm)	Seawater (mg/l)	Carbonate type zone (I) Aver. (mg/l)	Carbonate type zone (I) No	Na sulfate subtype zone (II) Aver. (mg/l)	Na sulfate subtype zone (II) No	Chlorite–Mg sulfate subtype zone (III) Aver. (mg/l)	Chlorite–Mg sulfate subtype zone (III) No	Mg sulfate–chloride type zone (IV) Aver. (mg/l)	Mg sulfate–chloride type zone (IV) No	Lakes on the whole plateau Aver. (mg/l)	Lakes on the whole plateau No
La	39	0.7	0.4	1.2×10^{-5}	0.08	5	0.01	1			0.36	5	0.20	11
Ce	43	1.1	0.7	6.2×10^{-6}	0.26	24	0.04	7	0.01	3	1.18	5	0.32	39
Nd	26	5	0.8	9.2×10^{-6}	0.53	25	0.28	11	0.07	6	1.04	5	0.47	47
Sm	6.7	1.3	0.3	1.7×10^{-6}	0.34	10	0.01	1	0.02	2	0.43	5	0.31	18
Gd	6.7	1.2	0.6	3.4×10^{-6}	0.24	24	0.02	9	0.01	3	0.63	5	0.22	41
Dy	4.1	0.5	0.05	2.9×10^{-6}	0.04	24	0.01	11	0.01	5	0.17	5	0.05	45
Yb	2.7	0.5	0.3	2.0×10^{-6}	0.05	6					0.04	5	0.05	11
Y	2.4	5	0.5	3×10^{-4}	0.04	8	0.02	2	0.003	1	0.05	5	0.04	16
$\sum$REE	165.35	17.70	4.63	7.4×10^{-5}	1.58		0.40		0.11		3.91		1.74	
$\sum$Ce	121.6	9.4	2.31	3.1×10^{-5}	1.21		0.35		0.09		3.02		1.39	
$\sum$Y	43.75	8.3	2.32	4.3×10^{-5}	0.37		0.06		0.02		0.89		0.35	
$\dfrac{\sum\text{Ce}}{\sum\text{Y}}$	2.78	1.13	1.00	0.72	3.29		6.33		3.91		3.38		3.97	
$\dfrac{\sum\text{Ce} \times 10^{6}}{\sum \text{salts}}$				8.86	8.90		2.29		2.58		18.18			
$\dfrac{\sum\text{Y} \times 10^{6}}{\sum \text{salts}}$				12.29	2.71		0.36		0.66		5.38			
$\dfrac{\sum\text{REE} \times 10^{6}}{\sum \text{salts}}$				21.14	11.62		2.65		3.24		23.56			
Data source	From Liu Yingjun et al., 1984			From Wu Ruiduan, 1982	This paper		This paper		This paper		This paper			

Note: Data of this paper measured by Jue Songqiao, Institute of Rock and Mineral Analysis.

areconsistent with precipitation of large amounts of Mg and Ca in carbonates and borates. This is also accompanied by increases in the Li content of the sediments, suggesting uptake of Li from the lake waters. B, Li, Rb, Cs and K in waters of various types also show positive correlations with SO_4^{2-} and Cl^- (Figs. 1.16 and 1.17).

4.3.5. THE RARE EARTH ELEMENTS

The preliminary analyses listed in Table 1.10 suggest the following features of the rare earth elements (REE) distribution in saline lakes of the plateau.

(1) The light REE (LREE) abundances in the lake waters show Nd > Ce > Sm > La, compared to La > Ce in seawater. In the saline lakes Nd concentrations are generally 0.1–10 mg/l, reaching a maximum of 10.6 mg/l in the intercrystalline brine of Zabuye South Lake. In brackish lakes the Nd concentration is generally 10^{-6}–10^{-2} mg/l and is below 10^{-2} mg/l in the freshwater lakes. The heavy REE (HREE) show relative abundances of Gd > Dy > Yb > Y, With Gd reaching maximum concentrations of 3.0 mg/l in Chalaka.

(2) The LREE/HREE ratios of the majority of the lakes range from 2–10, with only a few lakes (e.g., Chalaka) having ratios less than 1. These contrast with the HREE enrichment seen in seawater, but are similar to the average LREE/HREE ratio of the crust (2.78). These data suggest that the REE in the lake waters are derived from the crust, with no significant input from mantle sources.

4.4. Hydrochemical Types and Their Corresponding Saline Mineral Associations

There are a great variety of saline minerals in the saline lakes on the Qinghai–Tibet Plateau. To data 53 mineral species have been identified (Fig. 1.18), of which 6 are chloride minerals: halite, carnallite, sylvite, bischoffite, hydrohalite, and antarcticite; arranged in order of their abundance in the lake elements; with the same arrangement herein thereafter); 19 sulphate minerals (gypsum, mirabilite, glauberite, astrakhanite, thenardite, partially hydrated gypsum, polyhalite, epsomite, glaserite, syngenite, schoenite, kainite, tychite, celestite, hexahydrite, hydroglauberite, tetrahydrite, uklonskovite and barite); 16 are carbonate minerals (calcite, dolomite, aragonite, magnesite, trona, hydromagnesite, northupite, gaylussite, soda, Li–bearing magnesite, nahcolite, zabuyelite, thermonatrite, strontianite, nesquehonite and Li–bearing dolomite); 12 are borate minerals (borax, pinnoite, kurnakovite, ulexite, hydroboracite, inderite, tincakonite, inyoite, macallisterite, hungtsaoite and carboborite). Among these minerals, caboborite, hungtsaoite and macallasterite are new borate minerals discovered in China (Xie et al., 1964, 1965; Qu et al., 1964, 1965), zabuyelite is a new lithium carbonate mineral discovered by the author in 1983 (Zheng et al., 1985) and Li−bearing magnesite and Li−bearing dolomite are also new varieties of minerals that were discovered in the early 1980's (Zheng et al., 1984).

Saline lakes of different hydrochemical types and water compositions have different saline mineral associations and show different metallogenic characters (Zheng et al., 1974). With the change of hydrochemical types of saline lakes from carbonate type (strong, moderate and weak subtypes)to sodium sulphate subtype to magnesium sulphate subtype to chloride subtype, their saline mineral associations change from alkaline to weakly alkaline to neutral to slightly acid. As carbonate type saline lakes change from strong to medium to weak subtypes, their sodium carbonate mineralogy changes from a nahcolite−trona association and soda−trona−thermonatrite association to a mineral association dominated by gaylussite. Large amounts of lithium carbonate and borate (large borax, with smaller amounts of macallisterite) are also found in the moderate and strong carbonate subtypes. The moderate carbonate subtype lakes are characterized by the presence of the complex salt tychite with sodium carbonate, together with large amounts of mirabilite (e.g in Bangkog Co). Chlorides minerals consist mainly of sodium− or potassium−bearing chlorides, but in the strong carbonate saline lake subtype (e.g. Dujiali Lake) and some of the moderate subtype (e.g., Bangkog Co and Chakala Lake) halite has a relatively dispersed distribution. In parts of the moderate, and especially in the weak carbonate subtype saline lakes there are relatively pure and thick halite layers (e.g. in Bangyu Caka and Xia Caka).

In the sodium sulphate and magnesium sulphate subtypes, the carbonate minerals are dominated by calcium, magnesium and magnesium−calcium carbonates (e.g., in Yiliping Playa, Gas Hure Lake (Gao, 1984), Da and Xiao Qaidam Lakes, Chagcam Caka II and III, etc.). The borates are composed mainly of sodium borate, weakly alkaline sodium−calcium, calcium−magnesium and magnesium borates (e.g. in Chagcam Caka and Da and Xiao Qaidam Lakes). The sulphate minerals are characterised by the occurrence of sodium−calcium, sodium−magnesium, potassium−calcium, or magnesium sulphates, in addition to large amounts of sodium sulphate and gypsum (e.g., in Da and Xiao Qaidam Lakes, Chagcam Caka II and III and Dalangtan Playa). There are large amounts of chloride present in the two sulphate subtypes, together with appreciable amounts of patassium, magnesium or potassium−magnesium chlorides (e.g., in Dalangtan Playa, Dabsan Lake, and East and West Taijnar Lakes).

In the chloride type saline lakes, the carbonate mineral assemblage is similar to that in the sulphate type. The sulphate mineral association is dominated by patassium−magnesium sulphate, while sodium sulphate is almost absent (e.g., in Kunteyi Lake and North Huobuxun Lake). The chloride mineral association is the same as that of the magnesium sulphate subtype, but antarcticite has been found on the margins of Kunteyi Lake (Chen, 1984).

HYDROCHEMISTRY AND MINERAL ASSEMBLAGES

	Mineral name	Hydrochemical types	Carbonate type Strong	Carbonate type Moderate	Carbonate type Weak (Sub type)	Sulphate type Na	Sulphate type Mg	Chloride type
Chlorides	Antarcticite	$CaCl_2 \cdot 6H_2O$						÷ ×
	Carnalite	$KCl \cdot MgCl_2 \cdot 6H_2O$					H ○	H ○
	Bischofite	$MgCl_2 \cdot 6H_2O$					H ▲	H ▲
	Hydrohalite	$NaCl \cdot 2H_2O$				÷ ○	÷ ▲	
	Sylvite	KCl		▲			○	○
	Halite	$NaCl$	×	▲	○	○	○	○
Sulphates	Kainite	$KCl \cdot MgSO_4 \cdot 2.75H_2O$					▲	
	Schoenite	$K_2SO_4 \cdot MgSO_4 \cdot 6H_2O$					▲	▲
	Polyhalite	$K_2SO_4 \cdot MgSO_4 \cdot 2CaSO_4 \cdot 2H_2O$					▲	▲
	Syngenite	$K_2SO_4 \cdot CaSO_4 \cdot H_2O$					▲	
	Uklonskovite	$NaMg(SO_4)(OH) \cdot 2H_2O$					▲	
	Astrakhanite	$Na_2SO_4 \cdot MgSO_4 \cdot 4H_2O$				÷ ○	▲	
	Glauberite	$Na_2SO_4 \cdot CaSO_4$				▲	▲	
	Hydroglauberite	$5Na_2SO_4 \cdot 3CaSO_4 \cdot 6H_2O$				▲	▲	
	Epsomite	$MgSO_4 \cdot 7H_2O$				▲	▲	
	Hexahydrite	$MgSO_4 \cdot 6H_2O$				▲	▲	
	Tetrahydrite	$MgSO_4 \cdot 4H_2O$				▲	▲	
	Barite	$BaSO_4$					▲	
	Celestite	$SrSO_4$				▲		
	Gypsum	$CaSO_4 \cdot 2H_2O$		×	×	▲	○	○
	Half water Gypsum	$CaSO_4 \cdot 1/2H_2O$				▲	▲	
	Tychite	$2MgCO_3 \cdot 2Na_2CO_3 \cdot Na_2SO_4$		▲				
	Mirabilite	$Na_2SO_4 \cdot 10H_2O$	÷ ○	÷ ○	÷ ○	÷ ○	÷ ▲	
	Thenardite	Na_2SO_4	H ▲	H ▲	H ▲	H ▲	H ▲	
	Glaserite	$K_3Na(SO_4)_2$	▲	▲				
Carbonates	Strontianite	$SrCO_3$				▲		▲
	Calcite	$CaCO_3$	○	○	○	○	○	▲
	Aragonite	$CaCO_3$	○	○	▲	▲	○	▲
	Magnesite	$MgCO_3$	▲	○	▲	▲	○	○
	Dolomite	$MgCO_3 \cdot CaCO_3$	○	○	○	○	○	○
	Lithium bearing magnesite	Li_2MgCO_3		▲				
	Lithium bearing dolomite	$Li_2CaMg(CO_3)_2$		▲				
	Zabuyelite	Li_2CO_3		▲				
	Gaylussite	$CaCO_3 \cdot Na_2CO_3 \cdot 5H_2O$		○	○			
	Nesquehonite	$MgCO_3 \cdot 3H_2O$		▲				
	Northupite	$MgCO_3 \cdot Na_2CO_3 \cdot NaCl$	▲	○	▲			
	Hydromagnesite	$4MgCO_3 \cdot Mg(OH)_2 \cdot 4H_2O$	○	○		▲	▲	
	Trona	$Na_3H(CO_3)_2 \cdot 2H_2O$	H ▲	H ○				
	Soda	$Na_2CO_3 \cdot 10H_2O$	÷ ▲	÷ ○				
	Thermonatrite	$Na_2CO_3 \cdot H_2O$	H ▲	H ▲				
	Nahcolite	$NaHCO_3$	÷ ○	÷ ▲				
Borates	Carboborite	$MgCO_3(CO_3 \cdot B_2O_3) \cdot 10H_2O$					×	
	Hungtsaoite	$MgB_4O_7 \cdot 9H_2O$					÷ ×	
	Macallisterite	$MgB_6O_{10} \cdot 7.5H_2O$					÷ ×	
	Kurnakovite	$MgB_6O_{11} \cdot 15H_2O$				○	÷ ▲	
	Inderite	$MgB_6O_{11} \cdot 15H_2O$				÷ ▲	▲	
	Pinnoite	$MgB_2O_4 \cdot 3H_2O$				○	○	
	Hydroboracite	$CaMg_6O_{11} \cdot 6H_2O$					○	
	Inyoite	$CaB_6O_{11} \cdot 13H_2O$				▲		
	Ulexite	$NaCaB_5O_9 \cdot 8H_2O$			▲	○	▲	
	Borax	$Na_2B_4O_7 \cdot 10H_2O$	÷ ○	÷ ○	÷ ○	×		
	Tincakonite	$Na_2B_4O_7 \cdot 5H_2O$	H ▲	H ▲	H ×			
	Kevnite	$Na_2B_4O_7 \cdot 4H_2O$	×	×				

○ Major minerals ▲ Minor minerals × Trace mineral or minerals of marginal facies
H Hot phase mineral ÷ Cold phase mineral

Figure 1.18. Classification of saline minerals in saline lakes on the Qinghai–Tibet Plateau.

The mineral assemblages of various types of saline lakes on the Qinghai—Tibet Plateau have features of the cold-phase association. Mirabilite ($Na_2SO_4 \cdot 10H_2O$) that is generally precipitated at temperatures below 0 ℃, is widespread and present in large amounts in all the types of lakes, except the chloride type. Other salts that are precipitated at low temperatures include; soda and nahcoalite (e.g., in Dujiali Lake and Bangkog Lake) in the carbonate types; hexahydrite and schoenite in the sulphate types (e.g., in Dalangtan Playa and Da Qaidam Lake); and hydrohalite, bischofite and antarcticite in the chloride types (e.g., in Kunteyi Lake).

CHAPTER 5 CLASSIFICATION OF SALINE LAKES AND TYPES OF MINERAL DEPOSIT

There are many types of saline lakes on the Qinghai–Tibet Plateau and the different types of saline lakes contain many kinds of useful components, hence it is important to classify the saline lakes. Different disciplines will advocate different schemes of classification according to their objectives. In the fields of geology, physics and chemistry the classification is commonly based on the salinity of lake waters, hydrochemical types, saline mineral associations, mode of occurrence of lakes, relation and volume ratio of solid phase to liquid phase, origin of lake basins and useful minerals within the saline lakes (Jones et al., 1965; Eugster et al., 1978). The criteria of classification used in this paper are given in Table 1.11, and a preliminary scheme of classification for the mineral deposits is also proposed (Table 1.12).

In Table 1.11 the saline lakes on the plateau are classified into three types: potassium– magnesium– (lithium–boron–strontium) saline lakes, special saline lakes, and ordinary saline lakes. This classification is based on a number of factors. These include: consideration of the present state and trends of development and utilisation by Chinese and foreign geological and industrial groups; the requirements by industry for useful components in saline lake waters; and mineral features of saline mineral resources on the Qinghai–Tibet Plateau.Each lake type can be further divided into commercial deposits and potential deposits. The former can be mined according to the current industrial indices of China, i.e the mineral resources of the saline lakes can be exploited and utilized by industrial departments at the present time①. The latter category refers to the saline lake mineral resources that can be utilised for general purposes at the moment, or have been considered to be of commercial grade by foreign companies, but do not meet the industrial indices of China today. With the rise in the level of exploitation and utilisation of resources in China they may be included in the list of China's important saline lake mineral resources.

As is commonly known, whether or not a mineral deposits can be exploited and utilised not only depends on its grade, but is also related to its size, geographical position, communications network and market economics. This is particularly true for saline lake resources. Saline lakes contain deposits that form a solid–liquid

---

① Reference Manual of Industrial Requirements for Mineral Resources (revised version), 1987. Geological Publishing House, Beijing.

TABLE 1.11. General Classification of Saline Lakes on the Qinghai–Tibet Plateau

Principle	Name			Criteria or main features	Examples
Based on salinity of lake water	Freshwater lake			Salinity <0.1%	Mapan Yumco, Gyaring Lake
	Brackish Lake			Salinity 0.1%–3.5%	Qinghai Lake, Nam Co
	Saline lake			Salinity >3.5%	Zabuye Salt Lake, Caka
Based on mode of occurrence of lakes	Surface brine lake			Dominated by lake surface brine	Jingyu Lake
	Saline lake			Contains surface brine, intercrystalline brine, sludge brine, silt pore brine and salt beds	Zabuye South Lake, Da and Xiao Qaidam, Gas Hure Lake
	Playa			Salt lake basin with dried-up surface, generally containing salt bed intercrystalline brine, sludge brine or silt pore brine	Qarhan Playa, Dalangtan Playa
	Subsand lake			Surface covered with silts; below there occur salt beds and possibly pore brine (intercrystalline brine), which may have either dried up or migrated	Yiligou Mirabilite Lake

			Kc (%)	Kn_1	Kn_2	Kn_3	Kn_4	KI	KII^1	KII^2	KIII
Based on chemical composition of lake water	Carbonate type	Strong	>29	>>1	>>1	>>1	>>1	≧1	∞	—	—
		Moderate	8~29	>>1	>>1	>>1	>>1	≧1	∞	—	—
		Weak	0.1~8	>>1	>>1	>>1	>>1	≦1	∞	—	—
	Sulfate type	Na sulfate	0~0.1	≧1	≧1	>>1	>/<1	0	n	∞	—
		Mg sulfate	—	≦1	≦1	>>1	>/<1	—	0	∞	0
	Chloride type		—	<<1	<<1	≦1	<1	—	—	0	n

* $Kc = \dfrac{NaHCO_3 + Na_2CO_3}{\text{total salts}} \times 100\%$ (calculated with wt.% or mg/l of salt); $Kn_1 = \dfrac{CO_3 + HCO_3}{Ca + Mg}$; Kn_2

$= \dfrac{CO_3 + HCO_3 + SO_4}{Ca + Mg}$, $Kn_3 = \dfrac{SO_4}{Ca}$ and $Kn_4 = \dfrac{CO_3 + HCO_3}{Ca}$ (calculated with ion equivalents); KI

$= \dfrac{Na_2CO_3 \cdot NaHCO_3}{Na_2SO_4}$; $KII^1 = \dfrac{Na_2SO_4}{MgSO_4}$; $KII^2 = \dfrac{MgSO_4}{MgCl_2}$; and $KIII = \dfrac{MgCl_2}{CaCl_2}$ (Calculated with wt.% or mg/l of salt).

Table 1.12. Preliminary scheme of commercial classification of saline lakes on the Qinghai–Tibet plateau (liquid ore (mg/l) / solid ore (%))

	Special saline lake								Common saline lake		
Types	K–Mg–(Li–B–Sr) saline lake	Boron (saline) lake	Lithium (saline) lake	B–Li saline lake	Li–K saline lake	B–Li–K–(Br) saline lake	B–Li–K–Mg–(Cs–Rb–Sr–Br) saline lake	B–Li–K–Cs–Rb–Br–(Ge–W) saline lake	Halite–(sulfate) saline lake	Mirabilite–(halite) saline lake	Sada lake
Formation	1. KCl–K–Mg–B–Li–Sr brine 2. K₂SO₄–K–Mg–(Li–B) brine	Borax–boron brine	Lithium brine	1. Borax–B–Li brine 2. Ulexite–B–Li brine	Li–K brine	1. Borax–B–Li–K brine 2. Ulexite–borax–B–Li–K brine	1. Ulexite–pinnoite–kurnakovite 2. ulexite–pinnoite–mixing brine 3. ulexite–hydroboracite 4. B–Li–K–Mg brine	1. Borax (macallisterite)–mixing brine 2. Li and potash salt–borax–mixing brine	Sulfate–halite and Na–Mg mirabilite brine	Halite–mirabilite	1. Soda water 2. Soda and trona

Category Item	Comm. Pros.	Comm. Pros.	Comm. Pros.	Comm. Pros.	Comm. Pros.	Comm. Pros.	Comm. Pros.	Comm. Pros.	Comm. Pros.	Comm. Pros.	Comm. Pros.
KCl	$\dfrac{12000}{6} \dfrac{5000}{2}$				$\dfrac{12000}{6} \dfrac{5000}{2}$	$\dfrac{12000}{6} \dfrac{5000}{2}$①	$\dfrac{12000}{6} \dfrac{5000}{2}$	$\dfrac{12000}{6} \dfrac{5000}{2}$			

—continued

	Main mineral association	Example of mineral deposit
	1. Sylvite, carnallite, bischofite, halite (of chlorite type) and their brine 2. Sylvite, carnallite, potassium sulfate, Na–Mg–(Ca) sulfate, halite and their brine (Mg sulfate subtype)	Qarhan Lake, Xiezhu Lake, South Huobuxung Lake, Dalangtan Playa, Tuanji Lake, Gas Hure Lake
	1. Borax, mirabilite and B–Li brine 2. Ulexite and B–Li brine	Yinhu Lake, Jiabo Co
	Mirabilite–halite and Li brine	Xiao Caka (LiCl 450 mg/l)
	Epsomite, halite, mirabilite and B–Li brine	Chalaka II Lake (LiCl 300–4900 mg/l); B$_2$O$_3$ 1100–1800 mg/l; KCl not up to commercial grade), A'ong Co, Terang Punco
	1. Borax, halite, mirabilite and their B–Li–K brine 2. Ulexite, pinnoite, halite and their B–Li–K brine 3. B–Li–K brine	Yurba Lake (LiCl about 1000 mg/l; B$_2$O$_3$ 10300 mg/l; not up to commercial grade), Tuoba Lake
	1. Ulexite, pinnoite, kurnakovite, hydromagnesite (mirabilite) 2. Ulexite, pinnoite, halite, mirabilite, gypsum and their B–Li–K–(Cs–Rb–Br) brine 3. Ulexite, hydroboracite, pinnoite, gypsum	Xia Caka, Pongjin Co, Mami Co, Aqqiktol Lake, Riwan ces of the Da Qaidam Caka (I)
	1. Borax (tincakonite), mirabilite, soda and their Li–K–(Rb–Cs–Br) brine 2. Li carbonate, glaserite (sylvite), borax, soda, halite, mirabilite and their B–Li–K–Rb–Cs–(Br–Ge–Sr) brine	Terrace–I of Chagcam Caka, bottoms of the Da and Xiao Qaidam lakes, bottom of Chagcam Caka, terraces of the Da Qaidam Lake, Yiliping Playa (B–Li–K–Mg–Rb) brine
	Gypsum; mirabilite (Mg sulfate), halite and their brine (Mg may reach 26000–27000 mg/l; KCl 8000 mg/l)	Bangkog Lake, Dujiali Lake, Zabuye Salt Lake
	Halite (gypsum, Mg sulfate), mirabilite	Hoh Salt Lake, Caka
	1. Soda-bearing lake water type 2. Soda, trona (Mirabilite), halite	Yiligou Playa, terrace–I of Dong Co
	Mud–flats and soda pits of Pung Co, Dung Co (brackish lake) and Hatu	

—continued

MgCl₂	25000/5	10000/2.5					25000/5	10000/2.5	25000/5	10000/2.5
B₂O₃		400	1000/2	400①/1.5		1000/2	400/1.5	1000/2	400/1.5	1000/2	400/1.5
Carbonate type (low–Mg)				300	150/0.114	200/0.344	150/0.114	200/0.344	150/0.114
Li Sulfate– Cl chloride type (high–Mg)	400⑤	150③		500	150	400	150	400	150	400	150
Sr	9000④/40	300④					9000	300		
Br	300	100⑥				300	100	300	100	300	100⑥
Cs							10④	1④	10	1
Rb							60④	3④	60	3
WO₃									70	5
Ge									0.1	
NaCl									100000⑥/50	50000/30	100000/50	50000/30
Na₂SO₄									50000/60	30000⑥/15	50000⑥/60	30000/15
Na₂CO₃+ NaHCO₃											100000⑦/25	20000/20

CLASSIFICATION OF SALINE LAKES

This table was mainly prepared on the basis of the Reference Mannual of Industrial Requirements for Mineral Resources"(1972), Geological Publishing House. Besides the following notation is made.

① This type of solid ore consists mainly of magnesium borate and is similar to B—Mg ore in Northeast China, so is used for reference.

② When lithium is mined independently, the upper limit of the LiCl industrial index 200—300 mg / l proposed in the "Mannual" (P_{111}) is used as the commercial grade.

③ Lithium is difficult to extract from lithium—bearing brine of sulfate type and chloride type because of high magnesium content, so its grade is raised correspondingly here.

④ The industrial indices proposed in КЛИМЕНКО И. А. др, 1981, Состояные и перспективы к о мплексной Утилизации ненных компонентов природных и Техногенных минерализовонных Вод (Москва) are reduced to the prospective indices properly.

⑤ It is set according to the Br commercial grade (100 mg / l) for multi—purpose utilization of brines in Qijiaojing, Xinjiang; the index for Br extraction from oil waste water and sewage at home is 50 mg / l.

⑥ The index of liquid sodium sulfate is set temporarily according to the industrial indices.

⑦ The index of liquid sodium carbonate is set by reference to the mining grade of the upper brines of Searles Lake of the United States, while its prospective is temporarily set according to the fact that alkalis can be produced from lake waters of Bong Co etc. in various seasons even if they contain less than 20000 mg / l $Na_2CO_3+NaHCO_3$.

⑧ Common indices of saline lakes are used for ordinary salts (NaCl, Na_2SO_4 and alkalis) of various types of saline lakes.

Abbreviations: comm. = commercial; pros. = prospective.

phase dynamic balance that changes considerably with different seasons and evaporation conditions. It may also be evaporated by local solar energy to raise its grade. Hence, the definitions of the various types of saline lake mineral resources should be regarded as a preliminary guide that can be used as broad grounds for the evaluation, exploitation and utilisation of saline lake mineral resources. It should thus, be used in a flexible and appropriate way in the light of the economic situations of different regions. For example, the grade of KCl brine used in China and abroad is presently commonly lower than 1%, e.g., the KCl grade in brines mined in Zhigong, Sichuan, is 0.4%—0.7%, the KCl grade in exploitable brines in the production area of salt flats of Lake Bonneville is 0.8%—1.2% and may be as low as 0.4%, and the KCl grade in brines in Erton Salt Lake in the former U.S.S.R., is 0.5%—0.8%. Hence, the industrial index of KCl listed here (12000 mg / l) is that currently used in China, while an index of 500 mg / l is considered to be an index for propsective resources. In terms of the economic feasibility for mining and utilisation, a KCl grade of 5000 mg / l may be regarded as the commercial grade.

CHAPTER 6 TECTONOGEOCHEMISTRY AND REGIONAL GEOCHEMISTRY

6.1. Tectonogeochemical Characteristics

Tectonogeochemistry is a newly emerging cross disciplinary study that is still to be perfected. It deals with the behaviors and results of material components in tectonic processes (Chen et al., 1983). As far as saline lake mineral deposits are concerned, it is mainly concerned with the study of the controls of tectonics over geochemical fields of saline lake districts. Conversely, the geochemical processes within saline lakes may be studied as the effects of tectonic effects. The unusual tectonic conditions of the Qinghai–Tibet Plateau gave rise to various types of tectonic basins, created space for water and salt accumulation, controlled sedimentation, influenced the climate to some extent and provided the fundamental conditions for the formation of the saline lakes.

The existence of the Qinghai–Tibet Plateau as a newly created independent tectonic unit, is an activated structure that formed on the foundation of older fault systems and through the suturing that occurred along the Indus–Yarlung Zangbo in the Cenozoic. The important features of this structure are as follows:

(1) Activation in the Himalayan Stage resulted in the formation of lofty mountains, such as the Kuntun, Qilian, western Qinling, Tanggula, Gangdise and Himalaya mountains. It also produced a very thick piedmont molasse, and overthrusting generated by the Himalayan movement gave rise to a fan–shaped fold belt (Huang, 1984). It was bounded on all sides by a series of deep faults, forming the latest, highest fault block plateau with an average elevation of more than 4500 m above sea level.

(2) Since the Eocene, northward pushing of the Indian plate produced strong compression and shear, which caused N–S shortening and E–W extension of the plateau. This formed a series of tectonic basins, bringing about new sedimentation and magmatism in different fault–fold basins and blocks of the plateau, leading to new sedimentary sequences. For example, the very thick Cenozoic evaporite and petroleum–bearing sediments in the Qaidam Basin is the result of a large amplitude fall in the activated fault block. The intrusion of B–Li–F–P–bearing granites in the Higher Himalayas in the Meso–Cenozoic and the second retrograde metamorphism of Precambrian metamorphic rocks during the Himalayan Stage (Mao et al., 1984) are the result of addition of large amounts of new components

into the original geologic bodies.

(3) Recent studies of the deep geology of the plateau (Gao et al., 1982; Chen et al., 1986) indicate that the crustal structure of the modern Qinghai–Tibet Plateau forms a complete and independent unit up to 50–70 km thick (some 25–35 km thicker than normal continental crust). The asthenosphere underlying the plateau is about 120 km thicker than in the adjacent regions. The top of the asthenosphere is 80–90 km beneath the Indian block and up to 120–140 km deep below Tibet. There is a low velocity layer and a low–Q layer at depths of 20–50 km in the upper–middle crust, and there is an anomalous upper mantle layer at the bottom of the lower crust (Li, 1983; Yang, 1985; Sun et al., 1985; Shen, 1985). There are interpreted as the manifestations of a high–temperature molten layer. Various components of these deep–level molten materials have infiltrated into the surface of the upper crust in different forms at various times during the Tertiary–Quaternary. This has influenced the compositions of various geologic units on the plateau to varying extents. Under some conditions, these deep molten bodies have emerged on the surface to form large volcanic sheets. For example, in the Hoh Xil and western Kunlun fault–fold belts there are Tertiary and Quaternary volcanic rocks covering an area of a few hundred square kilometers. In some areas partially molten pockets were formed as a result of shearing between the upper and lower crust. Areas of strong heat anomalies have also produced areas of geothermal activity on the plateau. For example, the geothermal fluids in the Yarlung Zangbo area carry large amounts of B, Li, Cs, As, F, K, Sn, and Hg to the surface from depth.

6.2. B, Li, Cs and Rb Contents

Limited regional geochemical studies have been made on the distribution of B, Li, Rb, Cs and K on the plateau. In the early 1960's, the author carried out geochemical surveys on the distribution of B in northern Tibet and the Institute of Saline Lakes undertook a relatively detailed regional geochemical surveys of B and Li in the Qaidam Basin. Since the 1970's–1980's, there have been a large number of studies on other geochemical and petrological aspects, there is very little new data concerning the above mentioned elements. This is especially true for southern Tibet, Karakorum–Hoh Xil and southern Qinghai–eastern Tibet, where there are almost no geochemical data available. Hence, it is rather difficult to discuss the regional geochemistry of B, Li etc. on the whole plateau. Here we have only attempted to discuss the data obtained through many years of investigation and data provided by the Geological Survey Party of the Qinghai–Tibet Plateau and related organisations① since the 1980's, coupled with relevant data provided by their predecessors.

6.2.1. BORON CONTENT IN ROCKS (TABLE 1.13)

According to preliminary analysis of the data many types of rocks of various ages on the Qinghai–Tibet Plateau show high boron potentials. Of these rocks, Precambrian metamorphic rocks in some areas, Late Cretaceous igneous rocks and Cenozoic rhyolite, granite and sedimentary rocks have high B contents, and there is a tendency for various Cenozoic rocks to have higher B contents. For example, the average boron content in Cenozoic granite, rhyolite, dacite, travertine, argillaceous rocks and sandstone is several tens of times higher than its average abundance in the earth's crust. This is particularly the case for the widespread Cretaceous–Eocene volcanic rocks in the Gangdise Mountains and the B–bearing leucogranite in the Higher Himalayas where the boron contents are 130 and 17 times higher, respectively, than the global average boron abundance in these types of rocks.

The boron contents are not high in Precambrian igneous rocks (granite and diorite) and various types of metamorphic rocks in the Kunlun, Gangdise and Nyainqêntanglha Mountains, and only significantly increase in metamorphic rocks on the northern and southern sides of the plateau. For example, the boron contents of gneiss in the Himalayas and Qilian Mountains are about 20 times higher than those in the central part of the plateau. The boron contents in other metamorphic rocks of the Precambrian (Nyalam Group) also increase to different extents (Table 1.13). According to isotope age studies (Mao et al., 1984), there were two phases of regional metamorphism in the Nyalam Group at 640–660 Ma and 10–20 Ma. The more recent data corresponds to the phase of B–, Li and Rb–rich leucogranite activity of the Himalayan Stage. Hence, it appears that the second phase of metamorphism introduced dispersed components (such as B, Li, Cs and Rb) into the Precambrian rocks. The Precambrian (and Lower Paleozoic) metamorphic rocks in the Qilian Mountains also have higher boron contents, of which the Lower Paleozoic marble in Lengquangou contains up to 467 ppm boron (average of three

---

① Feng Binggui et al. of the Geological Survey of the Qinghai–Tibet Plateau provided Li and Rb data for the Karakorum–Kunlun mountains; Guo Tieying et al. provided samples from a section of the Qiangtang area; Fan Yongnian et al. of the Party provided Carboniferous samples from northern Tibet and Precambrian samples from the Himalayas; Wang of the Sino–French Scientific Expedition provided a part of the analytical data for samples of volcanic rocks of various ages from southern Tibet; Mei of the Institute of Geochemistry, Academia Sinica, provided a part of the samples and data from the western sector of the Gangdise Mountains.

CHAPTER 6

TABLE 1.13. Statistics of average Boron contents (ppm) in various rock types and herbs on the Qinghai–Tibet Plateau

Name		Age (area)	B aver. ①	B Aver. ②	T and W ③	B ④	①/③	①/④	②/③	②/④
Ultrabasic rock		Caledonian	28 (2)	28.5	3	1	9.33	28	9.5	28.50
		Late Jurassic	29 (4)				9.66	29		
Basic rock		Caledonian	5 (1)	662	5	5	1	1	132.4	132.4
		Terminal Cretaceous	2575 (2)				515	515		
		Tertiary	54 (1)				10.8	10.8		
		Quaternary	14 (2)				2.8	2.8		
Intermediate rock	Diorite	Caledonian	28 (2)	34	9 (syenite)	15 (diorite)	3.11	1.87	3.78	2.27
		Hercynian	8 (1)				0.89	0.53		
		Early Yanshanian	25 (2)				2.78	1.67		
		Late Yanshanian	102 (3)				11.33	6.80		
		Early Permian	8.5 (1)	29			0.94	0.57	3.22	1.93
		Early Cretaceous	19 (3)				2.11	1.27	2.56	1.53
	Andesite	Late Cretaceous	30 (11)	23			3.33	2.00		
		Tertiary	34 (3)				3.78	2.27		
		Quaternary (Kunlun)	9 (5)				1	0.60		
Acid rock	Granite	Caledonian	2.5 (2)	68	10 (low-Ca)	15	0.25	0.17	6.80	4.53
		Hercyian	29 (14)				2.90	1.93		
		Yanshanian	35 (14)				3.50	2.33		
		Himalayan (E₂)	1 (2)	378			0.1	0.07	37.8	25
		Himalayan (N₁)	274 (7)				27.4	18.27		
	Rhyolite	Early Cretaceous	11.5 (4)	688			1.15	0.77	68.8	46
		Late Cretaceous–Eogene	2000 (23)				200	133.33		
		Tertiary	52 (2)				5.2	3.47		

—continued

Clays and argillaceous rock	Shale	Carboniferous	26 (2)				0.26	0.26		
	Mudstone	Early Cretaceous	34 (4)				0.34	0.34		
	Mudstone	Eogene	170 (20)				1.70	1.70		
	Mudstone	Miocene	112 (28)	100	100		1.12	1.12		
	Mudstone	Pliocene	160 (10)				1.60	1.60	1	1
	Salt–bearing clay	Late Pleistocene	1203 (39)	547		100	5.23	12.03		
	Carbonate clay	Early Holocene	338 (42)		230		1.47	3.38	4.33	
	Ooze	Modern brackish lake	452 (5)	995			1.97	4.52		5.47
	Salt–bearing argillaceous material	Holocene (Tibet)	~2530 (1075)				~11.00	~25.30	3.25	9.95
		Holocene (Qaidam)	451 (188)				1.96	4.51		
	Clay	Modern weathering crust	90 (17)	130	22–40 *		2.25	0.90		
	Soil	Holocene	170 (140)				4.25	1.70		
Sandstone	Sandstone	Paleozoic	76 (5)				2.17			
	Sandstone	Carboniferous	6 (1)				0.17			
	Sandstone	Triassic	53 (1)				1.51			
	Sandstone	Jurassic	48 (2)				1.57			
	Sandstone	Early Cretaceous	44 (6)	71	35		1.26		2.63	
	Sandstone	Eogene	92 (7)				2.62			
	Sandstone	Neogene	134 (18)				3.83			
	Sandstone	E.–M. Pleistocene	125 (3)				3.57			
	Sandstone	Late Pleistocene	61 (3)				1.74			
	Sandstone	Holocene	926 (2)				26.46			
Carbonate	Limestone	Paleozoic	8 (1)				0.40			
	Limestone	Early Cretaceous	24 (16)				1.20			
	Limestone	Eogene	79 (2)	51	20		3.95		2.55	
	Lms+marl	Neogene	35 (2)				1.75			
	Fresh–water Lms	E.–M. Pleistocene	1.8 (9)				0.09			
	Travertine	Holocene	155 (23)				7.75			
Evaporite	Gypsum	Mid Jurassic	1.3 (3)	~1						
		Eogene	<1 (2)							
Metamorphic rock	Gneiss	Pre-cambrian	Qilian	109.2 (25)			>5.46			
			Kunlun	5 (1)	61	<20 *				
			Nyainqêntaglha	6 (3)						
			Himalayas	123 (7)			>6.15			
	Schist	Pre-cambrian	Qilian	168 (3)						
			Altun	36 (10)						
			Kunlun	32 (15)	57					
			Nyainqêntaglha	8 (1)						
			Himalayas	39.5 (6)						
Plant (herbs); salinized withered grass	Ash		Recent	120 (99)	120					
			Recent	31 (1)						

Notes: T and W: from K.K. Turekiun and Wedepohl, 1961; B: from A.P. Vinogradov, 1962; * from H. Aubert et al., 1977, comparison on the basis of 40 ppm; * * from J.J. Connor et al., 1975 (Samples from Missouri, U.S.A.

TABLE 1.14. Statistics of average Lithium contents (ppm) of various rock types of the Qinghai–Tibet Plateau

Name		Age	This paper Li aver. ①	Li Aver. ②	T and W ③	B.L. ④	①/③	①/④	②/③	②/④
Ultrabasic rock		Sinian (Xitieshan)	18 (2)	11.67	0.n	0.5	36.00 18.00 16.00			23.33
		Hercynian	9 (1)							
		Terminal Cretaceous	8 (1)							
Basic rock		Caledonian	5 (1)	16.4	17	15	0.29	0.33	0.97	1.09
		Terminal Cretaceous	18.6 (2)				1.09	1.24		
		Tertiary	6 (1)				0.35	0.40		
		Quaternary	36 (2)				2.12	2.4		
Intermediate rock	Diorite	Sinian	22 (2)	32.67	28 (syenite)	20	0.79	1.1	1.17	1.63
		Hercynian	33 (5)				1.18	1.65		
		Indosinian	37 (2)				1.32	1.85		
		Early Yanshanian	5 (2)				0.18	0.25		
		Late Yanshanian	73 (5)				2.61	3.65		
		Himalayan (E₂)	26 (7)	33.96 *			0.93	1.30	1.21	1.70
	Andesite	Early Cretaceous	41 (3)	35.25			1.46	2.05	1.26	1.76
		Late Cretaceous	35 (1)				1.25	1.75		
		Neogene	29 (3)				1.04	1.45		
		Quaternary	36 (5)				1.29	1.80		
Acid rock	Granite	Caledonian	27 (4)	65	40 (W– poor type)	40	0.68	0.68	1.63	1.63
		Hercynian	156 (2)				3.90	3.90		
		Indosinian	19 (1)				0.75	0.75		
		Yanshanian	42 (1)				1.05	1.05		
		Himalayan (E₂)	48 (2)	57 *			1.20	1.20	1.43	1.43
		Himalayan (N₁)	98 (7)				2.43	2.43	0.28	0.28
	Rhyolite	Late Cretaceous	10 (3)	11			0.28	0.28		
		Tertiary	12 (1)				0.30	0.30		

* Average value of extrusive and intrusive rocks.

Clays and argillaceous rock	Mudstone	Devonian	20 (1)	73	60		0.33	0.33	1.22	2.35
	Mudstone	Carboniferous	125 (2)				2.08	2.08		
	Mudstone	Early Cretaceous	20 (3)	141		60	0.33	0.33		
	Mudstone	Neogene	127 (48)				2.12	2.12		
	Loess	L. Pleistocene	227 (21)				3.98	3.78		
	Carbonate clay	Early Holocene	98 (16)	210	57		1.72	1.60	3.68	
	Salt–bearing clay	Holocene Tibet	524 (139) 304				5.33	5.07		
		Qaidam	83 (48)							
Sandstone	Sandstone	Precambrian	24 (4)	57.75	15		1.60		3.85	
	Sandstone	Carboniferous	37 (1)				2.47			
	Sandstone	Jurassic	17 (1)				1.13			
	Sandstone	Early Cretaceous	13 (2)				0.87			
	Sandstone	Cretaceous–Eogene	27 (6)				1.80			
	Sandstone	Neogene	132 (19)				8.8			
	Coarse–fine sand	L. Pleistocene	114 (29)				7.60			
	Sands	Early Holocene	98 (16)				6.53			
	Sands	Late Holocene								
Carbonate	Limestone	Precambrian	15 (2)	121.8	5		3.00		24.36	
	Limestone	Early Cretaceous	22(7)				4.4			
	Marl	Neogene	219 (5)				43.80			
	Travertine	Holocene	120 (2)				24.00			
	Carbonate	Holocene	233 (3)				46.60			
Metamorphic rock	Gneiss	Pre–Sinian	35.7 (26)	44.20						
		Precambrian	52.7 (6)							
	Schist	Pre–Sinian	135 (37)	83						
		Precambrian	31 (6)							
	Slate	Precambrian	298 (5)	166.50	66	40 (slate+ clay)	4.51	7.45	2.52	4.16
		E. Carboniferous	35 (1)				0.53	0.88		
	Granulilite	Precambrian	33 (7)							
	Quartzite	Precambrian	40 (5)							
	Phyllite	Precambrian	177 (1)							
	Marble	E.Carboniferous	29 (1)							
		Precambrian	107 (2)							
		L. Paleozoic	26 (1)							

samples)①. It is uncertain whether the rock was affected by late-stage hydrothermal processes.

The boron content increases markedly in terminal Cretaceous plateau volcanic rocks. Tourmaline basalt (Mei, 1984) of this age has been found in Dayemagou in the western sector of the Gangdise Mountains, where they contain up to 2600 ppm boron.

The rock types with higher than average boron contents on the plateau include rhyolite, clays, basic rocks, mudstone–shale, granite and sandstone. The boron contents show a comparatively small range in magmatic rocks of the same type and age. For example, the intermediate extrusive volcanics intercalated with the Upper Cretaceous red beds at Ngari and Nagqu have boron contents of about 30 ppm. In contrast, the boron contents of limestones of varying ages are generally low, and are only elevated where the argillaceous or organic content is higher. Due to the limited number of samples and the greater diversity of rocks on the plateau, it is difficult to calculate the average boron content of rocks in the whole region, but comparison of the data with similar rocks from elsewhere in the world the rocks of the Qinghai–Tibet Plateau have higher boron contents, except for basic rocks, which generally have similar low boron contents (Table 1.13).

6.2.2. LITHIUM CONTENTS IN ROCKS

The variation in lithium contents in various types of rocks in the region shows the following features.

(1) The Li content increases in the transition from basic to acid rocks: the average Li content is 13 ppm in ultrabasic rocks and up to 43 ppm in acid rocks. This is in accord with the incompatible nature of Li that leads to its concentration in the mineral biotite (Liu et al., 1984). Compared to similar rocks from elsewhere in the world the Li content of volcanic rocks on the plateau are slightly elevated (Table 1.14).

(2) For sedimentary rocks, the Li content is highest in argillaceous rocks and lowest in sandstones with intermediate values in carbonate rocks (Table 1.14). The global average Li content of marine limestone is less than 20 ppm, which is less half that in sandstones from the plateau. In contrast, the Li content in continental carbonate rocks (including Quaternary travertine and Neogene marls) is up to 3 times higher than that in sandstone. It is generally thought that Li is mobilised under supergene conditions and absorbed onto clay minerals. Because the ionic radius of Li (6.8 mμ) is very similar to that of Mg (6.6 mμ), Li substitutes for Mg in carbonate rocks. This feature is especially conspicuous in some arid and semi–arid depositional environments and it has been suggested that the increase in the Li con-

---

① Institute of Saline Lakes of Qinghai, Academia Sinica, 1978.

tent in the black marly deposits in saline lakes of Tibet is very likely related to their carbonate content (Zheng and Liu, 1974). This hypothesis has been verified by the discovery of Li-bearing magnesite in Bangkog Lake and Zabuye Salt Lake and natural lithium carbonate in Zabuye Salt Lake (see Chapter 3 for details). Hence, in Neogene deposits of the Qaidam Basin, the Li content in carbonate-bearing mudstone and marl tends to be higher than in pure mudstone. For example, in Kaitemilike, Neogene mudstone contains 25 ppm Li, while marl contains up to 240 ppm Li (Institute of Saline Lakes, 1978). A similar trend of increasing Li content in Mg-bearing deposits in closed sedimentary basins has been seen elsewhere in the world. For example, magnesite deposits with relatively high Li contents are found in the Tertiary Overton Basin of Nevada, USA (Vine, 1979); the Li content in the Unterbraichbakha magnesite deposit in the former USSR is up to 0.09%; and an increase in Li contents has been noted in Permian marine magnesite-anhydrite and sedimentary dolomites from the Caspian Sea area in the Ural depression (Kornevsky, 1973). In the halite stage of the Mengyejing potash deposit (China), the Li content of the magnesite is 0.0225-0.025% (Xu et al., 1978). Hence, it is apparent that Li may occur as isomorphs of Mg in magnesite in both endogenic and exogenic conditions. This geochemical behaviour of Li is not restricted to saline lake deposits, but can also occur in magnesian rocks (including magnesian clay) in restricted basins Li-Mg exchange, leading to deposition of Li-rich sediments.

(3) The content of Li is high in some metamorphic rocks on the plateau (e.g. slate, phyllite and schist), but relatively low in others (e.g., quartzite and marble). The distribution of Li in metamorphic rocks largely reflects the Li-bearing characteristics of their protoliths.

(4) Some rocks on the plateau show complex variations in their average Li contents as a function of age. In general, the Li content is higher in Cenozoic sedimentary rocks. For example, in clays the average Li content is ~ 72 ppm in the Devonian-Carboniferous, ~ 20 ppm in the Lower Cretaceous, up to 127 ppm in the Neogene, and up to 209 ppm in the Quaternary (Table 1.14).

The Li content of ultrabasic rocks is highest in Precambrian examples, with average Li levels of ~ 20 ppm in ultrabasic rocks across the plateau. Both Li and B contents are higher in ultrabasic rocks on the plateau, considering their largely lithophile behaviour. This may reflect mixing reactions with local crustal rocks.

In intermediate-acid rocks of the plateau Li levels are higher in both late Yanshanian and Quaternary rocks. In comparison, Li is higher in basic rocks of Quaternary volcanic and the Li content tends to increase in terminal Late Yanshanian to Himalayan rocks. In acid (granite) rocks Li contents increase markedly in Hercynian and Himalayan (E_2-N_1) rocks and is relatively low in Caledonian and Indosinian granites (Table 1.14).

TABLE 1.15. Statistics of the average Cesium contents (ppm) of various rock types on the Qinghai–Tibet Plateau

Name		Age	This paper Cs aver.①	Cs aver.②	T and W ③	B ④	①/③	①/④	②/③	②/④
Ultrabasic rock		L.Jurassic	6.7 (3)	4.35	0.n	0.1	67 / 20		43.50	
		L.Cretaceous	2 (1)							
Basic rock	Basalt	L.Jurassic	7 (3)	17 (4)	1.1	1	6.36	7	16	17
		Term.L.Cret.	55 (2)				50	55		
		Tertiary	~0.5 (1)				0.46	20.5		
		Quaternary	5 (1)				4.54	5		
Intermediate rock	Diorite	M.–L.Yanshanian	8 (4)	6.5	0.6 (syenite)		13.33		10.83	
		Himalayan (E₂)	5 (5)	7.08 **			8.33			
	Andesite	E.Cretaceous	4 (12)				6.67		11.8	
		L.Cretaceous–Eogene	14 (9)	7.6			23.33		12.66	
		Tertiary	5 (2)				8.33			
Acid rock	Granite	Caledonian	10 (2)	18.50	4 (Ca-poor type)	5	2.50	2	4.63	3.70
		M.–L.Yanshanian (K₁)	8 (4)				2	1.60		
		M.–L.Yanshanian (K₂–E)	6 (4)				1.50	1.50		
		Himalayan (E₂)	29 (1)	22.50 **			7.25	5.80	5.63	4.50
		Himalayan (N₁)	21 (7)				5.25	4.20	10	7.90
	Rhyolite	Cretaceous	8 (4)	39.50			2	1.60		
		Tertiary	71 (1)				17.75	14.2		
Clays	Mudstone	E.Cretaceous	5.7 (7)	22.85	5	12	1.14	0.48	4.57	1.9
	Carbonate clay	Holocene	40 (2)				6.67	3.33		
Sandstone	Quartz sandstone	L.Triassic	4 (1)	19	0.n		~n.10		~n.10²	
	Sandstone	E.Cretaceous	8 (5)				~n.10			
	Calcareous gravels	E.–M.Pleistocene	17 (1)				~n.10²			
	Sandstone	L.Pleistocene	26 (2)				~n.10²			
	Sandstone	Holocene	40 (1)				~n.10²			
Carbonate	Limestone	E.Cretaceous	6 (7)	7 (9)	0.n		~n.10		~n.10²	
	Travertine	Holocene	10 (2)				~n.10²			
Siliceous sinter		L.Pleistocene–Holocene	515 (22)	515 (22)	(12) *		43		43	

* Replaced by the Cs content of clays; ** The average content off extrusive and intrusive rocks.

6.2.3. CAESIUM CONTENT IN ROCKS

There are scanty data of Cs distribution in rocks on the plateau. The samples discussed here are all taken from the plateau region at elevations greater than 4200 m above sea level in the Kunlun–Himalaya mountains. This preliminary study indicates that the variation in Cs distribution across the plateau has the following features.

(1) Cs contents increase from basic to acidic igneous rocks. The average Cs content in 4 ppm in ultrabasic rocks, 4.7 ppm in basic rocks, 7 ppm in intermediate rocks, and 22.5 ppm in acid rocks. Among the various types of igneous rocks, rhyolite has the highest Cs concentrations of up to 26.5 ppm (Table 1.15).

(2) Sedimentary rocks the Cs contents decrease from clays to sandstones to carbonates. The higher Cs contents in clays reflects absorption of Cs on clays and the presence of Cs in mica minerals. Within sandstones, Cs is largely present in dispersed clay minerals and feldspar grains. The sedimentary rock with the highest Cs contents is siliceous sinter, which contains up to 515 ppm. This is a reflection of the large ionic radius and ionic potential of Cs that enhance its uptake by hydrated silica.

(3) The average Cs content of sedimentary rocks also has a tendency to increase in Cenozoic rocks. For example, Upper Cretaceous clays contain an average of 5.7 ppm Cs, whereas Holocene clays contain up to 40 ppm Cs. In igneous rocks Cs contents are generally relatively high in Late Cretaceous–Early Eogene and Tertiary intermediate–acid rocks. For example, in Himalayan (Tertiary) acid rocks there is a marked increase in Cs contents and the average Cs content reaches 45 ppm in Tertiary rhyolite in the southern part of Coqên.

TABLE 1.16. Rb / Cs ratios of various rock types on the Qinghai–Tibet Plateau (south of the Kunlun Mountains)

Type	Ultra–basic rock	Basic rock	Inter–mediate rock	Acid rock	Shale+ clay	Clay	Sandstone	Carbonate
Qinghai–Tibet Plateau (south of Kunlun)	5.28	4.28	15	5.69	2.36	0.54	4.24	3.21
Crust	20	40	500	400	16.67	66.67	600	30–3.33

(4) The Cs contents of all the rocks in the region are a few to tens of times higher than average Cs in abundances in equivalent rocks from elsewhere in the world (Table 1.15). The rocks with particularly high Cs contents include ultrabasic rocks, Tertiary rhyolites and Quaternary sedimentary rocks. The exception is provided by Early Cretaceous mudstone whose Cs contents are relatively low (average

TABLE 1.17. Statistics of the average Rb contents (ppm) in various rock types on the Qinghai–Tibet Plateau

Name		Age	This paper Rb aver. ①	Rb aver. ②	T and W ③	B ④	①/③	①/④	②/③	②/④
Ultrabasic rock		Hercynian	22 (1)	23	0.2	2	110	11	115	11.50
		L. Jurassic	38 (3)				190	19		
		L. Cretaceous	9 (1)				45	45		
Basic rock		Term. L. Cretaceous	< 20 (4)	30	45		0.66	0.44	0.66	0.44
Inter-mediate	Diorite	Hercynian	100 (4)	100.33	110 (syenite)	100	0.91	1	0.91	1
		Indosinian	79 (3)				0.72	0.79		
		E.–M. Yanshanian (J–K₁)	103 (3)				0.94	1.03		
		L. Yanshanian	92 (4)				0.84	0.92		
		Himalayan	115 (7)	106.67			1.05	1.15	0.97	1.07
		Quaternary	113 (1)	*			1.03	1.13		
	Andesite	E. Cretaceous	50 (12)	113			0.45	0.50	1.03	1.13
	Andesite-dacite	L. Cretaceous –Eogene	41 (15)				0.37	0.41		
	Andesite	Eogene	183 (1)				1.66	1.83		
	Andesite	Quaternary	178 (5)				1.62	1.78		
Acid rock	Granite	Caledonian	233 (2)	192	170	200	1.37	1.17	1.13	0.96
		Hercynian	129 (4)				0.76	0.65		
		Indosinian	97 (4)				0.57	0.49		
		M.–L.Yan. (K₁)	243 (4)				1.43	1.22		
		M.–L. Yan. (K₂)	180 (6)				1.06	0.90	0.75	0.64
		Himalayan (E₂)	279 (2)	128 *			1.64	1.40		
		Himalayan (N₁)	284 (7)				1.67	1.42		
	Rhyolite	E. Cretaceous	82 (3)	54			0.48	0.41	0.32	0.27
		L. Cretaceous –Neogene	37 (3)				0.22	0.19		
		Tertiary	43 (1)				0.25	0.22		
Clays	Mudstone	E.Cretaceous	103 (3)	54.00	140	200	0.74	0.52	0.39	0.27
	Mudstone	Miocene	90 (1)				0.64	0.45		
	Carbonate clay	Holocene Tibet	21.5 (2)		110		0.20	0.11	0.49	
		Holocene Qaidam	1.6 (2)				0.02	0.01		
Sand-stone	Quartz sandstone	L.Triassic	17 (1)	80.50	60		0.28		1.34	
	Sandstone+ conglomerate	E. Cretaceous	75 (3)				1.25			
	Conglomerate	E.–M. Pleistocene	65 (1)				1.08			
	Sandstone	L. Pleistocene	165 (2)				2.05			
Carbo-nate	Limestone	E. Cretaceous	31 (6)	22.5	3		10.33		7.5	
	Travertine	Holocene	14 (2)				4.67			

* The average content of extrusive and intrusive rocks

Cs abundance of 0.48 ppm).

(5) The majority of different rock types on the earth have Rb / Cs ratios in the range 16–30, with sandstones having higher ratios up to 600. However, equivalent rock types on the plateau generally have relatively low Rb / Cs ratios, i.e., rocks from this area are comparatively enriched in Cs (Table 1.15). This enrichment of Cs is a geochemical feature of rocks from the surface of the plateau with elevations over 4200 m above sea level (Table 1.16).

6.2.4. RUBIDIUM CONTENTS IN ROCKS

The samples discussed here are all taken from the plateau region at elevations greater than 4200 m above sea level in the Kunlun–Himalaya mountains. This preliminary study indicates that the variation in Cs distribution across the plateau has the following features.

(1) Rb contents have a tendency to increase from basic to acidic igneous rocks. The average Rb content in ultrabasic rocks is about 23 ppm, and reaches up to 128 ppm in acid rocks. The various types of igneous rocks also show different Rb contents. The highest Rb contents, up to 192 ppm, and seen in granites, followed by andesite, whith Rb contents of up to 113 ppm (Table 1.17).

(2) In sedimentary rocks, Rb contents decreases in the order of mudstones to sandstone to clays to carbonates, i.e., a similar pattern to that observed for Cs distributions.

(3) The Rb contents in diorite, andesite and granite of this region have a tendency to increase slightly as the rocks become younger, although Upper Jurassic ultrabasic rocks and Early Cretaceous rhyolites also have relatively high Rb contents. The Rb contents of sedimentary rocks also have a tendency to increase in younger rocks, although an opposite trend is observed in clay minerals. It is possible that the replacement of Rb by Na and K in slightly alkaline waters on the plateau may account for the recent depletion in Rb contents, but further study is required.

(4) In general, the Rb contents of rocks on the plateau are similar to those seen in similar rocks from elsewhere on the earth. The only exception is argillaceous rocks whose Rb contents is notably low relative to average global Rb abundances and ultrabasic rocks, which have relatively high Rb contents. Again, this feature is similar to that seen in Cs, Li and boron distributions.

Rb, Cs, Li and boron are typical lithophile elements whose abundances are all higher in the earth's crust than in the upper mantle. However, the average contents of these elements in ultrabasic rocks on the plateau are anomalously high, approaching those seen in the crust (Table 1.18). This enrichment is difficult to account for on a petrologic basic. The location of geothermal springs on the margins of ultrabasic bodies (e.g., the Zigedan Lake spring group) suggests that geothermal activity may be responsible for enriching these rocks in Rb, Cs, Li and boron.

TABLE 1.18. Correlation of Li, Cs, Rb and B contents (ppm) between ultrabasic rocks of the Qinghai–Tibet Plateau and the crust and mantle of the earth

	Name	Li	Cs	Rb	B	Data source
Ultrabasic rock	Qinghai–Tibet Plateau	11.67	4.35	23	28.5	This paper
	Global	0.5	0.1	2	1	A.L. Vinogradov (1962)
Crust	Continental	21 $\frac{32}{7}$	1.4 $\frac{1.1}{2.8}$	78 $\frac{113}{3.5}$	13	From Liu Yingjun (1979, 1984)
	Marine					
Mantle	Upper	4.1	0.3	2.6	2	
	Lower	0.5	0.1	2	1	
Ratio	$\frac{\text{Qinghai – Tibet Plateau}}{\text{Global}}$	23.34	43.50	11.50	28.50	
	$\frac{\text{Qinghai – Tibet Plateau}}{\text{Crust}}$	0.56	3.11	0.30	2.19	
	$\frac{\text{Qinghai – Tibet Plateau}}{\text{Upper mantle}}$	2.85	14.50	8.85	14.25	
	$\frac{\text{Qinghai – Tibet Plateau}}{\text{Lower mantle}}$	23.34	43.50	11.50	28.50	

6.3. Boron Content in Plants

Boron contents were measured in 97 plant samples from the northern Tibet lake district. The sampling localities were generally selected from the desert–steppe area at elevations of 4350–4600 m above sea level. The boron within the plants was extracted into solution and the results of the analyses are presented in Table 1.19.

Plant ash in the district has an average boron content of 115 ppm. This is approximately an order of magnitude higher than the global average and indicates that there is strong absorption of boron by the plants from the high boron contents in the Tibet lake district. Boron has an important role in physiology of plants (Vinogladov, 19XX) and this is reflected in the physiology of the plants that are growing in the Tibet lake district. The boron–rich areas are generally barren of vegetation, but relatively rare species of plants are found within these area. On the margins of the boron–rich areas and some of the magnesium borate layers there are usually sparse growths of such plants as *Polygonum sibricum,* saline alkali grass and *Potentilla* sp. Of these, *Polygonum sibricum* is generally present in a morbid state with its leaves turning red, while the saline–alkali grass is apt to turn yellow. Lush

TABLE 1.19. Boron content in herbs (Ash) in Northern Tibet

Serial No.	Name	No. of samples	B content (%)	Aver. B content (%)
1	Sisymbium humila	1	0.053	0.053
2	Polygonum sibiricum	10	0.095–0.00	0.032
3	Thlrmopsis sp.	1	0.031	0.031
4	Hiracllum sp.	1	0.028	0.028
5	Leontopodium alpiuum	9	0.061–0.003	0.020
6	Aster bowerii	1	0.017	0.017
7	Potentilla sp.	18	0.032–0.008	0.016
8	Iris cf. dutherii	1	0.016	0.016
9	Morina parvitlova	1	0.015	0.015
10	Artemisia stracheyi	2	0.015–0.013	0.014
11	Hlmistepts	1	0.010	0.010
12	Lagotis brachyslachys	8	0.013–0.004	0.007
13	Saussurea sp.	1	0.007	0.007
14	Christolea sp.	1	0.006	0.006
15	Androsace tapete	2	0.009–0.004	0.006
16	Oxytropis chiliophylla	8	0.014–0.000	0.005
17	Oxytropis mollis	1	0.005	0.005
18	Pldicularis alaschanica	4	0.005–0.001	0.003
19	Stipa aliena	6	0.005–0.000	0.002
20	Stipa purpurea	16	0.004–0.000	0.001
21	Allium przlwals kianum	3	0.000	0.000
22	Cragana sp.	1	0.000	0.000

growths of plants can only be seen in swamp and spring areas that are relatively far from boron–rich areas. These plant species all have relatively high boron contents; for example, the boron content of ash from *Polygonum sibricum* reaches 320 ppm, while the ash of *Potentilla libura* contains up to 160 ppm. Hence, these plants are best described as "boron–enduring" plants. (The boron content in halophilic algae will be discussed in Chapter 8).

The strong absorption of boron by organisms is seen in the sedimentary records. Tertiary oily, grayish—black shale from Lunpola contains up to 300 ppm boron and the organic mud—lime from the bottom of Bangkog Lake contains 500–5,000 ppm boron. It appears that plants on the plateau may have played an important role in boron concentration in saline lakes during the Cenozoic biotic evolution, but further studies of the biogeochemical cycle of boron await further study.

6.4. Regional Hydrochemistry

Inland drainages (rivers, springs and groundwater) usually ultimately converge in lakes and become the main source of lake waters and solutes, which are genetically associated with the chemical and physical composition of lake waters and sediments. Hence, if we are understand the composition of lake waters it is necessary to study the composition of the hydrochemical features of these natural waters. Systematic hydrochemical studies have not been carried out on river waters and ground water on the whole plateau and the relevant data are scarce. There are only a few hydrographic stations and there are few long—term observations and even less information concerning concentrations of boron, lithium and potassium. Hence, only a preliminary discussion is possible here, based on analyses of waters collected during many years of reconnaissance by the author and co—workers, together with some relevant data collected previously.

6.4.1. HYDROCHEMICAL FEATURES OF RIVER WATERS

Analyses of more than 100 samples from 70 different rivers on the plateau indicate that there is some zonation in the hydrochemistry of river waters. The distribution of river hydrochemical zones are similar to those of lake waters on the plateau described in Chapter 4. Some rivers also show anomalous boron and lithium concentrations. The whole plateau can be divided into six hydrochemical zones (Fig. 1.19).

Zone I.   This zone is dominated by Ca—Na bicarbonate waters (on the basis of the Shokalev classification). Water samples were taken from 45 rivers within this zone, of which 31 rivers belong to the bicarbonate ($HCO_3^-+CO_3^{2-}$) group of waters, 2 rivers are from the sodium chloride—bicarbonate group, 1 is a sodium sulphate—chloride water and the other belongs to the sodium chloride—sulphate type. The Gangdise—Nyainqêntanglha mountains from the boundary of this zone. The southern part of the zone (subzone I_1) is an the inflow—outflow area in which the rivers have an average salinity of about 150 mg/l and are, hence, belong to the weakly mineralised class of water①. The northern part of the zone (subzone I$_2$) is an inflow area in which the rivers have an average salinity of about 1100 mg/l, and thus belong to the group of highly mineralised waters (Table 1.20).

TABLE 1.20. Salinities (mg/l) and B, Li and K contents of river waters on the Qinghai–Tibet Plateau

Hydro-chemical zone	I I$_1$	I$_2$	II	III	IV	V	VI	Aver.	World river water	World sea water
Salinity	901 (35)①	150 (7) / 110 (28)	1010 (9)	1134 (21)	576 (2)	324 (11)	150 ± ②	682.5	–	35000
B$_2$O$_3$	17.32 (28)		2.7 (9)	7.68 (10)	11.18 (2)	–	–	9.72 (I–IV)	0.032199	14.33
B	5.38 (28)		0.84 (9)	2.39 (10)	3.47 (2)	–	–	3.02 (I–IV)	0.01 (0.119)	4.45
Li	0.48 (18)		0.14 (4)	0.56 (7)	–	–	–	0.39 (I–III)	0.0011	0.17
K	10.58 (32)		5.29 (14)	28.30 (12)	8.1 (1)	8.95 (2)	–	12.24 (I–V)	6.5	392
B/Cl	0.0483		0.0062	0.001	0.0065	–	–	0.0155 (I–IV)	0.0013	0.00023
Li · 10³ / Σ salts	0.53		0.14	0.082	–	–	–	0.25 (I–III)	–	0.003
Li · 10³ / Σ salts	11.74		5.24	25.04	14.06	27.62	–	16.74 (I–V)	–	11.2

① Number in parentheses refers to the average number of samples (rivers); all data from the present party except that noted. ② Estimated from Guan Zhihua et al. (1985).

Zone II. This zone is dominated by Ca–Na sulphate and Na chloride waters. Among the 16 rivers selected, 10 are chloride waters (of which 5 are Na chloride waters and 5 are Na bicarbonate–chloride waters) and 6 are Ca–(Na) sulphate waters. The entire zone belongs to an inflow area and the river waters have an average salinity of about 1000 mg/l, i.e., at the lower limit of the group of highly mineralised water.

---

① The river waters are classified in terms of salinity as follows: weakly mineralised water (<200 mg/l), moderately mineralised water (200–500 mg/l), strongly mineralised water (500–1000 mg/l), and highly mineralised water (>1000 mg/l).

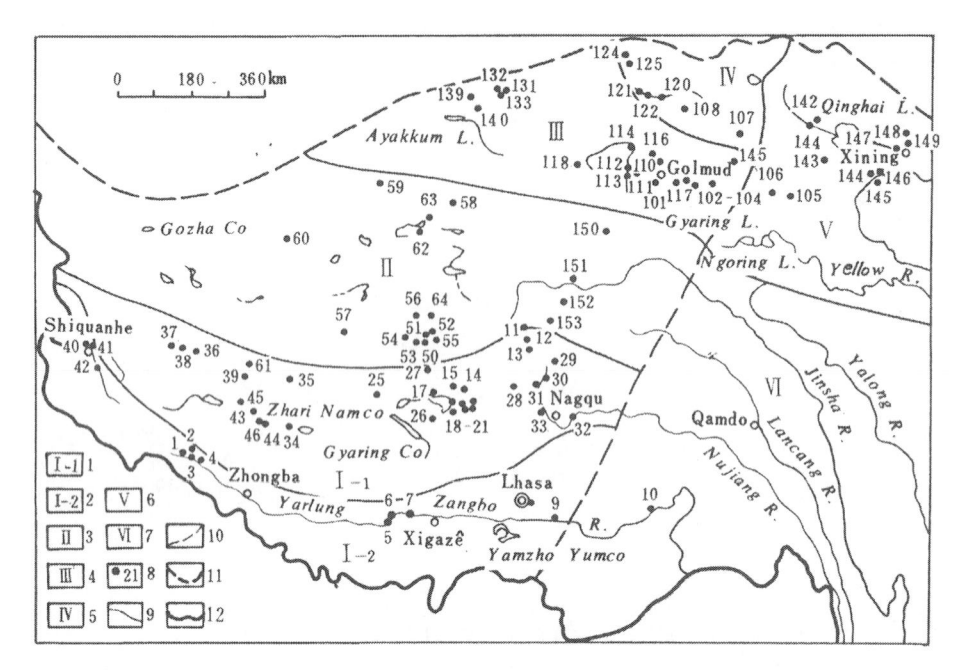

Figure 1.19. Sketch map showing the hydrochemical zones of river waters on the Qinghai–Tibet Plateau. 1. Ca–Na–HCO$_3$ water: salinity < 200 mg/l; 2. Ca–Na–HCO$_3$ water: salinity 500–1000 mg/l; 3. Ca–Na–Cl–SO$_4$ water: salinity > 1000 mg/l; 4. Na–(Ca)–SO$_4$–HCO$_3$–Cl water: salinity > 1000 mg/l; 5. Ca–Na–HCO$_3$–SO$_4$ water: salinity 200–1000 mg/l; 6. Na–Ca–Cl–HCO$_3$ water: salinity 200–500 mg/l; 7. Ca–(Mg)–HCO$_3$–(SO$_4$) water: salinity < 200 mg/l; 8. Sampling locality; 9. Boundary of zones (including subzones); 10. Inferred boundary; 11. Northern boundary of the Qinghai–Tibet Plateau; 12. National boundary.

Zone Ⅲ. Na–(Ca) chloride waters are dominant in this zone. Among the 18 rivers in this zone, 13 are chloride waters (of which 7 are Na bicarbonate–chloride waters, 5 are Na–(Ca) sulphate–chloride waters and 1 is a Na chloride water (the lower reaches of the Urt Moron River). In addition, 4 rivers belong to the Na–(Ca) chloride–bicarbonate group of waters and 1 belongs to the Na chloride–carbonate group (e.g. the Golmud River).

Zone Ⅳ. In this zone, Ca–Na bicarbonate–sulphate waters predominate. Of the 7 rivers in this zone, 3 are Na sulphate waters, 1 is a Na sulphate–chloride water and 2 are Ca bicarbonate rivers. This zone is an inflow area and the average salinity of rivers is 570 mg/l, which exceeds the lower limit of the strongly mineralised group of river waters.

Zone Ⅴ. This zone contains Ca–(Na) chloride–bicarbonate waters. Among the 10 rivers in this zone, 3 are Ca–(Na) chloride–bicarbonate waters, 4 are Ca–(Na) bicarbonate waters, and there is one each in the Ca sulphate–bicarbonate,

Na bicarbonate–sulphate and Na bicarbonate (Xiangride River, No. 100 in Fig. 1.19) groups of waters. There are high levels of alkali metals in some of the fluvial swampy areas of this zone. This zone contains both inflow and outflow areas (Fig. 1.10). The inflow area include the Qinghai Lake and Huobuxun Lake districts, in which rivers of the Ca–(Na) chloride–bicarbonate types predominate (Nos. 141 and 105 in Fig. 1.19), and there are also Na bicarbonate–sulphate (No. 143) and Na bicarbonate waters. The rivers of this zone have average salinities of 257 mg / l, and belong to the moderately mineralised group of waters.

Zone VI. According to Guan et al. (1984) the river waters in this zone mostly belong to the Ca–(Mg) bicarbonate type, although some belong to the Ca–(Mg) sulphate type, such as those found in the Nyingchi reach of the Nyang Qu (River) (No. 10 in Fig. 1.19). Most parts of the zone belong to the outflow area, and the river waters have low salinities (< 200 mg / l) (Guan et al., 1984), and are classified as weakly mineralized river water.

6.4.2. B, Li AND K CONTENTS

There are few data for B, Li and K contents in river waters on the plateau and no separate analyses for Na and K have been made in hydrogeological surveys, hence the hydrogeological distribution of K in river waters is presently unknown.

The boron content of global river waters is generally low, around 0.01 ppm (Liu et al., 1984), although Biqiyehua (1981) suggested that the boron content in river waters range from 0.2 to 0.038 ppm, with a mean of 0.12 ppm. The average B / Cl ratio in global river waters is about 0.0013, which is about four times higher than in seawater. The average boron concentration in river waters of the plateau (zones I–V) is 3 mg / l, which is about 25–300 times higher than the average value for world river waters. Their B / Cl ratio are 12.5 times higher than that of world river waters and 70 times higher than that of seawater. Among the various hydrochemical zones of the plateau the highest boron contents are seen in river waters zone I, with those from zones IV and III having the next highest and the rivers of zone II having the lowest boron contents (Table 1.20).

The average Li content in world river waters is much lower. The Li content of seawater is 0.17 ppm with $Li \times 10^3 / \sum_{salts} = 0.00049$. The average Li content of river waters on the Qinghai–Tibet Plateau is about 0.4 mg / l, which is 36 times higher than that in seawater (Table 1.20), while the $Li \times 10^3 / \sum_{salts}$ ratio being 125 times higher. The relative Li conentrations in the various hydrochemical zones are similar to those of boron.

The average K content in river waters of the plateau is about 9.9 mg / l, which is slightly higher than that for world river waters (6.5 mg / l), but 38 times lower than that in seawater. The $K \times 10^3 / \sum_{salts}$ ratio is slightly higher than that for seawater (Table 1.20). K concentrations are highest in zone III with the next highest concentrations being in zone I.

These preliminary data suggest that the river waters of the plateau have distinct boron and Li anomalies, with less conspicuous K anomalies.

6.4.3. HYDROGEOCHEMISTRY OF GROUNDWATER

On the basis of their hydrogeochemical origin, groundwaters may be classified into three basic types (Shen, 1986): (1) lixiviation—infiltration water, which is of meteoric origin and formed by the interaction of water and rock; (2) sedimentation—buried water, known as connate water; (3) juvenile water, which is of volcanic, magmatic, metamorphic or hot spring origin. The first two types are known as exogenous water, while the latter type is called endogenous water. In general, the endogenous water is usually mixed with exogenous water. In particular, the origin of the hot spring waters is complex. In the lake waters of the Qinghai—Tibet Plateau, the waters that are most importance in recharging the saline lakes are lixiviation—infiltration waters (roughly equivalent to phreatic water and a protion of interstitial water), with geothermal waters being of secondary importance. Geothermal waters are widespread on the plateau and there are large discharges in some areas, hence they have important effects on the trace and major element chemistry of some saline lakes. Connate waters, including sedimentary porewaters and oil field waters is only of major recharge significance in local areas, e.g., in some saline lakes of Qaidam, particularly in Late Quaternary saline lakes in oil field areas in the western part of Qaidam.

Lixiviation—infiltration waters
Normal temperature freshwater springs[①] are common on the plateau, and freshwater layers are often encountered in drill holes in the lake districts. These types of groundwaters largely fall in the lixiviation—infiltration category of waters.

This type of spring is generally seen in clusters on the ground surface, particularly in alternating sand and mud zones on lake edges and in locations where fault zones intersect. So far, 121 springs and 26 drill holes have been identified containing freshwater interlayers. These waters may be grouped into four hydrogeochemical zones (Fig. 1.20).

a. *Na—(Ca) bicarbonate water zone* (I).   On the basis their salinity this zone can be divided into two subzone I_1 and I_2, with the Gangdise—Nyainqêntanghla mountains as the boundary. The salinity of subzone I_1 is < 200 mg/l, while that of subzone and I_2 is about 1260 mg/l. Na—(Ca—Mg) bicarbonate waters predominate in this zone.

[①] Including rising springs (interlayer freshwater) and gravity springs (commonly phreatic water).

b. Ca–(Na–Mg) chloride–sulphate–bicarbonate water zone (II). In this zone bicarbonate waters are slightly more abundant than sulphate waters. This entire zone is an inflow area, with an average salinity of about 1100 mg / l.

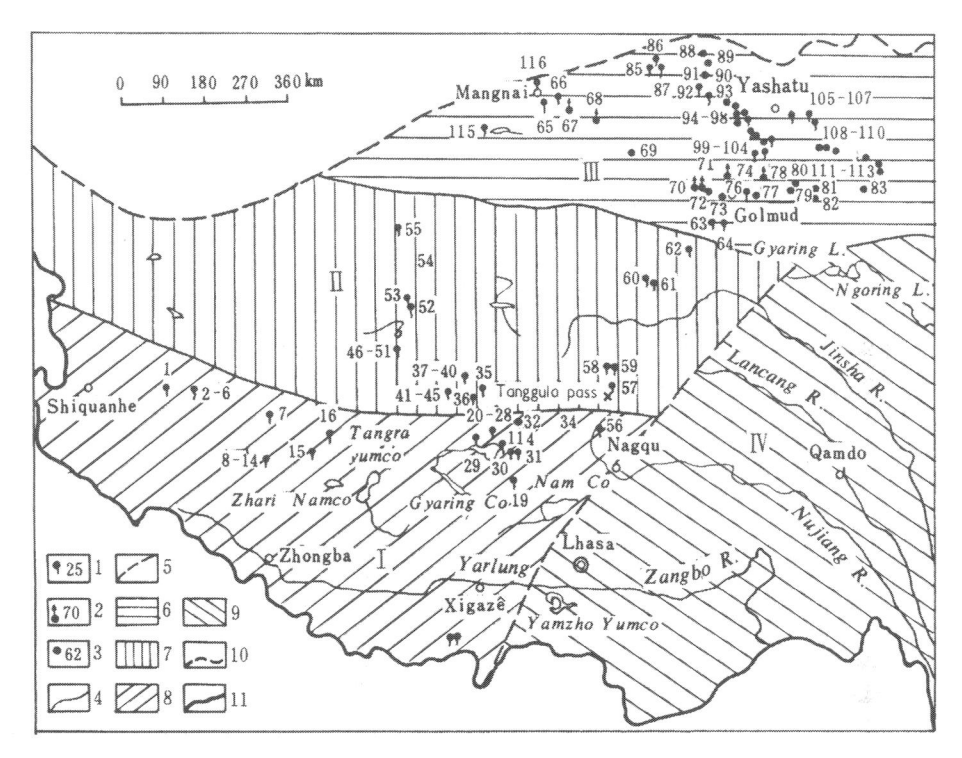

Figure 1.20. Sketch map showing hydrochemical zones of spring waters and lixiviation–infiltration waters on the Qinghai–Tibet Plateau. 1. Spring water; 2. Artesian water; 3. Non–artesian water; 4. Boundary of zones; 5. Inferred boundary of zones; 6. Na–Ca–SO$_4$–Cl–water (average salinity about 4700 mg / l); 7. Ca–(Na–Mg)–Cl–SO$_4$–HCO$_3$ water (average salinity 1080 mg / l); 8. Na–(Ca)–HCO$_3$ water (average salinity about 1260 mg / l); 9. Ca–(Na–Mg)–SO$_4$–HCO$_3$ water, inferred (average salinity less than 200 mg / l); 10. Northern boundary of the Qinghai–Tibet Plateau; 11. National boundary.

c. Na–(Ca) sulphate–bicarbonate–chloride water zone (III). Chloride waters predominate in this zone. Among the 47 springs studied in this area 33 are chloride waters, of which 29 are Na chloride waters. There are 3 Ca–chloride springs and 1 Mg–chloride spring in opposite corners of the basin, while the others are all distributed in the west–central to southeast parts of the basin. Na–(Ca–Mg) bicarbonate waters are distributed near the outer margins of the lake basin and Ca sulphate waters are found on the northern margin. The salinity of springs in this zone reaches as high as 4700 mg / l.

d. *Inferred Ca–(Na–Mg) sulphate–bicarbonate water zone* (IV). This zone is part of an outflow area, but there are only few data available. The basis of these data, the waters have been preliminary classified as Ca–(Na–Mg) sulphate–bicarbonate waters and are of relatively low salinity.

TABLE 1.21. Salinities (mg/l) and B, Li and K contents of spring waters (lixiviation–infiltration waters) on the Qinghai–Tibet Plateau

Item	Qinghai–Tibet spring water (interlayer freshwater) I I_1 / I_2	II	III	Aver. ①	Aver. of plateau river water ②	World river water ③	World sea water ④	①/②	①/④	②/③
Salinity	1258 (40) / 34(2) / 1573 (39)	1080 (28)	4720 (79)	2353	683	202	35000	3.45	0.067	3.38
B_2O_3	21.33 (31)	3.14(8)	11.82 (8)	12.10		1.38	14.33	1.24	0.84	7.04
B	6.59 (31)	0.98(8)	3.67 (8)	3.75	9.72	0.43	4.45	1.24	0.84	7.02
Li	1.15 (21)		0.4 (1)	0.78 (I,III)	3.02	0.17	0.17	2.00	4.59	2.29
K	17.98 (37)	23.74 (14)	95.83 (16)	45.85	0.39	6.2	392	3.75	0.12	1.91
Cl	187 (33)	195.41 (8)	134.29 (10)	172.23	12.24	7.8	19000	0.214	0.009	103
$Li \cdot 10^3 / \sum salts$	0.914		0.085	0.4995 (I,III)	803 / 0.25	0.84	0.003	1.99	166.5	0.30
$K \cdot 10^3 / \sum salts$	14.29	21.98	5.03	13.77	16.74	30.69	11.2	0.82	1.23	0.55
$B_2O_3 \cdot 10^3 / \sum salts$	16.88	2.9	2.5	7.43	6.83	0.40	0.30	12.83	2.50

B, Li and K have been measured in waters from zone I–III, and the analyses show the following features:

(i) The average boron content in the spring waters and a portion of the interlayer freshwaters on the plateau is slightly higher than in the nearby rivers, while the K and Li concentrations of the springs are between 2 and 3.5 times higher those of the rivers (Table 1.21). In comparison with seawaters, the boron contents are slightly lower, while the Li contents are 3.5 times higher and the K contents are 8 times lower.

(ii) The B, Li and K concentrations in the spring waters in the various hydrochemical zones show the same variation trneds as those in the river and lake waters, i.e., the B and Li concentrations in zone I are higher than those in zones II and III, but the K content increases from zone I to zones II and III (Table 1.21).

Thermal water and mud–volcano water

The Qinghai–Tibet Plateau is one of the world's most active geothermal areas. There are more than 600 groups of hot springs that are widespread across the plateau. In addition, there are also local mud–volcano waters. According to ^{14}C datas and stratigraphic relationships all these features were formed after the middle and late Late Pleistocene. They were most likely formed as the result of intensification of thrusting of the plateau accompanied by crustal thermal expansion in the fourth phase of collision that caused large thermal flows at deeper levels to upwell at the ground surface along the fracture system created by the collision. On the basis of an analysis of the relation between the distribution of hot springs and fracture systems on the plateau it is apparent that most hot springs formed along N–S–, NE– and NW–WNW–trending active fractures, especially at loci where two or more fracture zones intersected. The hot springs (locally mud volcanoes) are distributed in groups and clusters and commonly concentrated near sutures on the plateau. From north to south the plateau can be divided into five geothermal zones.

TABLE 1.22. Frequencies (%) of hydrochemical types in geothermal zones I–V on the Qinghai–Tibet Plateau

Zone \ Type	B-rich-Na-Cl water	B-bearing Na-Cl water	Na-HCO$_3$-Cl water	Na-HCO$_3$-SO$_4$ water	Na-HCO$_3$-SO$_4$ water	Na-HCO$_3$ water	Ca-HCO$_3$ water	SO$_4$-HCO$_3$ water	Na-SO$_4$-Cl water	Na-SO$_4$-HCO$_3$ water	Mg-HCO$_3$ water	Na-Cl-SO$_4$ water
I	20	–	32.3	22.8	7.4	3.4	4.7	5.4	3.4	–	–	–
II	4.7	2.8	23.3	24.3	22.4	6.5	5.6	8.4	1.9	–	–	–
III	–	11.21	–	22.2		22.2	–	22.2	–	11.1	–	11.1
IV	25(mud volcanos)	12.5	–	–	12.5	–	6.25	–	25	–	6.25	12.5
V	–	–	10	10	–	10	40	–	–	–	10	20

a. Shiquanhe–Yarlung Zangbo geothermal zone (I). This zone is one of the world's most important geothermal zones. Trending WNW, it lies largely along the Yarlung Zangbu suture zone. Its western part trends towards the NW from Shiquanhe and enters Kashmir to link with the geothermal zones of Afghanistan, Iran and Turkey. Its eastern sector swings SSE and passes Tengchong to link with the geothermal zones of Burma and Thailand. This zone, especially in its part west of eastern Tibet, is characterised by the presence of some special hydrochemical types, high temperatures, enrichments in B, Li, Cs, As and Cl, and the production of abundant geyserite.

TABLE 1.23. Frequencies (%) of trace elements and temperatures in geothermal zones I–V on the Qinghai–Tibet Plateau

Zone	% in various $\frac{HBO_2 \cdot 10^3}{\sum salts}$ intervals					% in various $\frac{Li \cdot 10^3}{\sum salts}$ intervals						% in various $\frac{Cs \cdot 10^3}{\sum salts}$ intervals				
	10–40	80–120	160–200	280	360–440	0–1	2–3	4–5	6–7	8–9	10–11	0–1	2	3	4–6	10–13
I	41.01	33.33	12.99	5.13	7.63	48.6	24.32	10.8	5.4	8.1	2.7	60.1	15.8	2.6	10.5	10.5
II	97.26	2.74				77.8	13.8	7		1.4		93.2	6.8			
III	100					100						100				
IV	60	20	20			–						–				
V	100					100						–				

Zone	% in various $\frac{As \cdot 10^3}{\sum salts}$ intervals			% in various $\frac{F \cdot 10^3}{\sum salts}$ intervals					% in various temperature (°C) intervals				
	0–2.5	5	7.5	1–3	6	9	12–18	21–27	36	<20	20–40	40–60	60–85	>85
I	82	10.26	5.74	48.72	28.21	7.69	10.25	5.13		10	10	30	20	30
II	100			85.91	9.96	1.41	1.41	1.41		15.1	36.6	24.4	22.3	1.2
III	100			80	20					33.3	11	55.6		
IV	–			–						–	14.9	14.3	42.9	
V	–			25	25	25		25		40	30	30		

(i) Hydrochemical types. Na bicarbonate–chloride, B–rich–Na chloride and Na bicarbonate–sulphate waters predominate in the zone, accounting for about 75% of all the hot springs of this zone (Table 1.22). These water types are generally considered to be related to magmatic activity (Truesdell, 1975; Dong et al., 1981).

(ii) Water temperature. This zone is dominated by high–temperature and super–heated waters (low–temperature hot springs <20 °C; low temperature thermal water 20–40 °C; medium–temperature thermal water 40–60 °C; high temperature water 60–85 °C; super heated thermal water >85 °C –boiling point). The waters with temperatures of 60–>85 °C account for 50% of the geothermal waters (Table 1.23). Hence, there are large amounts of geyserite and Cs–bearing geyserite

TECTONOGEOCHEMISTRY AND REGIONAL GEOCHEMISTRY

being precipitated from the super-heated waters.

(iii) Features of trace element contents. The spring with the highest B content is the Sogdoi Hot Spring in the Moincêr area, Gar county, which has a boron content of up to 1917 mg/l. This is followed by the Lhabu Lang hot spring in the Dobê area, Ngamring county, which has a boron content of up to 1750 mg/l. The spring with the highest Cs content is the Lhabu Lang hot spring and Semi hot spring in the same area, with Cs contents up to 51 mg/l. The Sogdoi and Semi hot springs have the highest Li contents of 50 and 35 mg/l, respectively. The Sogdoi hot spring also has the highest As content of up to 126 mg/l. followed by the Semi hot spring, with an As content of 31 mg/l. The highest F content is present in the Dagejia geyser in the Sangsang area, Ngamring county, with 25 mg/l of fluorine. Lhabu Lang hot spring also has high F contents of 20 mg/l. This zone has the highest concentrations of trace elements of all the zones on the plateau. Enrichments of Br (10.9 mg/l in Qiwulan hot spring, Barga area, Burang county) and Rb (2.5 mg/l in Semi hot spring) are also seen in this zone (Dong et al., 1981). The ratio of B, Li, Cs, As and F total salts are also highest in the Shiquanhe-Yarlung Zangbo geothermal zone (Table 1.23).

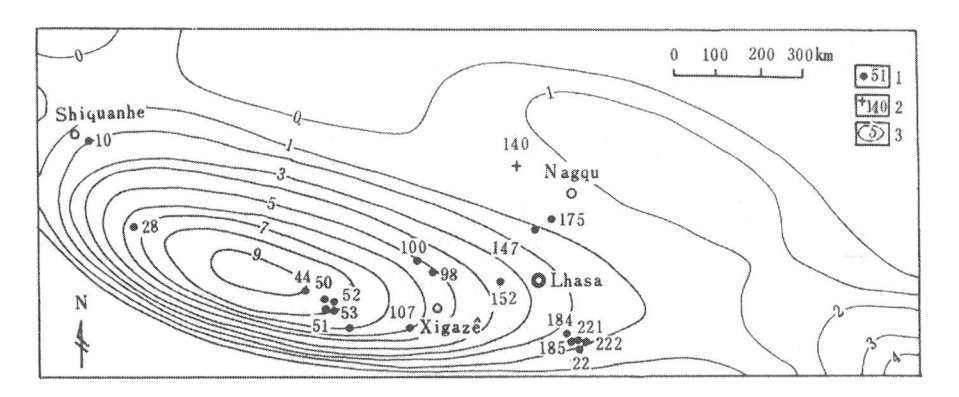

Figure 1.21. Trend surface analysis of Li in geothermal waters on the Tibet Plateau. 1. Li contour (in ppm).

We have also studies the trend in Li and Cs concentrations in geothermal waters on the Tibet Plateau①. Three different examples of these trend are shown in Figures I-21, 22 and 23. The Li and Cs contours are equispaced with units of mg/l. The contours with the highest contents are 7-8 ppm for Li, 9 ppm for Cs and 200-300 ppm for boron.

---

① Most of the analytical data is from Dong et al. (1981), with some from the author's laboratory.

CHAPTER 6

b. Bangong Lake—Nujiang River geothermal zone (Ⅱ). This zone lies north of the Shiquanhe—Yarlung Zangbo geothermal zone (I). The bulk of zone Ⅱ tends WNW and is largely located along the Bangong Lake—Nujiang River suture, although some hot spring are also found along its northern and southern sides. This zone is characterised by abundant low— to high—temperature thermal waters and travertines, with enrichments of B, Li, Cs, Rb, As and F.

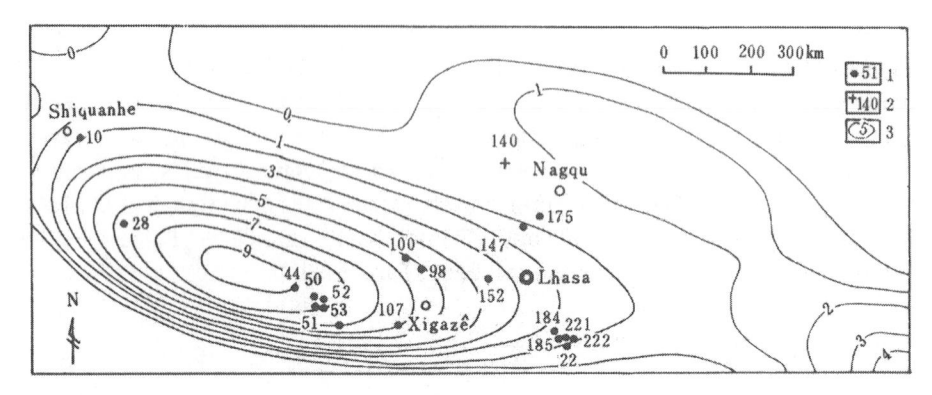

Figure 1.22. Trend surface analysis of Cs in geothermal waters on the Tibet Plateau. 1. Hot spring with older and younger geyserite and its No.; 2. Hot spring with older geyserite and its No.; 3. Contour of Cs (in ppm).

(i) Hydrochemical types. In this zone Na bicarbonate—chloride waters and Na bicarbonate—sulphate waters account for 48% of the total thermal waters. Ca bicarbonate—sulphate waters account for a higher proportion (23%) than in observed in zone I. There are also a small number of B—bearing Na chloride waters (Table 1.22).

(ii) Water temperatures. Low— high—temperature thermal waters predominate (83% of the waters), although some superheated waters are also present. Low—temperature springs are also more abundant, compared with zone I (Table 1.23). There is a large amount of travertine in this zone, and geyserite is only present locally (e.g., in the Zigedan Lake and Nau Co area). These observations are compatible with the water types and temperatures found in this zone.

(iii) Features of trace element contents. The spring with the highest contents of Rb (11 mg / l), B (308 mg / l), Li (15.4 mg / l) and Br (7.25 mg / l) in this zone is Quduowa hot spring at Luoma Township, Xungba, Gêgyai. The springs with the highest contents of F (17.8 mg / l) and As (6.9 mg / l) are Burong hot spring, Damxung county and Jipu hot spring, Dongqiao, Amdo county, respectively. Among all the hot springs on the Qinghai—Tibet Plateau, Quduowa hot spring has the highest Rb content, while its Li, Br and As contents rank second. The corresponding ratios of Li, Cs, As and F to total salts rank second, while the ratio of boron to total salts rank third (Table 1.23).

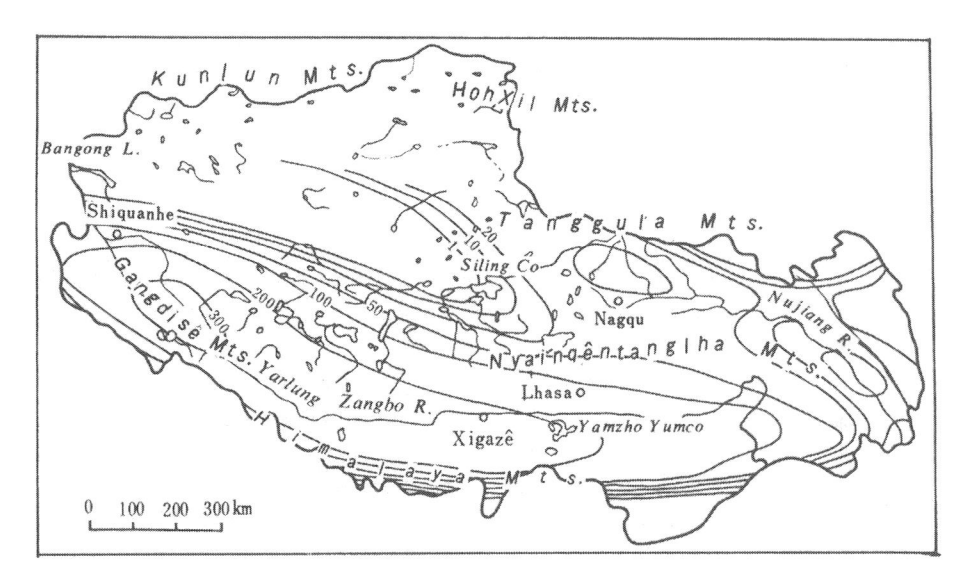

Figure 1.23. Trend surface analysis of HBO₃ in geothermal waters on the Tibet Plateau (from Dong Wei, Zhang Zhifei et al., 1981).

c. North Qiangtang–Jinsha River geothermal zone (III). The bulk of this zone lies in the NNW–trending Lungmu Co–Jinsha river suture. Some hot spring groups are also scattered along NW–, NE– and even N–S–trending active faults that splay out from both sides of the suture. This zone is largely unpopulated, hence data are scarce. According to data from nine hot springs they are low–temperature hot springs and medium– low–temperature thermal waters. The hydrochemical types are dominated by sulphate–bicarbonate, Na bicarbonate–sulphate and Na bicarbonate waters (accounting for 6 of the hot springs), with subordinate boron–bearing Na chloride type springs (Table 1.22). The concentations of B, Li, Cs, Rb, As, F and Br are relatively low. In this zone the hot springs at East Co Nyi lake has the highest boron (403 mg / l), Li (43.5 mg / l) and fluorine (13.2 mg / l) contents. The West Juheqiao hot spring in the eastern part of Markam has the highest Cs content (0.4 mg / l). The contours of the surface trends of these elements are shown in Figs. 1.21, 22 and 23.

d. Guide–South Qilian geothermal zone (IV). The main part of this zone is distributed along the WNW–trending South Qilian suture and the Guide fault zone. A small portion of this zone is also located in the area of Xiangride and Xinghai. There are 11 hot springs in this zone, with low– to high–temperature (5 springs) thermal waters predominating (Table 1.23). The hydrochemical types are dominated by Na sulphate—chloride waters with subordinate B—bearing Na chloride waters. Mud volcanoes are also well developed in Yashatu, South Qilian, where all the waters are of the B—bearing Na chloride type (Table 1.23). The level of trace elements in this zone are moderate. The highest boron concentrations (189 mg / l) are in the Beishan hot spring of Da Qaidam; the highest Li concentrations (20 mg / l) are in Qunaihai, Guide County; and the highest fluorine concentration (18 mg / l) are in the Xinjie hot spring, Guide county. The average boron concentrations in this zone (88 mg / l) are lower than in zones I and II, ranking third on the plateau. In addition to the hot springs, groups of small mud volcanoes are exposed along the fault zone in the Yashatu area of South Qilian. Of the mud volcano waters the Northeast " Dazhaisuluntuoluohai" spring has the highest boron content (478 mg / l). The average boron concentration of five mud volcano water is 235 mg / l.

e. North Qilian—Xining geothermal zone (V). The main part of this zone is scattered along the WNW—trending Lower Paleozoic Central Qilian suture up to the fault zone north of Xining. There are ten hot springs known in the area, with low—temperature hot springs and low— to medium—temperature thermal waters. There are four Ca bicarbonate waters, with subordinate Na chloride—sulphate waters. The scanty data from this zone indicate that the boron content ranges from 7—58 mg / l, with the highest Li content being only 4 mg / l (Wangjiazhuang, Yingsheng Township, Ledu county). The fluorine content is in the range 0.4—20 mg / l. Overall, trace element contents in the hot spring waters in this zone are the lowest on the plateau.

Sedimentation—buried waters
The sedimentation—buried waters in the region include oil—field waters and saline lake brines. The brines include interlayer brines of ancient and modern saline lakes. In this section a brief discussion will be presented concerning the oil—field waters and entrapped brines of ancient saline lakes. These two kinds of water are mainly distributed in the western part of the Qaidam basin, with some connate water found within the oil—bearing structure of the Lunpula Basin.

There are abundant oil—field waters in the widespread, high—yield, Tertiary oil—bearing structures in the western part of Qaidam. The structure outcrops along fractures to form ulexite and halite veins. Hence, oil—field waters are important in recharging modern saline lakes in some structural settings. The oil—field waters are mostly of the Ca chloride type, with some Mg chloride (Na chloride) waters. They all have relatively high B, Li, Br and I contents. For example, the oil field waters of the Kaitemilik oil field contain up to 51 mg / l of Li, whereas those of the Youquanzi oil field contain up to 92 mg / l Br and 1800 mg / l boron (Table 1.24). The boron content of oil field waters from this structure may reach a maximum of 2370 mg / l (Table 1.24). The confined sedimentary waters encountered during

TABLE 1.24. Chemical composition of connate waters on the Qinghai–Tibet plateau

Location	Age and water type	Sampling depth (m)	Salinity (g/l)	Na	K	Li	Rb	Cs	Ca	Mg	Cl	SO$_4$	CO$_3$	HCO$_3$	B$_2$O$_3$	Br	I	B/Cl	B$_2$O$_3 \cdot 10^3$/$\sum$ salts	Li$\cdot 10^3$/$\sum$ salts	Br$\cdot 10^3$/$\sum$ salts	Remarks
Well Zhong-1 of Kaitemilik structure	Neogene oilfield water	190	145	51500	220	51	–	–	1700	1500*	84900	2100	–	480	1600	83.7	–	5.9	11.0	0.35	0.58	Data from ①
Youquanzi structure	ditto	400	196	68300	1200	–	–	–	3600	1900	118000	–	–	28	1800	91.61	–	4.8	9.2	–	0.47	ditto
Huatu structure of Mangnai	ditto	–	207	77670	210	27.3	1.53	–	2410	930	123600	–	–	410	1570	80.75	0.81	4.0	7.6	0.13	0.39	Data from Zhang Changmei
Well Gao-1 of Lenghu No.5	ditto	–	43.8	14940	33	2.35	–	–	850	750	26600	–	–	180	120.75	10.68	23.27	1.4	2.8	0.05	0.24	ditto
Lunpula	Eogene sedimentary water	11.1	261	3.33	5.0	–	–	1.21	7.87	14.5	37.2	–	69.9	340.21	–	–	7.3	30.7	0.45	–	Data from Tibet Petroleum Party

① Zheng Mianping, 1958. Report on Scientific Investigation of the Qaidam Salt Lake. Sampling time: November 1957; analysts: Zhang Changmei et al.

drilling in the Lunpula basin have salinities of up to 11.1 g / l and area clearly distinct from the interlayer freshwater in the region. These waters also have higher boron and Li contents and their ratio of B and Li to total salts are up to 10 times higher than those in the Qaidam oil field waters. This suggests there is some additional source of B and Li in Lunpula compared to those in the western part of Qaidam (Table 1.24).

The sedimentary brines of ancient saline lakes in the western part of Qaidam mainly occur in Neogene and Quaternary salt-bearing rock series and show polycyclic features, such that there are four or more brine layers. They are mostly hypersaline brines and are commonly of the Na sulphate subtype and Mg chloride type, while the B and Li contents are relatively low.

6.5. Origins of Hydrochemical Types and Their Zonation

6.5.1. FACTORS CONTROLLING THE FORMATION OF DIFFERENT HYDROCHEMICAL LAKE TYPES

There are two different views regarding the formation and evolution of the hydrochemical saline lake types: one is related to climate and advocates that climatic changes result in the cyclic deposition of salts in lakes and the different hydrochemical types were formed when normal or reverse changes occurred in the lakes (Valiasko, 1954); the other view relates the formation and evolution of the salt lakes to their inflow sources and suggests that the different hydrochemical types occur in waters before they enter the lakes, rather than as a result of inter-lake processes (Dorov, 1948) In this latter scenario the "rock types in the water catchment areas control the hydrochemical types of lake basins in arid zones" (Stanov, 1962).

On the basis of the regional hydrochemical studies discussed above, especially the similarity of the hydrochemical zoning of river waters and luxiviation-infiltration waters to the hydrochemical zoning of lake waters (cf. Figs. 1.19 and 20 with Fig. 1.13) together with the evolution of inflow water, the author considers that the inflow source is the dominant factor. However, the inflow source conditions refer to the structural and hydrogeological conditions as well as the rock types of the catchment area. In general, the climatic conditions do not play the dominant role in the formation and conversion of lake water types, but they may be of significance under local geological-geographical conditions and in a few individual lakes. The reason why inflow is more important is that there is a high volume of periodic runoff which recharge and discharge the lakes at intervals, rather than a single injection of water, followed by closed basin evolution of the lake waters. Hence, the properties of inflow waters have a much greater impact on the formation and variation of lake water types than the climatic factor. Because the properties of the inflow waters are determined by the lithologic, structural and hydrologic conditions in the catchment of the lakes and these conditions (especially lithologic conditions) are not usually liable to undergo radical changes during the period of existence of a lake, the properties of the inflow waters tend to remain relatively constant during the period of the lakes existence. The model of a progression from carbonate to sulphate to chloride waters during closed basin evolution of the lake waters is relatively uncommon (Valiasko, 1954).

Changes in the water type of a saline lake can only take place when the inflow disappears or becomes very small, e.g. there may be salt–forming physico–chemical processes, exchange reactions between porewaters and sediments and biogeochemical processes in playa stage①. Some examples are given below.

The sources of plateau lakes largely comprise meltwater and rainwater–snowwater that are usually calcareous (magnesian) waters. When they contact different types of rocks they undergo reactions with those rocks that produce waters of different hydrochemical types. This is seen in the different compositions of the river waters and lixiviation–infiltration waters on the plateau. As these natural waters continuously feed the lakes, the lakes inherit the water types of these inflow waters before salts are deposited in the lakes. For example, in Silin Lake the river waters and freshwater lakes in its catchment are mostly of the Na bicarbonate or Ca (Mg) bicarbonate types. Although the lake has undergone long and continued evolution and become a brackish lake, its water type is still of the Na carbonate type, i.e., it has inherited the features of the water types of the inflow waters in its catchment. The hydrochemical types of the inflow waters on the plateau are generally consistent with the lake waters that they feed (Appended Map 1 and Figs. 1.19 and 20), i.e., the composition of the inflow waters is of greater importance than the physicochemical and biochemical factors that change lake water types. Hence, the water types of most saline lakes on the plateau remain consistent with the runoff of their catchment until the saline lakes die. Examples are provided by Bangkog Lake, Tujiali Lake and Xa Cake which all contained water of the carbonate type thoughout their histories even though they have developed into semi–playas and playas (for details, see Chapter 3). Elsewhere, e.g., in the East African rift valley, the soda lavas and the Si– and F–rich sodium carbonate thermal waters (the products of strong neovolcanic activity) resulted in the formation of a series of alkaline lakes. These range from freshwater lakes to semi–playas and and contain ususual

---

① The climate largely determines the salinity of the lakes and the stages of deposition of salts from saline lakes, i.e., carbonate to sulphate to chloride stage, which are very different from the above mentioned hydrochemical lakes of types. Waters with different chemical compositions may experience the three stages, but their hydrochemical types and corresponding mineral compositions are different. In some papers the deposition stage of saline lakes has been confused with the water types of lakes, leading to errors in interpretation.

F−, Si−, and alkali−bearing saline minerals (Eugster, 1979).

In only a few cases when the weakening of the influence of the chemical composition of of the catchment runoff do changes in intra−lake physicochemical and biochemical processes cause changes in the water types of the lakes. For example, in the Dalangtan saline lake of Qaidam, the saline lake water was of the Na sulphace subtype during its early development, but at a later stage the saline lake water type was transformed into a Mg sulphate subtype as the inflow from the catchment was reduced and large amounts of Na sulphate and Na chloride were deposited in the saline lake.

6.5.2. FORMATION OF VARIOUS WATER TYPES AND THE HYDROCHEMICAL ZONES

Studies indicate that the formation of the water types of most of the lakes on the plateau (and the river waters and luxiviation−infiltration waters within their catchments) is closely related to the lithological and geochemical conditions. The main factors controlling the lithological and geochemical conditions and the nature of water−rock interactions are the structural and hydrogeological conditions. They are described as follows (Table 1.25; Fig. 1.24).

Formation of the carbonate type and its hydrochemical zone
As noted above, the formation of the carbonate lake water type (Na−HCO$_3$ water) is largely a matter of the formation of its recharge water−Na bicarbonate water. The results of a large mumber of studies have suggested that the main factors responsible for the formation of Na bicarbonate waters are: 1) weathering and hydrolysis of rocks composed of sodium silicates (especially feldspars); 2) substitution of Ca^{2+}, Mg^{2+} and Na$^+$ in gels such as clays; 3) biochemical reduction of sodium sulphates; 4) hydrolysis of H$^+$ in waters and Na$^+$ in rocks and minerals; 5) metamorphism of organic matter; 6) recharge of sodium carbonate−bearing waters from depth. The presence of fold mountains and fault zones both favor the enhancement of the reaction of CO$_2$ in air and water with rocks. Studies of natural waters of the former U.S.S.R. (Bichiyewa, 1981) indicate that bicarbonate waters are mainly formed through hydrolysis of feldspar−bearing rocks in recharge areas. The reaction formula is;

$$NaAlSi_3O_8 + H_2O + CO_2 - - \rightarrow HAlSi_3O_8 + Na^+ + HCO_3^-$$

The hydrolysis of feldspars generally shows a wide regional distribution pattern, while other processes have a more restricted range.

From an analysis of the distribution of carbonate type waters on the plateau and the associated structural and lithological−geochemical features, it is apparent that there are two main factors controlling the formation of carbonate−type saline

TECTONOGEOCHEMISTRY AND REGIONAL GEOCHEMISTRY 117

lakes. The first of these is the hydrolysis of feldspar—rich rocks in provenances and the second is the association with recent strong geothermal and volcanic activity. There are two local carbonate saline lake (or salt—marsh) districts on the plateau, i.e., the salt—marsh facies trona districts in the vicinity of Zhongjia—Hatu southeast of Qaidam, and the carbonate lakes in the southern part of the Kunlun Mountains.

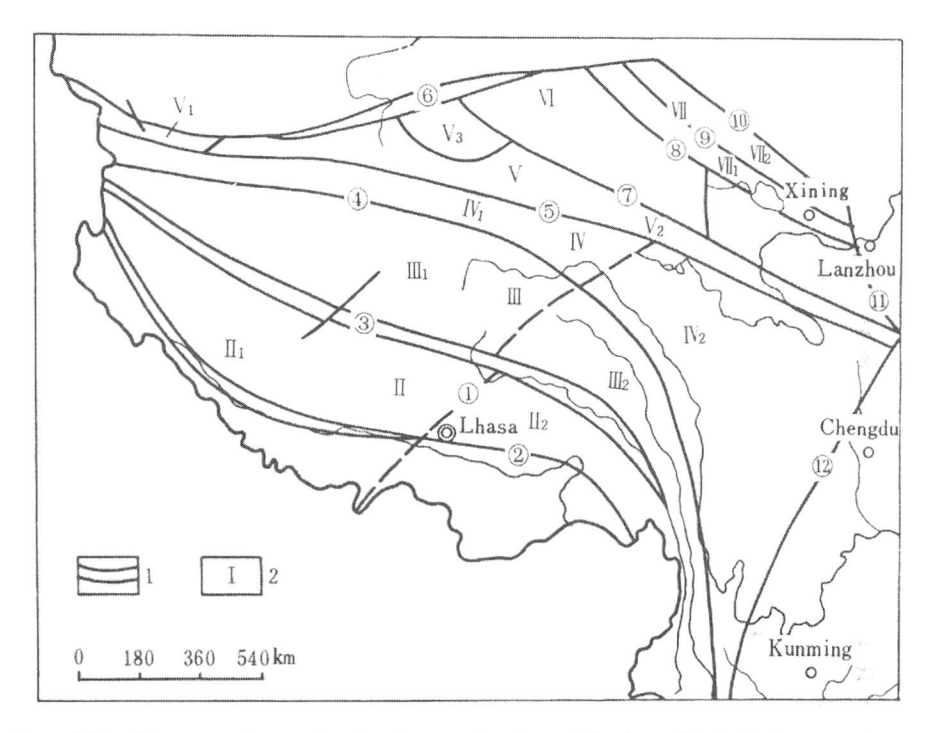

Figure 1.24. Diagrammatic map showing the tectonic units and dominant lithological—geochemical association areas on the Qinghai—Tibet Plateau. 1. Suture zone or deep fault and their No. (for explanations, see Table I—25). 2. Tectonic unit and dominant lithological—geochemical association area and their No. (for explantions, see Table 1.25).

The former is mainly related to the hydrolysis of feldspars of Hercynian granites and Paleozoic volcanic rocks on the plateau. Rivers in this district, such as the Xiangride River (No. 106 in Fig. 1.19), belong to the Na carbonate type and alkalis are extracted from the catchment area sediments in the lower reaches of the rivers. The latter are an unusual, nearly N—S—trending carbonate lake group (from Aru Co (no. 307) through Luoto Lake (no. 305) up to Shugu Lake north of the Kunlun Mountains in Appended Map 1) appearing locally in an E—W—trending chloride—bearing Mg— sulphate sub—type zone. Its formation was intimately associated with the recent alkaline volcanism in a N—S trending fault zone in the southern part of the western Kunlun Mountains.

The main carbonate–type saline lakes on the plateau are concentrated in the southern part. These form two carbonate hydrochemical subtype zones (I_1 and I_2; see Appended Map 1). Their origin is related to the two factors described above. The two carbonate subzones are both characterised by feldspar–rich igneous rocks and clastic rocks (associations I and II_1 and II_2 in Table 1.25). There are generally fold mountains around the lake district, which is favorable for the promotion of lixiviation of rocks and soils, resulting in desalinization of soils. Hence, river waters and ground waters on the plateau consist dominantly of bicarbonate waters. In addition, evolution of these water types can also be observed. For example, at the source of the Za'gya Zangbo River on the southern slopes of the Tanggula Mountains, there are Jurassic and Eocene clastic and carbonate rocks that result in Ca–Mg–HCO$_3$ type of water (no. 106 in Fig. 1.19). In its lower reaches the river runs through Quaternary lacustrine sediments (clayey and locally salinised) and the waters change to the Ca–Na–HCO$_3$ type (no. 14 in Fig. 1.19) and Na–Cl–HCO$_3$ type (no. 17 in Fig. 1.19) as a result of exchange of Ca and Mg in the waters for Na in the clays. The two carbonate subtype zones also include two geothermal zones. The waters in these zones are characterised by weakly to strongly alkaline, low salinity Na waters. Hence, this area is one of the four major accumulation zones of CO$_2$ fluids in the world (the others being the Mediterranean Sea, Lesser Asia and California) (Shi, 1980) In addition to the dominant lithological–geochemical associations of the plateau, the formation of carbonate type waters is also due to the high crustal stress that brings about the accumulation of CO$_2$ at varying depths (including thermal metamorphism of magmatic and carbonate rocks). These geothermal waters are abundant and tend to converge to form lakes. This constitutes an important factor for the formation of some carbonate lake water types of the zone (for details, see the subsequent quantitative estimate of material sources).

Formation of sulphate type waters and their hydrochemical zone
The distribution of sulphate–type saline lakes on the plateau is wider than that of carbonate–type saline lakes. The sulphate type predominates in lithological–geochemical association areas II_2, III, IV, V, VI, VII and VIII. The common feature of the lithological–geochemical associations of these hydrochemical zones is the occurrence of gypsum beds, or relatively abundant sulphides (IV–VII in Table 1.25), or alternate occurrence of hydrolysis of feldspars and the dissolution of sodium chlorides in places. Gypsum is soluble and can form sulphate waters, while sulphides under oxidation to form sulphuric acid, which in turn reacts with Ca and Ca–Mg carbonates that are common. Hence, the concentration of Ca–Mg and sulphate ions can build up in these zones.

TABLE 1.25. Tectonic units and dominant lithological–geochemical associations on the Qinghai–Tibet Plateau

Tectonic unit ①		Dominant lithologic–geochemical association ②	Boundary of system or zones
System	Zone		
Himalaya fault–fold system	I. Higher Himalaya fault–fold zone (block)	Meso–Cenozoic clastic rocks and volcanic rocks and Precambrian metamorphic rocks locally with Tertiary tourmaline leucogranite and Upper Paleozoic limestone Hydrolysis of feldspars and intracrustal CO_2 concentration area	Yangbajain–Dulan fault Indus–Yarlung Zangbo suture zone $(K_2–E_1)$ ③
Gangdise–Gaolishan fault–fold system	II$_1$ Gangdise fault–fold zone (block)	Early Cretaceous limestone with volcanic rocks and clastic rocks, locally granite (γ_5^3) and Tertiary clastic rocks (north part); Late Cretaceous–Tertiary volcanic rocks with clastic rocks and locally Upper Paleozoic sandy shale, limestone and granite (γ_5^3) Hydrolysis of feldspars and intracrustal CO_2 concentration area	
	II$_2$ Lhasa–Bomi fault–fold zone (block)	Triassic sandy slate, Mesozoic gypseous clastic rocks, Permo–Carboniferous limestone, sandy shale and volcanic rocks, and locally Eogene evaporite–bearing red beds and granite (γ_5^3) Mixed sulphate leaching, sulfide oxidation and feldspar hydrolysis area	Bangong Co–Nujiang River suture zone (J_3)
Qiangtang–Sanjiang fault–fold system	III$_1$ Qiangtang fault–fold zone (block)	Jurassic gypseous carbonate rocks with clastic rocks and Paleozoic metamorphic rocks and locally Cenozoic clastic rocks and Triassic and Triassic sandy shale Dominant sulphate leaching and sulfide oxidation area	
	III$_2$ Sanjiang fault–fold zone (Block)	Triassic sandy slate with limestone, coal measures and granite $(\gamma_5^1–\gamma_5^3)$, Mixed sulfide oxidation and feldspar hydrolysis area	Lungmu Co–Jinsha River suture zone (P in east and T in west)
Hoh Xil–Bayan Har fault–fold system	IV$_1$ Hoh Xil fault–fold zone (block)	Jurassic gypseous Ca–Mg carbonate rocks with clastic rocks, Cenozoic evaporite–bearing clastic rocks and locally Cenozoic volcanic rocks Area of sulphate leaching, sulfide oxidation and locally chloride (NaCl etc.) leaching	
	IV$_2$ Bayan Har fault–fold zone (block)	Triassic sandstone, magnesian limestone and volcanic rocks in western part; sandy slate with limestone and coal measures and locally granite (γ_5^2) and Early Permian phyllite or sandy slate and crystalline limestone in eastern part Dominant sulfide oxidation and feldspar hydrolysis area	Kunlun Range Southern–Margin parasuture zone (C_3) ④

Kunlun fault–fold system (India)	V_1 west Kunlun fault–fold zone (block)	Proterozoic high– and low–grade metamorphic series, Paleozoic magnesian limestone and sandy shale with volcanic rocks and siliceous rocks, and locally granite (γ_4^2) and Quaternary volcanic rocks Area of sulfide oxidation, feldspar hydrolysis and local CO_2 concentration in Quaternary volcanic rocks	Kunlun Range Southern– Margin parasuture zone (Cs)
	V_2 East Kunlun fault–fold zone (block)	Upper Paleozoic arenaceous rocks and magnesian limestone with volcanic rocks, Triassic sandstone and slate with volcanic rocks, Jurassic clastic rocks, and locally Tertiary evaporite–bearing red beds Dominant sulphate leaching and feldspar hydrolysis area	
	V_3 Kumkol block	Tertiary evaporite–bearing clastic rocks and Mesozoic Ca–Mg carbonate rocks and clastic rocks Dominant sulphate and chloride leaching area	Altun fault North Kunlun fault
Qaidam block (terminal Indosinian)	VI Qaidam Block	Tertiary–Quaternary evaporite clastic rocks, locally Mesozoic clastic rocks with coal measures and Paleozoic–Sinian carbonate rocks and clastic rocks and Proterozoic–Paleozoic high– and low–grade metamorphic series Dominant sulphate and chloride leaching area	South Qilian fault
Qilian fault–fold system	VII$_1$ South Qilian fault–fold zone (block)	Proterozoic–Lower Paleozoic metamorphic rocks and granite (γ_3) and locally Triassic sandy slate with limestone and Tertiary gypseous clastic beds Mixed feldspar hydrolysis and sulphate leaching area	Central Qilian suture zone ($\in_2, O_2$)
	VII$_2$ North Qilian fault–fold zone (block)	Paleozoic volcanic rocks, limestone and sandy shale, Sinian sandstone, slate and crystalline limestone, and locally granite (γ_3) Mixed feldspar hydrolysis and sulfide oxidation area	
Gonghe– Minxian fault–fold system	VIII Gonghe– Minxina block	Triassic sandstone and slate with limestone and marl and locally Tertiary red beds, Quaternary gypseous lacustrine deposits and granite (γ_5^1) Mixed feldspar hydrolysis and sulfate leaching area	Hot spring– Yangkang West (western part) Dangchang fault (eastern part)

① After Liu Zengqi et al., 1984, with slight supplements and modifications. ② After Liu Zengqi (ed.), 1980. Geological Map of the Qinghai–Tibet Map (1:500,000). ③ Codes in parentheses refer to the age of ophiolites. ④ After Jiang Chunfa, 1986. New data and new idea about the Kunlun structure (Report Outline).

$$(1) 2H^+ + SO_4^{2-} + CaCO_3 - - \to CaSO_4 + H_2O + CO_2$$
$$(2) 4H^+ + 2SO_4^{2-} + CaMg(CO_3)_2 - - \to CaMg(SO_4)_2 + 2H_2O + 2CO_2$$

There is more limestone in the lithological–geochemical association areas II$_2$, III, IV, VII and VIII, where the reaction (1) proceeds and Na–(Ca)–SO$_4$ waters are formed when the resulting Ca–SO$_4$ waters exchange with ions absorbed by clays. For example, the Pamei Zangbo river runs into Bienuoze Lake in hydrochemical zone II and takes its source in sulphide–bearing limestones. In its middle and lower reaches the river waters are of the Na–Ca–HCO$_3$–SO$_4$ type (No. 36 in Fig. 1.19), and when the river enters a Quaternary lacustrine sedimentary area the waters change to the Na–SO$_4$ type as HCO$_3^-$ in the waters interacts with Ca^{2+} in clays. As a result the waters of Bienuoze Lake are of the Na sulphate subtype (No. 312 in Appended Map 1). In lithological–geochemical areas IV$_1$ and V, the Mesozoic and Tertiary evaporite series are relatively thick and the Mg content in the carbonate rocks is higher. This favors reaction (2), leading to the formation of Mg–bearing waters. For example, inflow waters to lakes in the central part of hydrochemical zone III (the chloride–bearing Mg sulphate type zone) are mostly derived from Cenozoic and Mesozoic evaporites. Hence, the lake waters are largely of the Na–Mg–Cl or Na–(Mg–Ca)–SO$_4$–Cl–(HCO$_3$) type. The waters of the Tuotuo river (which is crossed by the Qinghai–Tibet highway) run through Early Quaternary evaporites and are of the Na–Mg–Cl–(SO$_4$) type and the waters of the Buqu river are of the Ca–(Na–Mg)–SO$_4$–(HCO$_3$)–Cl type. Due to their relatively high Mg content these kinds of inflow waters gives rise to many Mg sulphate saline lakes. Again, this suggests that the hydrochemical types of the saline lakes are intimately associated with the chemistry of the infow waters.

6.5.3. FORMATION OF THE CHLORIDE TYPE AND ITS HHDROCHEMICAL ZONE

Chloride type saline lakes are generally considered to be the products of the last stage of development of saline lakes, i.e., their chemistry is thought to be the result of interlake processes. Chloride–type lakes on the Qinghai–Tibet Plateau are largely distributed in Cenozoic evaporite basins of the Qaidam block, and less commonly in Tertiary evaporite basins of hydrochemical typee III. The source of chlorine ions in these lakes is closely related to ancient chloride deposits and brines (mainly sodium chloride with minor magnesium chloride and even calcium chloride). The structural–geological settings and hydrochemical features suggest that chloride type waters were not entirely formed during evolution of the saline lakes. For example, Changjing Lake, Xianhe Lake, Taiku Lake, Dogai Coring Lake, Cheyue Lake and Yanghu Lake (Nos. 415, 411, 413, 417, 418 and 402, respectively, in Appended Map I) are all chloride type lakes. They are all located in Tertiary evaporite–bearing basins, and some of them are still in the initial stage of formation of saline lakes and soluble salts have not been deposited. For example, the Cheyue Lake (No. 418) has a salinity of 60 g / l and a pH of 6.7 and belongs to the chloride type. Its water type is mainly related to selective dissolution of abundant chloride–bearing evaporites located around the peripheries of the lakes. The chloride–type waters of Qaidam basin are mainly distributed in its northern and southeastern corners. The long history of chemical differentiation of ancient saline lakes is important in defining the chemistry of the lake waters, but recharge by Na–Ca chloride waters flowing out along faults likely also play an important role.

CHAPTER 7 MATERIAL SOURCES AND MODEL OF SALT FORMATION OF SALINE LAKES

7.1. Material Sources of Saline Lakes

There are various views regarding the source of salts in the saline lakes on the Qinghai–Tibet Plateau. Special elements in the saline lakes might be derived from deep–seated thermal waters, marine sources and rock weathering. Zheng (1964) noted that Li, K, Rb and Cs have similar sources in the salt lakes to boron. In studies of Sambhar Lake in the Thar desert of India, near the plateau, Godbole (1972) suggested that this borax lake is either a residual lake of the Tethys or it has been produced from the weathering of trapped salts. Some Chinese geologists also believe that there are abundant marine strata near the saline lakes of Tibet and that the boron in the lakes is derived from these strata. The source of Li is thought to be from hot springs, weathering of rocks and decay of boron to lithium during cosmic radiation.

In this paper, tectonic–geochemical studies have been used to gain a knowledge of the various sources of salts to the different types of saline lakes.

7.1.1. WEATHERING OF ROCKS

The hydrochemical types of the lake waters and the origin of their zonation have been discussed in the preceding chapter. Studies of continental salt deposits in China and elsewhere in the world (Yuan et al., 1979; Smith, 1979) have suggested that the weathering product of rocks is undoubtedly a common source of salts in continental clastic type saline lakes. This is especially true for the major constituents; Ca, Mg, Na, S, Cl and HCO_3 (resulting from the interaction of CO_2 in air with rock).

The difference in the source rocks and the geochemical associations give rise to the variations in the composition of the source waters to the lakes, thus yielding the different hydrochemical zones discussed earlier. The rocks and natural waters on the plateau have anomalous contents of boron, Li etc. (Table 1.13 and 15), hence the concentrations of these elements are more variable in the lakes on the plateau than they are in neighbouring areas (e.g., Xinjiang, Gansu and Inner Mongolia). However, most of the saline lakes that only the sources from the weathering of crystalline and sedimentary rocks and lack enriched sources (e.g., B— and Li—rich geothermal waters, rocks or salt—bearing series) cannot form special boron and Li or K and Mg saline lakes of commercial value. In these cases the salt lake brines are mainly composed of common salts such as halite, mirabilite, gypsum, epsomite and bloedite, with lower levels of Li, boron, Cs, As and fluorine. These waters are usually of the sulphate type, as exemplified by Hoh saline lake and Xiligou lake (Nos. 629 and 603 in Appended Map 1) in hydrochemical zone IV of the lakes. The lakes are located on the margin of the Qinghai—Tibet tectonic body, and the rocks surrrounding the lakes are mostly low grade metamorphic rocks of the Changcheng system and Tertiary strata, with no boron— and Li—bearing rocks and geothermal waters in this area.

7.1.2. INHERITED SALT—BEARING ROCK SERIES

Regional geochemical and geological studies show that the average K content in rocks of the plateau is not high, e.g. it is 2% in the regional rocks of the Qaidam Basin (Chen, 1981), which is lower than the global average for crustal rocks. However, the largest K—Mg saline lakes on the plateau is located in Qaidam Basin. Its presence is not the result of special K—rich recharge sources in the region (e.g. ancient potash deposits and K—rich granites), but must be related to the special geologic history and geographic setting in the region. The geologic history relates to the fact that the basin inherits Neogene and Quaternary lake waters and large amounts of potash accumulated over a long period of time. The geographic features chiefly relate to the lack of vegetation and low surface run off in this high altitude, frigid zone. Potassium is lost from river waters during their course, so that recharge waters generally have higher K concentrations than average runoff. In addition, the continued salt formation in the Qaidam basin has let to concentration of the dispersed potassium. From the sedimentation history and geophysical setting of the Qarhan saline lake, the author has made the following conclusions regarding its origin and evolution. During the late stage of the evolution of the ancient Qaidam lake, the northwestern part of the basin was tilted and the residual ancient lake waters (that were rich in sodium and potassium) were concentrated in the low lying area of the Four Lake subsidence zone (especially Qarhan—Dabsan①). Another important source of dissolved salts is from the Tertiary and Early Quaternary salt—bearing rock series (for example, in the depression north of Kunteyi) and local oil field waters.

---

① Recently Zhu Yunzhu proposed that the Late Quaternary lake waters at the northern foot of the Kunlun Mountains migrated and concentrated towards the Four Lake subsidence zone.

Later, Zeng et al. (1980) investigated the main potash lake of Qaidam (the Qarhan salt lake playa) and estimated the composition of the ancient lake waters of the region from the composition of the sediments and lakes in the lake. The ancient lake waters of Qarhan has a K content of 7 kg of potassium per tonne of dry salts. This implies that the ancient lake waters carried a large amount of potassium when they migrated from the northwest to Qarhan lake. However, the K content of the ancient lake waters was not high (K / $\sum_{salts}$ = 7×10^{-3} and K / Cl = 13×10^{-3}) and relatively dispersed. Potassium and magnesium salts were precipitated during the playa stage as a result of continued evaporation and periodic dissolution events.

As the result of lixiviation experiments Chen et al. (1981) proved that large amounts of boron, Li and K can be leached from Neogene salt–bearing rock series of Qaidam. Drilling by the First Geological Party of Qinghai Province in the 1980's revealed that the Dalangtan K–Mg salt lake is situated directly on the Neogene–Early Quaternary salt basin (Tables 1.3 and 4) and that the late stage salt lake is more enriched in K–Mg salts than the early stage salt basin. These studies have all provided important evidence regarding the inherited source of potassium in the region.

As a result of the processes described above, the average K concentrations of the river waters and subsurface groundwaters in the Qaidam Basin are both higher than the world average. The K content in recharge river waters of the Qarhan salt lake is also relatively high, averaging up to 34 mg / l. The Na / K ratio of river waters in the drainage area of the Golmud river 140 km from its mouth is 19.5, this falls to 15.7 at the river mouth. A parallel fall in the K / $\sum_{salts}$ ratio is also seen, from 11 to 8.1×10^{-3} (Zheng et al., 1989). This sugests that there is only a small fall in potassium concentrations during the course of the river. According to Yang Ligang's estimate, the water recharge to Qarhan is 3×10^8 m^3 / a and the K$^+$ flux is 6547 tonnes / a. If this potassium flux had remained constant then about 1.6×10^8 tonnes of potassium have been delivered to the lake over the past 2.5 ka. This accounts for about 20%–30% of the total amount of potassium in the salt–bearing rock zone of this area, indicating that the recharge runoff is one of the important sources of potassium in this area.

7.1.3. DEEP–SEATED RECHARGE

Although saline lakes containing boron, Li, (K, Cs, As, F) are widespread on the plateau, their grades and reserves[1] are highest between the Yarlung Zangbo geothermal zone (I) and the Bangong Co–Nujiang river geothermal zone (zone II

---

[1] These refer to the bopron, Li, Cs, As, F and K grades and boron, Li, Cs, (As, F) reserves. The K reserves are highest in Qaidam because there are also K–Mg saline lake types there.

TABLE 1.26. Correlation of hydrochemical coefficients of B, Li, K, Cs, Rb, As and F between the geothermal zones and hydrochemical zones of lakes

Name		Coefficient	$\frac{Li \cdot 10^3}{\sum salts}$	$\frac{Rb \cdot 10^3}{\sum salts}$	$\frac{Cs \cdot 10^3}{\sum salts}$	$\frac{K \cdot 10^3}{\sum salts}$	$\frac{B_2O_3 \cdot 10^3}{\sum salts}$	$\frac{As \cdot 10^3}{\sum salts}$	$\frac{F \cdot 10^3}{\sum salts}$	$\frac{Rb}{Cs}$	$\frac{B \cdot 10^3}{\sum salts}$	Remarks
Qinghai–Tibet Plateau	Geo-thermal zone	I	2.14(37)	0.24(37)	2.25(37)	38.53(40)	73.49(39)	1.56(39)	4.84(39)	0.079(37)	250(19)	Calc. according to the data of Dong Wei et al.
		II	0.56(72)	0.126(73)	0.25(73)	11.35(73)	7.30(73)	0.46(71)	2.18(71)	0.56(73)	14(71)	
		III	0.41(9)	0.1(9)	0.13	9.88(9)	6.49(9)	0.09(9)	1.67(9)	0.48(9)	42(9)	Calc. according to the data of Qinghai Bureau of Geology and the authors
		IV	1.72(6)	–	–	17.25(1)	8.73(8)	–	1.52(5)	–	17(8)	
		V	–	–	–	–	9.66(2)	–	2.60(1)	–	85(2)	
	Lake hydro-chemical zone	I$_1$	0.725	0.00275	<0.01	73.275	40.92	0.346	0.224	70.3	185	
		I$_2$	2.798	0.1179	0.235	53.875	19.456	0.271	0.01	0.50	17	Corr. to geothermal zones I – II
		II	0.848	0.1125	0.0568	57.141	6.375	0.061	0.028	1.98	4	Corr. to geothermal zones II – III
		III	0.456	0.004174	0.043	15.335	7.515	–	0.02	2.83	4.4	Corr. to geothermal zones III
		IV$_1$	0.322	0.0158	0.026	8.126	7.072	0.0137	–	0.61	3.1	
		IV$_3$	0.327	0.0189	0.00013	18.595	1.95	0.00034	0.025	144.74	1.0	Partly Corr. to geothermal zones IV
		IV$_4$	0.059	0.0026	0	13.707	0.742	–	0.092	–	0.4	
		V	0.157	0.00818	0	18.38	0.057	–	0.091	–	0.04	Partly Corr. to geothermal zones V
Kizidere Hot Spring, Turkey			0.99	–	–	29.67	18.75	–	5.21	–	220	A. Tendam et al., 1971
Searles Lake (recharged by geothermal water)			0.24	0.003	–	77.38	32.35	–	0.05	–	27.93	E.W. Donald, 1963
Magadi Lake (East Africa Rift)			–	–	–	5.72	0.92	–	0.98	–	0.28	H.P. Eugster, 1980
Recharging Magadi Hot Spring			–	–	–	7.73	1.04	–	5.70	–	1.70	
Contin. sedim. brine in Qianjiang			0.45	0.45	–	18.91	0.11	–	–	–	0.03	Jianghan oil field, 1981
Lake Inder (recharged by marine salt deposits)			–	–	–	1.93	0.29	–	–	–	0.15	M.G. Valyashko, 1952
World seawater			0.0049	0.0034	0.00000008	11.02	0.43	0.000074	0.037	4000	0.24	Liu Yingjun, 1979
Illis oil-field water, U.S.			0.18	–	–	2.20	0.07	–	0.03	–	0.04	D.E. White, 1968

Note: Numbers in parentheses refer to the statistical number. Abbreviatios: calc. = calculated; coor. = cooresponding.

in the west—central sector) and the south Qilian geothermal zone (IV) on the northern margin of Qaidam, forming a distribution pattern with the Gangdise area as the main center of boron, Li, (K, Cs) salt lake resources and Qaidam (northern margin) as the secondary anomaly center (Figs. 1.16 and 17). There are three prominent tectono—geochemical features in these saline lake areas.

Strong tectonic mobilization
Compared to other natural waters, geothermal waters (mud—volcano waters) in fault zones are rich in boron, Li, (Cs, K, As, F) etc. The concentrations of these elements are up to 20 times higher than in river waters and lixiviation—infiltration waters of the plateau and in seawater. In addition, they are also higher than in the various special constituents in geothermal zones III and V of the plateau (Table 1.26).

Wide distribution of B, Li (Cs)—rich rocks
In the above mentioned areas, the rocks with the highest B, Li and Cs contents are mainly distributed from the Himalayas to the Bangong Co—Nujiang river geothermal zone. For example, terminal Cretaceous—Eocene intermediate—acid volcanic rocks distributed extensively in the central and western sectors of Gangdise—Bangong Co—Nujiang river geothermal zone have relatively high concentrations of B, Li etc., that are up to up to 130, 2 and 23 times higher, respectively, than global average concentrations of these elements. Similar enrichments in boron, Li and Cs are seen in terminal Cretaceous basic rocks in Moincêr of the western sector of the Gangdise mountains, the Tertiary rhyolite south of Coqên in the central sector of the Gangdise and the Miocene (tourmaline) granite in the Himalayas—Yarlung Zangbo. In addition, these rocks are also rich in other volatile elements, such as F and P (Searle, 1983) and most of the post—Cenozoic sedimentary rocks are enriched in boron, Li, Cs, etc., to different degrees (Tables 1.13 to 15). Boron—rich rocks have also been found on the northern margin of the Qaidam Basin. For example, Lower Paleozoic marble in Lengshuigou, Beishan, Da Qaidam has a boron concentration of up to 467 ppm, which is the highest boron content seen in Eocene rocks.

High B, Li, (K, Cs, Rb, F, As) anomalies in natural waters, especially in geothermal waters (mud—volcano waters), recharging saline lakes
In the major areas of saline lakes containing boron, Li, (K, Cs, As, F) reserves, the concentrations of boron, Li etc., in river waters and ordinary springs are higher than those in other saline lake areas of the plateau. For example, the boron and Li concentrations in the river waters are up to 5 times higher in these enriched areas (corresponding to riverwater hydrochemical zone I; Fig. 1.21) than those in secondary areas (zones II to V) (Table 1.27). When comparing river waters and springs

TABLE 1.27. Comparison of C values of B, Li and K in various hydrochemical zones of river waters on the Qinghai–Tibet Plateau

* C value Item	Zone I ②	Zone II ②	Zone III ②	Zone IV ②	Zone V ②	① ②
B₂O₃	537.9	83.85	231.06	347.2		403.7
B	538	84	231	347		300
K	1.63	0.814	2.62	1.25	1.38	1.54
Li	208.7	60.87				174.48
B/Cl	37.15	4.77	3	5		12.46

① Average value of river waters of Qinghai Tibet Plateau; ② Average value of world river waters; * C value = ion concentration in regional river waters / ion concentration in world river waters.

TABLE 1.28. Correlation of several components in natural waters on the Qinghai–Tibet Plateau

| Component | Plateau geothermal water | | | | | Qinghai–Tibet Plateau | | | World river waters | $\frac{1}{②}$ | $\frac{1}{③}$ | $\frac{①}{②}$ | $\frac{①}{③}$ | $\frac{①}{④}$ |
	Zone I	Zone II	Zone III	Zone IV	Zone V	Total (incl. interzonal area) ①	Aver. of river waters ②	Lixiviation infiltration freshwaters ③						
Li	7.16	1.16	0.9	3.34	4.0	3.49	0.4013	0.78	0.17	17.84	9.18	8.70	4.47	20.53
K	61	31.85	18.47	24.1	/	61.55	9.99	45.8	39.2	6.11	1.33	6.16	1.34	1.57
Rb	0.52	0.3	0.16	/	/	0.45	/	/	0.12	/	/	/	/	3.75
Cs	6.58	0.54	0.34	/	/	2.18	/	/	0.00003	/	/	/	/	72667
B₂O₃	248.1	16.6	11.4	18.35	3.15	81.51	13.00	12.08	1.38	19.09	20.54	6.27	6.75	59
As	5.63	0.21	0.07	/	/	1.35	/	/	0.0026	/	/	/	/	519
F	6.1	3.1	1.8	2.07	4	5.03		/	1.3	30.5	/	/	/	3.87

with geothermal waters of the plateau, geothermal waters show higher contents of B, Li and Cs. The boron and Li concentrations are comparable to other high boron and Li geothermal areas of the world, while the Cs contents are the highest observed in the world, reaching levels of 58 mg/l in the Labulang hot spring of the Yarlung Zangbo geothermal zone. The B, Li and Cs concentrations of the geothermal waters on the plateau are up to ten times higher than those observed in the river waters and interlayer waters (Table 1.28). Hence, the geothermal waters are an important source of the high B and Li concentrations in the river waters and springs in the areas of the special saline lakes.

For example, consider the Tatalin river, on the northern margin of Qaidam flowed from northeast to southwest into the Da Qaidam lake in the Late Pleistocene. In the Early Holocene, the alluvial fan in the eastern part of Da Qaidam arched up as a result of the east–directed stress produced by the continuous activity of the Altun sinistral fault. As a result the river turns to the south at a right angle after it leaves the mountain and flows into Xiao Qaidam lake. The middle and upper reaches of the Tatalin river lie largely within the South Qilian fault zone and receive inputs from the large amounts of high–boron–(Li) mud volcano waters in the area. The mud volcano waters are of the B–rich Na chloride and Ba carbonate types, containing up to 55–580 mg/l boron with the boron/$\sum$ salts reaching $11-84 \times 10^3$. In addition, this area also contains the Juhongtu and other sedimentary boron deposits. The tributaries of the river contain from 14 mg/l (in the Yamat river) to 333 mg/l (in the Heishui river) of boron (Fig. 1.25). The river waters are diluted with large amounts of melt water in the lower reaches of the river so the boron concentrations decrease to 3.6–24 mg/l. Hence, the Tatalin river became one of the main recharge sources of boron (and Li) of the Da Qaidam lake in its early stages and of the Xiao Qaidam lake in its later stages.

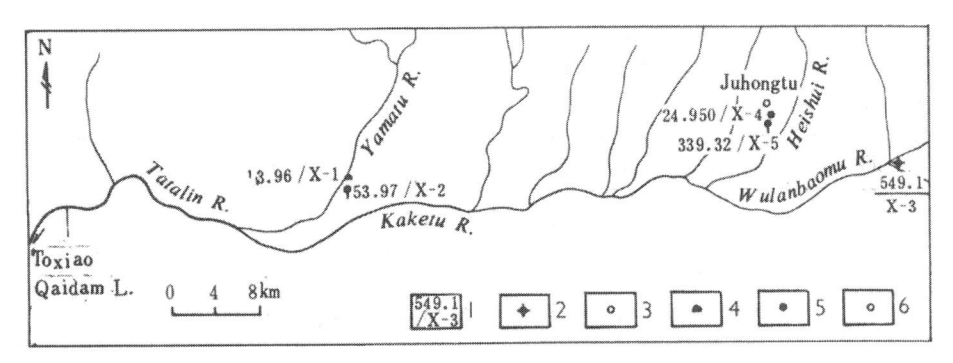

Figure 1.25. The B content in natural waters in the Yashatu area. 1. Numerator is the B_2O_3 content (mg/l) and demoninator is water sample No.; 2. Sampling site of mud–volcano spring waters; 3. Sampling site of spring waters; 4. Sampling site of river waters; 5. Sampling site of phreatic water; 6. Settlement.

Large amounts of dispersed elements, such as boron and Li, are leached from boron– and Li–rich rocks during interaction with natural waters. For example, in the Za'gya Zangbo (the largest river in northern Tibet) the boron and Li concentrations of the river waters increase markedly in the area of the outcrop of the Niubao rhyolite (north of Bangkog Co).

7.1.4. AFFINITY BETWEEN DISPERSED COMPONENTS IN SPECIAL SALINE LAKES AND GEOTHERMAL WATERS

Both the absolute and relative concentrations of a number of transition and volatile elements are higher in the saline lakes on the plateau (especially in those special lakes associated with boron and lithium deposits) than in global average saline lakes. Among these elements there are 31 (particularly boron, Li, Cs, As, U, Sb and F) which have relative concentrations that are higher than in seawater. The concentrations of these elements are also relatively high in geothermal waters on the plateau. Within the hydrochemical and geothermal zones there are good correlations between a number of the elements within the waters (the correlation coefficients are shown in Table 1.25).

(1) The characteristic coefficients of the ratio of boron, Li, Cs, Rb, As, F and K to total salts decrease from south to north in the lake hydrochemical and geothermal zones. The hydrochemical coefficients of lake waters and geothermal waters are similar to those of geothermal waters elsewhere in the world (Table 1.26), but are quite different from those of marine and average continental salt–deposit sedimentary waters and oil–field waters. For example, the B–Cl coefficient (B × 10^3 / Cl) is relatively constant in natural waters of differing origins and does not change with depth in seawater (i.e., about 0.24), but ranges from 14–250 in various geothermal waters on the plateau. In the lake hydrochemical zones on the plateau the B–Cl coefficient ranges from 17–185 in main special saline lake areas (geothermal zones I and II and main special saline lake areas) and from 1–4.4 in the secondary special saline lake areas. The B–Cl coefficients in the geothermal waters and special saline lake areas are similar to those of Kizildere hot spring and Searles Lake (220 and 28, respectively), but are very different from those of hot springs and lakes of other origins. For example, the B / Cl and B / $\sum$ salts coefficients are up to a hundred times higher in the special saline lakes on the Tibet–Qinghai Plateau than they are in the inflow hot springs to Lake Magadi, which is located in a rifting environment with mainly mantle source.

(2) The Rb / Cs ratio is low in the geothermal zones and lake hydrochemical zones I, II$_2$ and IV$_1$, which is in accord with the extensive distribution of special saline lakes in these zones, whereas the ratio is high (145) in lake hydrochemical subzone IV$_3$. This agrees with the observation that most lakes in this subzone are K–Mg saline lakes or ordinary lakes (there are special saline lakes at the northern margin of the zone) (Table 1.26). The average crustal abundance of Rb is 40 times higher than that of Cs. The Rb / Cs ratio is 4000 in oceanic waters and averages 22600 in global river waters. In contrast, the Rb / Cs ratio in rocks of various ages on the plateau is generally low, and is particularly low in Cretaceous–Tertiary basic volcanic rocks (2.9), andesite (2.9), Tertiary rhyolite (1.6) and Quaternary sedimentary rocks (0.6–1.4). The Rb / Cs ratio is relatively high in pre–Upper

Cretaceous rocks (4–13) as well as in Late Yanshanian and Himalayan diorite (12–30). The Rb / Cs ratios of the various rocks on the plateau reflect the generally high Cs and low Rb concentrations in these rocks and also indicate that Cs has a tendency for Cs to be concentrated in late–stage Cenozoic acid magmas. This Cs is leached and concentrated by geothermal fluids. The very high Cs concentrations and low Rb / Cs ratio in the geothermal waters and lake brines have not generally been found in similar environments elsewhere in the world. The similarity of the Rb / Cs ratios of the geothermal waters and the saline lakes on the plateau are another indication of the importance of this source of species in some saline lakes.

Study of isotopic composition (δD, $\delta^{18}O$ and $\delta^{34}S$) of natural waters in Tibet has also noted the similarity between the composition of the special saline lakes and geothermal waters, and these are distinct from marine saline lakes and ordinary inland saline lakes. The δD and $\delta^{18}O$ values of seawater and marine brines tend to be greater than zero, whereas those of continental saline lakes tend to be more negative than seawater (Delost et al., 1974). The brines in some of the special saline lakes on the plateau, e.g., Zabuye salt lake, have an average δD value of −59‰ and an average $\delta^{18}O$ value of −6.2‰ (for details, see Chapter 3). The Cs, boron, Li and As–rich geothermal waters have distinctly negative δD (−154‰) and $\delta^{18}O$ (−18.9‰) values. The values of Zabuye salt lake lie within the field of values seen in geothermal waters of iceland and the Salton Sea, although former lies at a much higher elevation above sea level than the latter. The $\delta^{34}S$ values of the geothermal waters on Tibet are also similar to the values of minerals in saline lakes.

7.2. Quantitative Estimate of Material Sources

In order to demonstrate the importance of the recharge of geothermal waters to the special saline lakes, an attempt has been made to estimate the output of boron, Li etc., from several geothermal zones and accumulation of these elements in the saline lakes.

7.2.1. OUTPUT OF B, LI ETC., FROM GEOTHERMAL WATERS

At the present time large fluxes of B, Li, K, Cs, As, F and Br are still being brought to the surface from depth in the five geothermal zones on the plateau. This is especially the case in zone I in the south (Dong et al., 1981), which is still at an "early" or "mature" stage in its development, whereas most of the other zones are in the "old" stage of their development. Here, the calculation is based on the outputs and analytic data of components from 273 hot springs in zones I–IV on the whole plateau. If their modern outputs and concentrations of the components are taken

TABLE 1.29. Output of characteristic elements from hot springs of geothermal zones I–V on the Qinghai–Tibet Plateau

Geothermal zone		Zone I–III		Zone IV–V		Total of zones I–V
		Tonnage	Percentage on the whole plateau	Tonnage	Percentage on the whole plateau	
Annual component output (t)	K	5753.58 (225)	/	384.92 (3)	/	6138.5 (228)
	Li	594.99 (224)	/	3.61 (4)	/	598.6 (228)
	Rb	104.89 (224)	/	/	/	104.89 (224)
	Cs	365 (224)	/	/	/	365.0 (224)
	B_2O_3	16992.53 (250)	/	373.73 (10)	/	17366.3 (260)
	As	161.02 (261)	/	/	/	161.02 (261)
	F	1105.68 (266)	/	0.91 (7)	/	1106.6 (273)
	Br	409.96 (56)	/	/	/	409.96 (56)
Component output since 19000 a B.P. (10 kt)	K	10931.8	93.7	731.3	6.27	11663.
	Li	1130.48	99.4	6.86	0.6	1137.3
	Rb	199.29	100	/	/	199.29
	Cs	693.5	100	/	/	693.5
	B_2O_3	32285.8	97.8	710.1	2.1	32996
	As	305.9	100	/	/	305.9
	F	2100.8	99.9	1.7	0.1	2102.5
	Br	778.9	100	/	/	778.9

Notes: (1) Original data from the Geothermal Group of the Comprehensive Investigation Party of the Qinghai–Tibet Plateau, Geological Survey Party of Tibet Bureau of Geology, the First and Second Hydrogeological Party of Qinghai Bureau of Geology and the present party. (2) Bracketed numbers refer to the number of hot springs in the statistics. (3) Zones IV–V are lacking in analytic data of Rb, Cs, As and Br or the flow data, so calculation is not able to be made.

TABLE 1.30. Annual output of components from the Yangbajain geothermal field

Outflow rate	(l / s)	980
	(m^3 / a)	30.9 × 10^6
B$_2$O$_3$	Concentration (mg / l)	163.2 (10)
	Annual total output (t)	5043.74
K	Concentration (mg / l)	42.5 (10)
	Annual total output (t)	1313.47
Li	Concentration (mg / l)	10.14 (10)
	Annual total output (t)	313.38
Cs	Concentration (mg / l)	6.45 (1)
	Annual total output (t)	199.34
Rb	Concentration (mg / l)	4.6 (1)
	Annual total output (t)	142.16
Na	Concentration (mg / l)	394.9 (10)
	Annual total output (t)	12204.5
HCO$_3$	Concentration (mg / l)	373.93 (10)
	Annual total output (t)	11556
CO$_3$	Concentration (mg / l)	11.9 (10)
	Annual total output (t)	367.77
Cl	Concentration (mg / l)	459.68 (10)
	Annual total output (t)	14206.5
SO$_4$	Concentration (mg / l)	70.18 (10)
	Annual total output (t)	2168.93
As	Concentration (mg / l)	2.15 (1)
	Annual total output (t)	66.45
F	Concentration (mg / l)	14.2 (1)
	Annual total output (t)	438.85

Note: Flow and water quality data from the literature (Dong Wei et al., 1981); bracketed numbers refer to the number of samples.

as constands① and the output of the geothermal field is assumed to have started 19 Ka ago (the ^{14}C age of the ancient travertines of the hot springs of Bangkog Co is 19430± 180 a), the five geothermal zones (which account for about 50% of the springs on the whole plateau) have outputs of 3.3×10^8 t B_2O_3, 1.14×10^7 t Li, 1.17×10^8 t K, 6.94×10^6 t Cs, 1.99×10^6 t Rb, 3.06×10^6 t As and 2.1×10^7 t F since 19 ka (Table 1.29). These outputs mainly stem from geothermal zones I–III (zone I in particular), while the outputs from geothermal zones IV and V only accounts for a few percent of the total output (Table 1.29). This evidently because these two geothermal zones have reached the late stage of geothermal activity. The small numbers and extent of hot springs in the two zones, however, may imply that the recharge from deep levels of the northern part of the plateau is less significant than that in the southern and central parts. The exception is the northern margin of Qaidam, where geothermal waters (mud–volcano waters) are of major significance. The outputs of B, Li and K from geothermal zones I, II and III, respectively account for 186%, 192% and 625% of the amounts of corresponding components in the lakes of lake hydrochemical zones I, II and III. For example, the still active geothermal springs of zone I could all form special saline lake deposits. Detailed estimates of the outputs of B, Li, Cs, Rb and K from ten hot springs in the Yangbajain geothermal field over the past 10 Ka show outputs 50×10^6 t B_2O_3, 3.1×10^6 t Li, 2×10^6 t Cs, 0.43×10^6 t Rb and 13×10^6 t K (Table 1.30). These data indicate that the deep–seated sources are of great importance in the development of special saline lake components in the central and southern parts of the plateau.

7.2.2. COMPARISON BETWEEN THE INPUT AND GEOCHEMICAL RESERVES OF LAKES

Taking the Mapam Yamco lake basin as an example. This lake is a rift lake (originally an outflow lake) in geothermal zone I. It divided into Mapam Yamco and La'nga Co (Nos. 001 and 002 in Appended Map 1) as a result of the climate becoming drier. The lake surface of the former lake is 15 m higher than that of the latter. The two lakes are connected by a water course. In recent times the climate has become drier, with an aridity index of up to 13②. So in the past few decades the watercourse has dried up (Ou (1978), resulting in La'nga Co and Mapam Yumco both becoming closed lakes. Hot springs are well developed to the northwest and southeast of the lake. A brief comparison is made between the data of 11 components of the inflow hot springs and surface runoff (lake–bottom hot springs excluded) and the geochemical reserves (Table 1.31). The input time is taken as the

---

① The extent of ancient sinter deposits in various geothermal areas suggest that ancient outputs in most geothermal areas were larger than those today.

②Calculated from data between 1973–1975 from the Burang Weather Observatory.

last 100 years to coincide with the increasing aridity of the climate. Only the geochemical reserves of the liquid phases are considered, as Mapam Co is a freshwater lake. In comparison, the outputs of boron, Li, K, SO_4, Ca and CO_3 from the geothermal springs are about 6.7% more than the geochemical reserves, with the excess of Ca and CO_3 being appreciable. The boron and K outputs is close to the geochemical reserves, while the Li outputs are close to double the reserves. The geothermal sources of Na, Mg, Cl and HCO_3 contribute 74% of their geochemical reserves, indicating that there must be an additional source from rock weathering. Cl and HCO_3 may also have a meteoric water source.

TABLE 1.31. Estimates of inputs of part of components and their geochemical reserves in the Mapam Yumco lake

Component	(1) Surface runoff Content (mg/l)	(1) Surface runoff Annual input (t)	(1) Surface runoff Input in the past 100 a (t)	(2) Hot springs SE of the lake Content (mg/l)	(2) Hot springs SE of the lake Annual input (t)	(2) Hot springs SE of the lake Input in the past 100 a (t)	Total input of (1)+(2) in past 100 a (t)	Geochemical reserves in mapam Yumco lake waters(t)	Input/geochemical reserves
HCO_3^-	57.21	26888.7	2688870	549.0–732.0	2277.09	227709	2916579	3948853.8	0.7386
CO_3^{2-}	11.25	5287.5	528750	18.2–123.0	255.92	25592	554342	182655	3.0349
Cl^-	3.09	1452.3	145230	112.0–119.0	425.14	42514	187744	223946	0.8383
SO_4^{2-}	8.17	3839.9	383990	66.9–82.4	266.27	26627	410617	369738	1.1106
Na	6.16	2895.2	289520	335.0–343.0	1254.23	125423	414943	645012	0.6433
K	1.60	752.0	75200	34.0–35.5	124.68	12468	87668	82287	1.0654
Ca	15.63	7346.1	734610	3.78–18.1	29.07	2907	737517	425272.5	1.7342
Mg	5.23	2458.1	245810	2.49–5.98	13.34	1334	247144	392616	0.6295
B_2O_3	4.18	1964.6	196460	9294–105.66	364.61	36461	232921	190588.5	1.2221
Li	0.1	47.0	4700	1.77–1.83	6.53	653	5353	2398.5	2.2318
F	Absent	–	–	9.0–11.0	37.42	3742	–	13284.0	–

Note: Estimated according to the hydrochemical data of the Comprehensive Scientific Investigation Party of the Qinghai–Tibet Plateau of the Chinese Academy of Sciences and the present party.

An additional example is provided by Bangkog Lake within geothermal zone II. Judging from the extent and the composition of travertine in the area, the extent of geothermal activity in this zone was far greater and the water temperatures and concentrations of elements such as boron, Li and Cs were far higher in the Late

Pleistocene—Early Holocene than today. For example, the ancient travertine in the Zigê Tangco hot spring is up to 10 km long with geyserite at the base. Extensive fossil travertine is also found in the Bujue hot spring south of Bangkog Lake. A comparison with the amount of travertine formed by modern, active hot spring suggests that the extent of ancient geothermal activity was several hundred times higher than today. The ancient hot springs on the plateau were formed in three stages, with the earliest stage having a ^{14}C age of 19.43 ± 0.18 ka[1]. Since the late Late Pleistocene there have been four inflow sources to Bangkog Lake: 1) hot spring waters like the Bujue hot spring, 2) the Lawa Zangbo River, 3) lixiviation—infiltration waters (springs), 4) the ancient Siling Lake waters. Because the composition of the ancient Siling lake waters is unknown and the lake was gradually separated from Bangkog lake after the Early Holocene, only the first three sources have been evaluated. In calculating the geochemical reserves of Bangkog lake and the flow of Lawa Zangbu river, we have also used drilling data and long-term observation data obtained by the original Northern Tibet Geological Party. Preliminary estimates indicate that the amounts of Li, boron, HCO_3, CO_3, K and Na supplied by the hot springs make up greater than 50% of the input (Table 1.32), demonstrating that this source is of importance in supplying these components to the lake. In contrast, hot springs account for less than 25% of the input of Mg, SO_4, Ca and Cl, indicating that they were largely derived from rock weathering. The ratio of the reserves of boron, K, HCO_3, CO_3, Cl and Na to their inputs vary from 0.71 and 0.53, indicating that they are essentially coincident. This calculation only includes the solid and liquid reserves in the lake, while the salts dispersed in clayey sediments around the lake have not been included. The ratios of the reserves to the inputs of Li, Ca, Mg and SO_4 are 1.38, 1.33, 6.04 and 3.50 respectively, suggesting that there are other sources of these elements that require further investigation.

7.2.3. SOURCES OF SPECIAL COMPONENTS

The discussions above suggest that geothermal waters and subsurface and surface waters leaching boron— and Li—bearing rocks are the main sources of these special components in saline lakes on the plateau. But these waters are only the carriers, not the primary sources. In order to ascertain the nature of the primary sources, it is necessary to study the origin of high boron and Li concentrations in geothermal waters. Comparison of the hot springs and saline lakes on the plateau with those elsewhere in the world suggest that there are some similarities (Zheng et al., 1964, 1982), particularly with hot spring waters in "continental collision belt" in the Mediterranean Sea (e.g. Kizildere hot spring in Turkey). However, they are di-

---

[1] ^{14}C samples were analyzed with the help of the Institute of Geography of the Chinese Academy of Geological Sciences.

MATERIAL SOURCES AND MODEL OF SALT FORMATION

TABLE 1.32. Estimates of the inputs of some components and geochemical amounts in the past 19400 years in Bangkog Co

Component	Lawa Zangbo Content (mg/l)	Lawa Zangbo Annual input (t)	Lawa Zangbo Input in the past 19400a (t) (1)	Bujue Hot spring south of Bangkog Co Content (mg/l)	Bujue Hot spring south of Bangkog Co Annual input (t)	Bujue Hot spring south of Bangkog Co Input in the past 19400a (t) (2)	Springs north of Bangkog Co (I–III) Content (mg/l)	Springs north of Bangkog Co (I–III) Annual input (t)	Springs north of Bangkog Co (I–III) Input in the past 19400a (t) (3)	Total input of (1)+(2)+(3) (t)	Geoch. reserv. of Bangkog (10^4t)	Geoch. reserv. / input	(2) / $\sum$ inputs
CO_2	183.7	7897.4	153209560	1024.5	20032	388620800	331.8–461.1	73.3	1421050	54325.1	32775.3	0.6	0.7
Cl	91.2	3920.9	76065460	84.8	1658	32165200	249.1–796.9	79.5	1541718	10977.2	5835.8	0.5	0.3
SO_4	107.8	4636.5	89947130	20.0	392	7604800	335.8–1013.6	110.9	2150684	9970.3	3484.8	3.5	0.1
Na	362.5	11288.0	218986424	649.5	12698	246341200	242.5	47.5	921500	46624.9	27047.4	0.6	0.5
K	22.0	946.0	18353176	52.1	1020	19788000	12.0	2.4	45784	3818.7	2722.6	0.7	0.5
Ca	30.3	1304.2	25302256	25.7	502	9738800	15.1–29.3	4.2	81092	3512.2	4664.2	63.3	0.3
Mg	29.0	1246.6	24184428	19.5	382	7410800	61.8–119.6	15.4	299536	3189.5	19262.8	6.0	0.2
B_2O_3	5.2	223.6	4338034	44.0	860	16684000	13.5–34.9	3.9	75854	2109.8	1444.8	0.7	0.8
Li	0.6	25.8	500520	1.5	28	543200	0.8	0.2	2910	104.7	144.0	1.4	0.5

Notes: (1) The input in springs (I–III) north of Bangkog Co is the total sum after separate calculation of the grades of various springs (I–III); the discharge of the hot spring was calculated according to its ancient sinters that are 200 times more than their modern ones. (2) The geochemical reserves of Bangkog Co refer to the total sum of the geochemical reserves of lake waters, intercrystalline brines and solide sediments of lakes I, II and III of Bangkog Co, excluding the early-stage hydromagnesite.(3) The component CO_2 includes HCO_3^- and CO_3^{2-}.

fferent from ordinary volcano and rift hot springs (e.g. those in Kamchatka and the East African Rift Valley). The boron / Cl ratios and the abundances of characteristic elements, such as fluorine, boron and Li of the plateau geothermal waters are distinct from those of hot brines, oil–field waters and marine waters (Table 1.26). Geothermic studies indicate that the Yarlung Zangbo geothermal zone is of magmatic origin (Dong et al., 1981; Huang 1986). In addition, the sulphur isotope composition of the geothermal waters show a relatively narrow range of less than ± 4‰, and the values are characteristic of a deep–seated origin of sulphur (Dong et al., 1981). Hence, it has been inferred that the origin of elements such as boron, Li, Cs, F, As, Rb, K and Cl is from the remelting of magmas. This may be related to the great thickness of continental crust on the plateau, which only allows volatile components of the magmas to make it all the way to the surface. The hot springs of the plateau are characterized by low salinity, intermediate alkalinity and high concentrations of the alkali metals and volatile components such as boron, CO_2, F, Cl, As and S. These compositions are distinct from the heavy metal–rich hyperthermal fluids and superheated brines characteristic of mid–ocean ridge and Red Sea type rifts. The presence of a low velocity zone at depths of 27–40 km in the crust of the plateau (Sun, 1983; Sun et al., 1985) suggest that the hyperthermal flow of the upwelling magma from deep levels may lead to remelting of sialic materials in the crust and the redistribution of lithophile elements. This would favour the concentration of volatile elements such as boron, Li, K, Cs, As and F. This hypothesis accords with some experimental data (Barsykov et al., 1972; Ellis et al., 1977). The volatile components noted above were probably an important factor in the upwelling of the magmas. The uplift of the plateau accelerated since the Late Quaternary. This has allowed the volatile components to diffuse up to the surface of the plateau in form of geothermal and lixiviation waters through a series of active fracture systems on the plateau centred on the Gangdise–Yarlung Zangbo convergence zone. This stage is related to the most important salt–forming stage of the Quaternary special lakes and hydrothermal deposits. The abundances of the special elements such as boron and Li in the plateau lake waters and geothermal waters have a general tendency to decrease progressively from the high–value center (the Bangkog–Nujiang to Yarlung Zangbo convergence zone) outward, and are markedly higher than those in the lakes in the surrounding areas of the plateau. These features are coincident with the high thermal flow in the interior of the plateau and large–scale concentration areas (Li, 1983; Yu et al., 1981; Shen, 1985). This strongly suggests that the sources of such special components as boron and Li on the plateau are related to the diffusion of heat energy from the deep seated molten mass of the Bangkog–Nujiang to Yarlung Zangbo convergence zone. Hence, although the saline materials of the plateau are derived from multiple sources, the boron, Li etc., in the special saline lakes are mainly derived from remelting of intermediate–acid magmas at deep levels in the crust, with the elements being brought to the surface by geothermal and lixiviation waters. These leached boron—rich rocks (mainly Cenozoic volcanic—sedimentary rocks, granite etc.) and concentrated the elements in multi—step saline lake sub—basins.

7.3. Salt Formation Models and Minerogenetic Series

A genetic model of the formation of saline lakes requires the intergration of a number of factors, including; the source of materials, the mode of salt migration, the conditions for salt concentration and deposition, as well as tectonic processes (such as uplift and generation of deposition basins). In light of the diversity of saline lake basins on the Qinghai—Tibet Plateau it is clear that several genetic models are possible and that it is not advisable to remain tied to a single simple genetic model.

7.3.1. TECTONO—GEOMORPHOLOGICAL FEATURES AND GENETIC MODELS

There are numerous saline lake basins on the plateau. On the basis of their tectono—geomorphological features, three genetic models may be proposed to their formation (Fig. 1.26).

Chain—like intermediately deep—shallow down faulted basins
These are mainly distributed on the first and second steps (with elevations of 4750—5100 m and 4200—4700 m above sea level) of the plateau and developed under compressional stresses of various blocks or fault—fold zones of the plateau. As the downfaulted basins of NW, NE and nearly N—S trends are arranged alternately, the lake basins have different amplitudes of deposition and degrees of differentiation. During the lake and river flooding period, these lake basins were separated by high watersheds to form lake systems, each covering an area of up to a few tens of thousands of square kilometers. The lake basins inside these lake systems were connected to form a network pattern. During the drying and shrinking stage, these lake systems disintegrated, but the sub—basins around the large lakes usually remained connected to form numerous chain—shaped or circular sub—basins. There were usaully fed by deep thermal waters along the fracture systems, as exemplified by the Bangkog—Siling lake group (district).

Deep intermontane fault—block basins
They are distributed on the third step (with elevations of 2670—3270 m above sea level), and belong to the basin style formed in the above mentioned structural basin—intermontane fault—block basin. They show large amplitudes of subsidence and a high degree of differentiation. During the flooding stage they were integrated into a large lake basin. During the drying and shrinking stage they became simple chain—like sub—basin groups, which were fed by waters leaching salt—bearing rock

CHAPTER 7

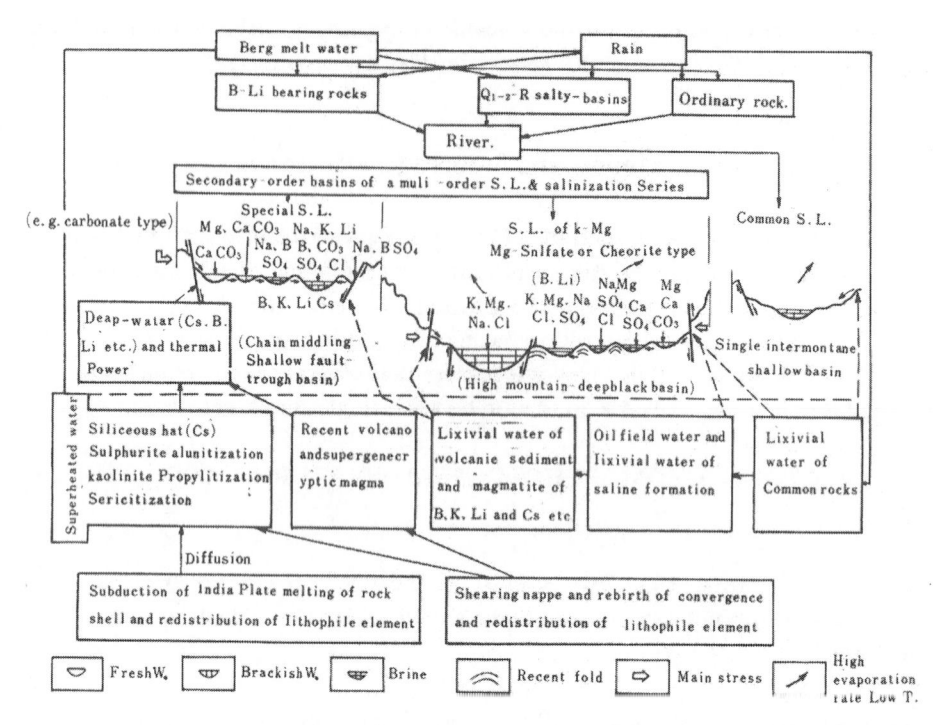

Figure 1.26. Diagram of mineralization for saline lakes on the Qinghai–Tibet Plateau. Abbreviations: S.L. = saline lake, Int. = intermediately, W. = water.

series and oil—field waters, and occasionally by deep—seated thermal waters. An example of this type is provided by the Qarhan lake group (district).

Isolated shallow intermontane basins
They are mainly distributed on the first step of the plateau and are occasionally found on the third step. This type of basin occurs as isolated intermontane basins and has a small downfaulting or downwarping amplitude, with a low degree of differentiation. Sub—basins are generally lacking from this type of basin, of which the Hoh salt basin on the third step is a good example.

7.3.2. SALT FORMATION MODELS AND MINERAGENETIC SERIES

On the basis of the arrangement of sub—basins of the saline lakes, the distribution patterns of salt sediments and the features of salt—forming processes the models of salt formation on the plateau may be divided into those of multi—step saline lake sub—basins and those of isolated salt—forming basins.

Multi—step saline lake sub—basins
The two above mentioned genetic models (i.e., chain like intermediately deep—shallow downfaulted basins and deep intermontane fault—block basins) are of common occurrence, but the former is marked by a combination of several multi—step saline lake sub—basin chains with a relatively small extent of salt formation, while the latter is usually represented by a single multi—step saline lake sub—basin chain with a relatively large extent of salt formation.

By using the saline lake salt formation model of multi—step sub—basins, we can break free from the conventional method of analysing only a single saline lake and, hence, study the genetic relation of the saline lake basin system and the processes of recharge and migration of waters and salts therein.

Formation of saline lake multi—step subbasins
On the basis of the evolution of lake basins and migration and recharge of waters and salts therein, multi—step sub—basins may be divided into lake systems, group, chains, rings, separated lakes and high and low stand lakes. These sub—basins are the result of the combined action of paleoclimatic and paleohydrologic changes and tectonic activity.

In a tectonically active region, inland lake basins are usually controlled by faults. A fault system that extends for a few hundred to a thousand kilometers controls the generation and distribution of basins, and secondary faults cut a lake basin system into several sub—basins. These sub—basins have different elevations above sea level and assume different deep or shallow forms, forming multi—step sub—basins.

Inland saline lakes were usually formed by long continued evolution of freshwater lakes. In the early humid stage, the freshwater lakes covered a relatively wide area, rivers and lakes were connected and various lake basins were linked in a network. They were often separated by high watersheds to form lake systems, each covering an area of ten thousand to a few tens of thousand square kilometers. This geologic period is called the river and lake flooding (or pan—lake) stage. The river flooding stage witnessed the most humid climate and most extensive lake area. Later, as the climate became arid and the tectonic, geomorphological and hydrogeological conditions changed, the pan—lake was dismembered, forming numerous intermontane sub—basins. This period may be called the dismembering stage in the evolution of the lake basin system. In this stage, appreciable differences in regional hydrogeological conditions in various sub—basins of the lake system resulted in the formation of brackish and saline lakes in co—existence with freshwater lakes. This situation continued for a relatively long period of time. A large lake basin and its surrounding sub—basins were generally connected to some extent. These lakes usually had different elevations and salinities and were still fed by surface and sub—surface waters, thus still remaining a lake group. A lake system was generally composed of one or more lake groups. The connection between various sub—basins in the same lake group often show chain— or ring—like patterns. Alternatively, the sub—basins may be arranged down a terrace, thus forming a lake—chain. Sometime sub—basins are present in a crescent form and arranged parallel to a large lake basin to form a lake—ring. Sub—basins with high elevations relative to those adjacent to them in the lake group are called highstand lakes, while those with relatively low elevations are called lowstand lakes. In the case of a sustained arid climate, the recharge was far less than evaporation loss and the lake basin evolved towards the extinction stage when the connection of surface waters and salts between the various sub—basins of a lake group were entirely interrupted and the lakes became shallow and were separated. Each sub—basin then became a local salt—forming center and may have further developed into a playa or an undersand lake covered with sediments.

Material sources
Clastic sediments in the multi—step saline lake sub—basin system are derived from weathering within the catchments, whereas saline materials come from multiple sources. (a) The weathering products of rocks within a catchment: This is a common source of salts, especially ordinary components such as Ca, Mg, Na, S, Cl and HCO_3, which enter the basin system by means of leaching by surface and subsurface waters and from aeolian sources. Salts of saline lakes developed in tectonically stable regions of the crust are usaully derived from weathering of rocks within the catchments. The salts in this kind of lake are composed mainly of ordinary salts, such as halite, mirabilite, trona, gypsum, epsomite and astrakhanite. The concentrations of dispersed elements, such as boron, Li, Cs, As and F in the brines is not high. (b) Deep—seated source: In tectonically active regions of the crust, ma-

terials from deep levels in the crust are a source of salts in multi–step sub–basins. They are an especially important source of such dispersed elements as boron, Li and Cs. The deep–seated materials are supplied to the saline lakes through the following channels: (i) Abundant geothermal waters formed by magmatism and faulting at shallow depths feed the basin in the form of hot or boiling springs. The concentrations of dispersed elements contained within these waters are up to 20 times higher than those in river waters and underground lixiviation waters of the plateau. Isotopic data and characteristic element ratios of components within the saline lakes in Tibet indicate that they have an affinity with the geothermal waters. (ii) Volcanic rocks and volcanic sedimentary rocks formed in the early–stages of volcanism contain abundant dispersed elements such as boron and Li. These enter the basin by means of weathering and leaching of the rocks. Surface and subsurface waters passing through such areas usually show higher concentrations of dispersed elements like boron and Li than in ordinary areas. (c) Ancient salt layers, occlusion brines and oil–field waters: studies of modern inland saline lakes suggests that ancient salt layer waters and oil–field waters can return to the surface and participate in the process of salt formation in some saline basins.

Water and salt migration
In multi–step sub–basins, waters (surface waters, groundwater and hydrothermal waters) are the most important transporting agents, while wind and biota also take part in the process of redistribution of materials to some extent. Clay–sized clastic materials are transported in suspension and by traction and are generally deposited in proximal basins due to gravitational differentiation, while some suspended clays can be transported in sub–basins. Owing to seasonal and secular periodic changes of dry and humid stages, structure of alternating salt beds and clastic beds tend to be formed in the sediments of inland saline lakes. Saline materials entering the basin system migrate as ions or ion complexes together with waters. Because of the action of gravity and selective chemical differentiation, salts always migrate from highstand to lowstand lakes. More soluble salts, i.e. components containing ionic bonds and univalent alkali metals, belong to strongly migrating associations (Cl, SO_4, Na, K, Li, Rb, Cs, Br, I, F and boron) and are liable to migrate to a lowstand lake, and the most chemically active components always migrate toward the lowest stand lake basin. The salinity of lake waters universally increase towards lowstand lakes. When the multi–step sub–basins enter the dismembering stage, they are still connected by surface and subsurface waters and salts. The migration of salts towards lowstand lakes is very conspicuous. Even in the early stages of lake basin desalination and extinction, when various subbasins have been entirely separated on the surface and become independent salt–forming centers, the most active components are still able to migrate to lower stand lakes through underground clastic layers or fractures. Within the separated lakes, the most active components tend to be transported by brines or intercrystalline brines to the lowest point within the

lake. In addition to the external factors, such as the hydrogeologic conditions, the chemical activity of ions, the stability of saline compounds and pH of the media all affect the migration of ions from highstand to lowstand lakes. The migrating abilities of ions in two media can be arranged in the following order: (i) the weakly acid–acid medium (Mg sulphate subtype–chloride type), $Cl > SO_4$ $(B(OH)_3$, $B(OH)_4^-) > HCO_3$ (CO_3) and $Mg > Ca > K >$ or $< Na$ (Li); (ii) in the weakly alkaline–strongly medium (carbonate type), $Cl > SO_4$ (polyborate ions) $> HCO_3$ (CO_3) and $Na >$ or $< K > Li > Mg > Ca$. In the latter medium, Mg and Ca are readily precipitated as Ca–Mg carbonates, and as K ions are readily absorbed by clays and plants they are usually more dispersed than Na.

Dynamic processes of waters and salts
When recharge becomes very small or ceases entirely, physico–chemical processes become the dominant factor in changing water types. This leads to the following model for the evolution in water types; carbonate→ sulphate→ chloride in a closed water mass. Solubility determines the order of removal of saline materials from brines. Multi–step sub–basins are open water masses fed by saline waters. Leached salt components carried by surface and subsurface waters feed the basin over a long period, causing extensive mixing; as a result, the basin waters reflect the character of the inflow waters and the long evolution. The evolution of water types does not always follow the simple model outlined above. For example, Bangkog and Xia Caka in Tibet have developed into the playa stage, but their waters are still of the carbonate type. Salt removal can result from evaporative concentration and crystallisation and / or cooling crystallisation caused by temperature changes and replacement of early deposited salts by brines. Selective adsorption of some ions by clay minerals and selective concentration of halophilic algae and bacteria also contribute to the processes of differentiation and concentration of saline materials. In addition, in areas where there is mixing of several water masses (lake shores, river mouths, springs, etc.), mixing crystallisation is a common occurrence. Long continued recharge of large amounts of some components can result in "overcompensation" crystallisation of some salts and can produce rich salt deposits.

Minerogenetic series
Due to multiple material sources and multiple salt–forming processes, the "ox–eye" and "tear–drop" zoning patterns are not completely applicable to inland basins. From highstand to lowstand lakes and from early to late stages of evolution of multi–step sub–basins, the following minerogenetic series of solid–liquid phases are formed:

1. carbonate type minerogenetic series: Ca–Mg carbonates, clay, diatom earth → hydromagnesite→ mirabilite (thenardite), borax→ halite, sylvite, glaserite, Li–bearing carbonate and K–Li–B–(Cs–Br) brines;

2. sulphate—type minerogenetic series: Ca—Mg carbonate, clay→ gypsum→ mirabilite (thenardite), Mg borate→ halite, schoenite, kainite (Mg borate) and K—Li—B—(Cs—Rb) brines;

3. chloride—type minerogenetic series: Ca—Mg carbonate, clay→ gypsum→ halite, gypsum (half—water gypsum)→ halite, carnalite (sylvite) and K—Mg—(B—Li) brines.

Conclusions

Mineralization in multi—step sub—basins of saline lakes is interpreted as follows: when salt components derived from multiple sources migrate in to multi—step sub—basins, extensive mixing occurs, saline materials are differentiated and regularly deposited, and more soluble salts migrate from highstand to lowstand lakes and are concentrated in the latter, thus forming several minerogenetic series (Table 1.33).

TABLE 1.33. Summary of the model of evolution of multi—step subbasins of saline lakes

Evolution stage	River—lake flooding stage	Dismembering stage	Extinction stage
Climate	Humid	Semihumid—sustained arid	Arid—extremely arid
Lake form	Pan—lake	Lake system—lake group lake chain, lake ring (highstand lake, lowstand lake)	Separated shallow basin
Salinity of waters	Freshwater lake	Freshwater lake, brackish lake, saline lake	Playa→undersand lake
Sedimentary formation	Clastic, clay and carbonate formation	Clay and carbonate formation, evaporite formation	Evaporite formation
Minerogenetic series		1. Carbonate type: Ca—Mg carbonate, clay→gypsum→mirabilite (thenardite), Mg borate→halite, schoenite, kainite (Mg borate) and K—Li—B—(Cs—Rb) brines 2. Sulfate type: Ca—Mg carbonate, clay→gypsum→mirabilite (thenardite), Mg borate→halite, schoenite, kainite (Mg borate) and K—Li—B—(Cs—Rb) brines 3. Chlrode type: Ca—Mg carbonate, clay→gypsum→halite, gypsum (half—water gypsum)→halite, carnalite (sylvite) and K—Mg—(B—Li) brines	

Saline lake sub—basins that are distributed as steps are common on the Qinghai—Tibet Plateau and in modern salt—forming areas of South and North America and Africa. In addition, they are also found in some ancient continental salt foming areas, such as those of the Cretaceous to Tertiary red beds in eastern China. Inland saline lakes are the inexorable products of a certain stage of evolution of a multi—step sub—basin system. The model of salt formation in multi—step sub—basins of saline lakes is representative of salt—forming models in arid inland areas. Using this theory, a number of economic deposits of boron, K, Li etc., have been discovered on the plateau. This theory is important in the study of mineralisation in ancient evaporite basins and mineral prospecting.

CHAPTER 8 SHORELINE DEPOSITS AND PALEOCLIMATIC ANALYSIS

In the Qinghai—Tibet Plateau saline lake region there are extensive shoreline deposits that reflect the paleoclimatic and paleoenvironmental variations in the Late Quaternary. These sites offer valuable sites for the study of the recent paleoclimatic evolution on the plateau. In a report written in 1959, Zheng Mianping, the chief author of this monograph, made a preliminary discussion of the lake terraces in the Bangkog—Siling lake area①. This report was studied by the late Professor Zhu Kezheng in the preparation of his important work (Zhu, 1973). He advised the author to pay attention to the great significance of lake terraces (including sand barriers, etc.) on the Tibetan Plateau in the study of paleoclimatic changes, and encouraged the author to prepare a scientific paper for publication. The writer would like to take this opportunity to express his deep respect to and cherish the memory of Professor Zhu, who was clearly a man of great foresight.

In light of the well preserved late Late Pleistocene lake shoreline geomorphology and deposits on the plateau, this chapter will make a preliminary analysis of the paleoclimatic evolution of the plateau since this time. This analysis will make use of geomorphological, sedimentological, geochronological, micropaleontological and stable isotope studies.

8.1. The Paleoclimatic Evolution of the Qinghai—Tibet Plateau Since the Late Part of the Late Pleistocene

This section discusses the regional paleoclimate of the plateau since 40 ka (Zheng, 1994). A total of 12 sites on the plateau and its adjacent areas were selected for the study (Fig. 1.27) (Du and Kong, 1983; Wasson et al., 1984; Chen and Bowler, 1985; Prell and Kutzbach, 1987; Sun and Wu, 1987; Zhang et al., 1988, 1994; Zheng, 1989; Zheng et al., 1989, 1992; Du et al., 1989; Chen et al., 1990; Li et al., 1991; Li and Li, 1991; Li, 1992; Xu, 1992; An et al., 1993; Gu et al., 1993; Shan et al., 1993; Shen et al., 1993; Fonte et al., 1993; Qi and Zheng, 1995).

---

① Zheng Mianping (1959). Preliminary report on the investigation of boron deposits in the eastern part of North Tibet; a report for internal use.

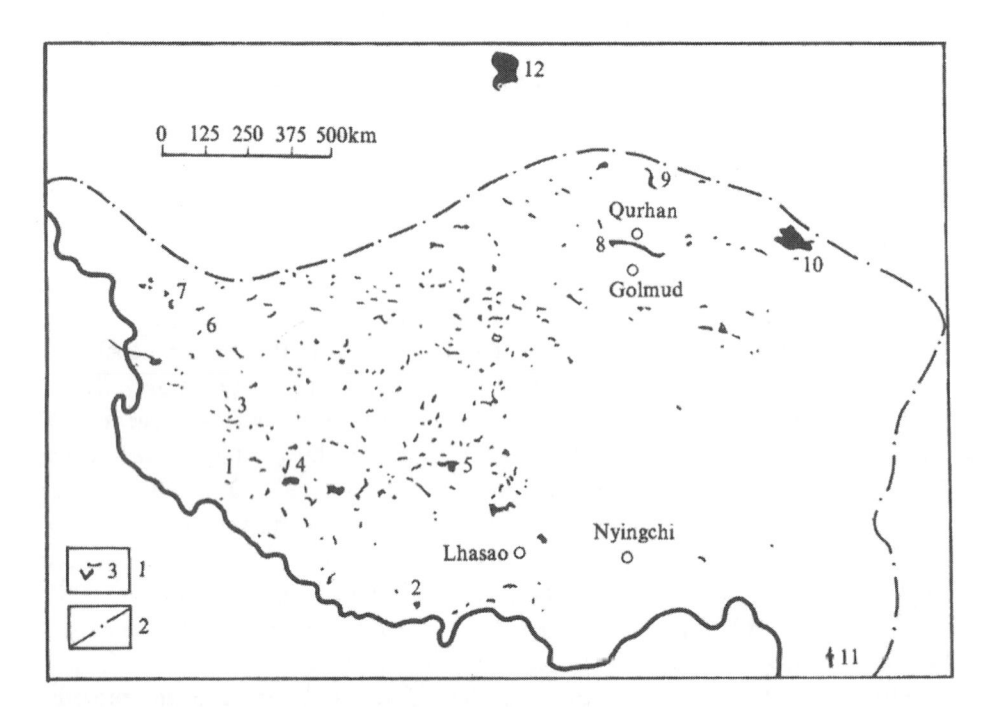

Figure 1.27. Lake areas of the Qinghai–Tibet Plateau and study localities. I. Modern lake (including playa); II. Plateau boundary. III. Lake area boundary; IV. Lake area No. 1. Qilian inflow area; 2. Qaidam inflow area; 3. Kumkol inflow area; 4. Hoh Xil inflow area; 5. Qiangtang inflow area; 6. Northern Tibet inflow area; 7. Southern Tibet inflow–outflow area; 8. Qinghai Lake–Gonghe outflow–iflow area; 9. Bayan Har outflow area; 10. Southeastern Tibet inflow–outflow area. Study localities: 1. Paiku Co; 2. Como Chamling; 3. Zhacang Caka; 4. Zabuye Lake; 5. Siling Co–Bangkog Co; 6. Lungmu Co; 7. Aksayqin Lake; 8. Qarhan; 9. Da Qaidam and Xiao Qaidam Lakes; 10. Qinghai Lake; 11. Erhai Lake; 12. Nop Nor Lake.

8.1.1. CLIMATIC EVOLUTION SINCE 40 KA

According to changes in the level of the lake surface and the nature of the lake deposits, the paleoclimatic evolution since 36 ka can broadly be divided into six stages;

Pan–lake stage during 40–28 ka (Zheng et al., 1983; Zheng, 1989; Li et al., 1991)
Between 40–28 ka there was a vast expanse of lakes on the plateau that was up to dozens times larger than the present surface area of lakes. The lake waters were fresh and tended to outflow. The lakes and rivers were connected to form pan–lakes. The climate was mainly cold and humid. At about 30 ka, the pan–lake system on the plateau started to disintegrate from north to south. First, the ancient

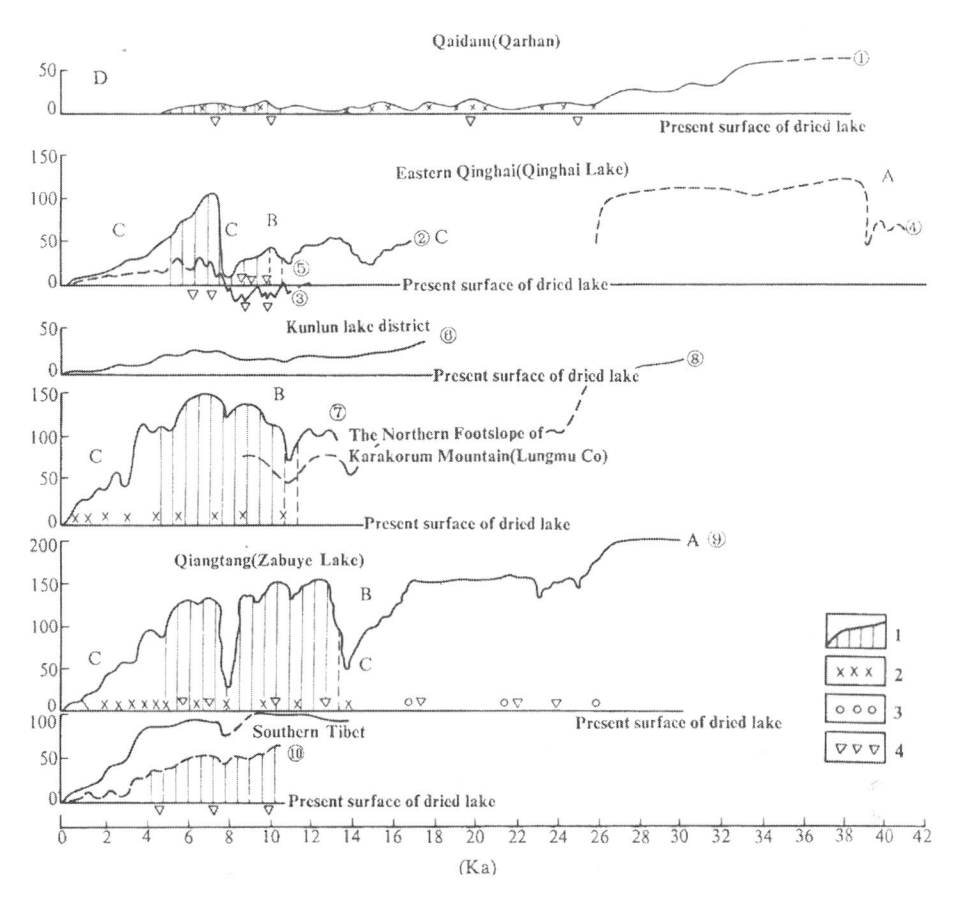

SHORELINE DEPOSITS AND PALEOCLIMATIC ANALYSIS 149

Figure 1.28. Sketch map showing the fluctuation of the ancient lake surfaces on the Qinghai–Tibet Plateau. A. High lake surface (overflow surface); B. Medium lake surface; C. Low lake surface; D. Playa surface. 1. Warm slightly humid–warm humid; 2. Salt–bearing segment; 3. *Ilyocypris* or *Pediastrum*; 4. High arbor sporopollen content segment. ① From Chen Kezao et al. (1985); ② From Xu Guowen (1992); ③ From Zhang Pangxi et al. (1994); ④ From Shan Fashou et al. (1993); ⑤ From Du Naiqin et al. (1983); ⑥ From Li Shijie et al. (1991) and Li Shuanke (1992); ⑦ From J. C. Fontes et al. (1993); ⑧ From Li Binyuan et al. (1991); ⑨ From Zheng Mianping et al. (1989); ⑩ From Li Binyuan et al. (1983).

Qarhan Lake in the Qaidam basin in the northern part of the plateau became a saline lake at 30 ka, and a self precipitating lake at about 25 ka (Chen et al., 1985). During the pan–lake stage, prior to 30 ka, the ancient lake surface was about 50–60 m above the modern level (Fig. 1.28) and its area was at least 25,000 km^2 (Fig. 1.29), compared to the modern area (including the playa) of 5850 km^2. The area ratio of the ancient lake to the modern one is about 4.3 and their volume ratio is 52.

150   CHAPTER 8

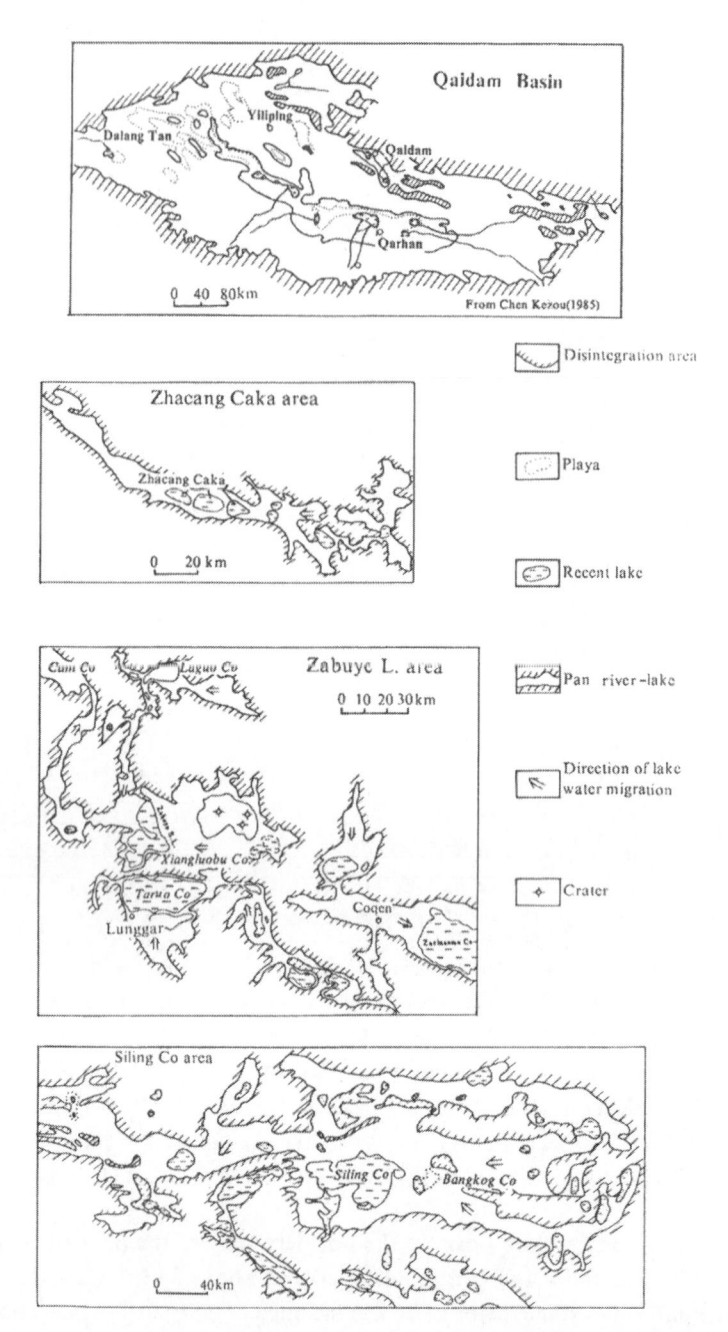

Figure 1.29. Distribution of the pan–lakes on the Qinghai–Tibet Plateau during Q_3^3 (40–28 ka). 1. Lake bank; 2. Playa; 3. Lake; 4. Pan–lake–river; 5. Direction of migration of lake waters; 6. Ancient volcanic crater.

SHORELINE DEPOSITS AND PALEOCLIMATIC ANALYSIS 151

The volume ratios of the ancient and modern Da and Xiao Qaidam lakes north of Qarhan have similar volume ratios of about 50 (Table 1.34). However, in the southwestern part of the plateau the pan–lake stage continued for a longer period. For example, pan–lakes were widespread in the northern Tibet lake area (inflow lake area) (Figs. 28 and 29) (Zheng, 1989). From the west to the east these lakes are: Chagcam–Dong Co with an area of up to 4500 km^2, compared to a modern area of 270 km^2; Zabuye–Mami Co with an area of 9700 km^2, compared to its modern area of 920 km^2 (Fig. 1.30); Bangkog–Siling Co had an area of 10 000 km^2, compared to its modern area of 3364 km^2 (Fig. 1.29). The ancient/modern area ratios of these lakes vary from 3–17 and their volume ratios are 8–40 (Table 1.34).

TABLE 1.34. Comparison of the present and ancient (Pan–river–lake Stage) areas and volumes of some lakes of the plateau

Pan–lake name	Modern lake area (km^2)	Area (km^2) of pan–lake stage	Area ratio	Modern lake volume (10^{8}m^3)	Volume (10^{8}m^3) of pan–lake stage	Volume ratio	Note
Qinghai L. (–Gonghe)	4432	14800	3.3	740	8883	12	Changed from an outflow lake to an inflow lake
Da and Xiao Qaidam	186	3750	20	1.86	94	50	
Qarhan	5850	26000	4.3	146	7500	52	
Aksayqin	182	1403	7.7	18.2	702	39	
Lungma Co	135	635	4.7	14	477	34	
Chagcam (–Dong Co)	270	4500	17	5.7	225	40	
Zabuye (–Mami Co)	927	>9780	10.6	409.7	11700	29	
Bangkog (–Siling Co)	3364	>10000	3	1334	10700	8	
Como Chamling	108	1800	17	5.1	2700	526	Modern lake waters mostly lose

The plateau lake surface remains are well preserved in the northern Tibet lake area. For example, before 28 ka the surface of Zabuye–Mami pan–lake was

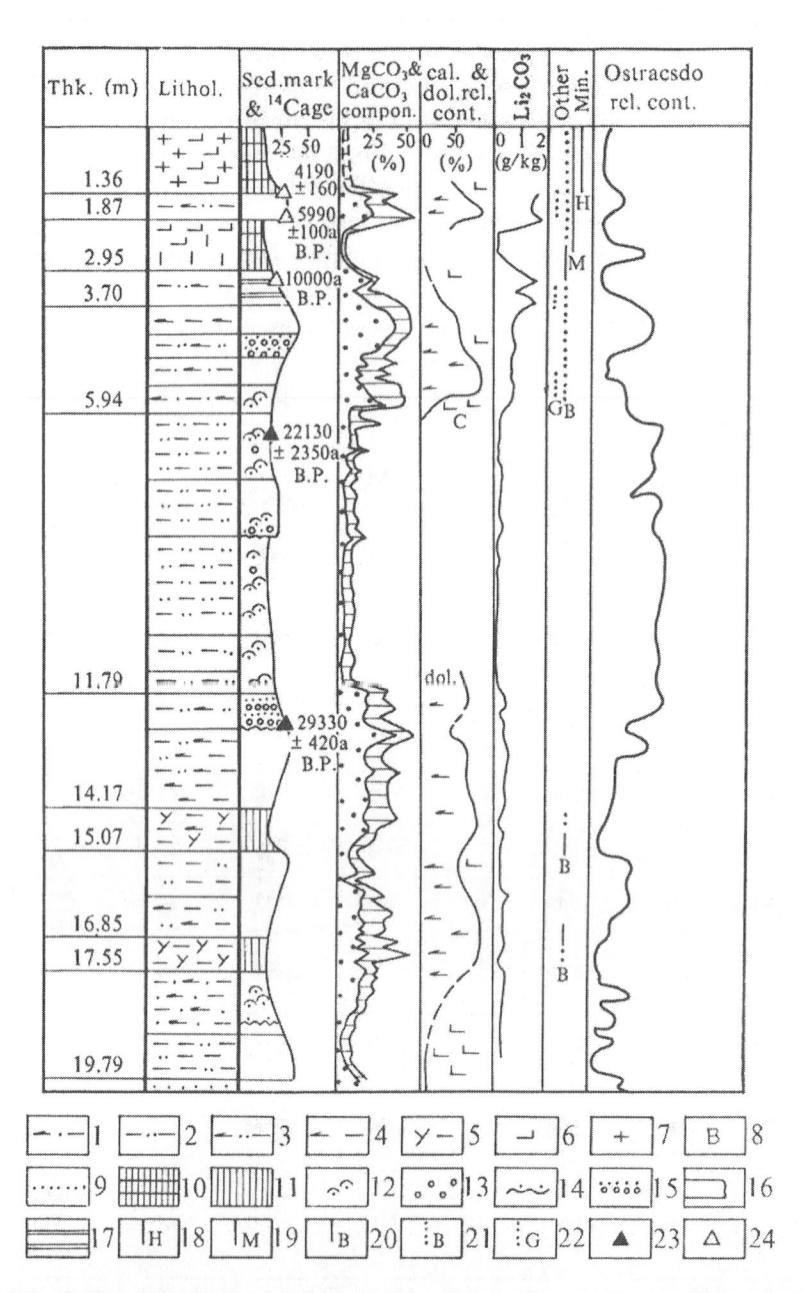

Figure 1.30. Sedimentary column of well ZK 91-2 in southern Zabuye, Tibet (modified from Qi and Zheng, 1995). 1. Silt and fine sand carbonate clay; 2. Silt-bearing clay; 3. Silt- and fine sand-bearing carbonate clay; 4. Carbonate clay; 5. Carbonate clay with ulexite; 6. Mirabilite; 7. Halite; 8. Borax; 9. Silt and fine sand; 10. Saline deposits; 11. Deposits intercalated with salts; 12. Ripple bedding; 13. Intrabasinal pebble; 14. Scour surface; 15. Graded bedding; 16. Massive bedding; 17. Horizontal bedding;

18. Halite—bearing bed; 19. Mirabilite—bearing bed; 20. Borax—bearing bed; 21. Dispersive borax; 22—23. ^{14}C age data of sporopollen; 24. Inferred age. Abbreviations: cal.=calcite; dol.=dolomite; rel. cont.=relative content.

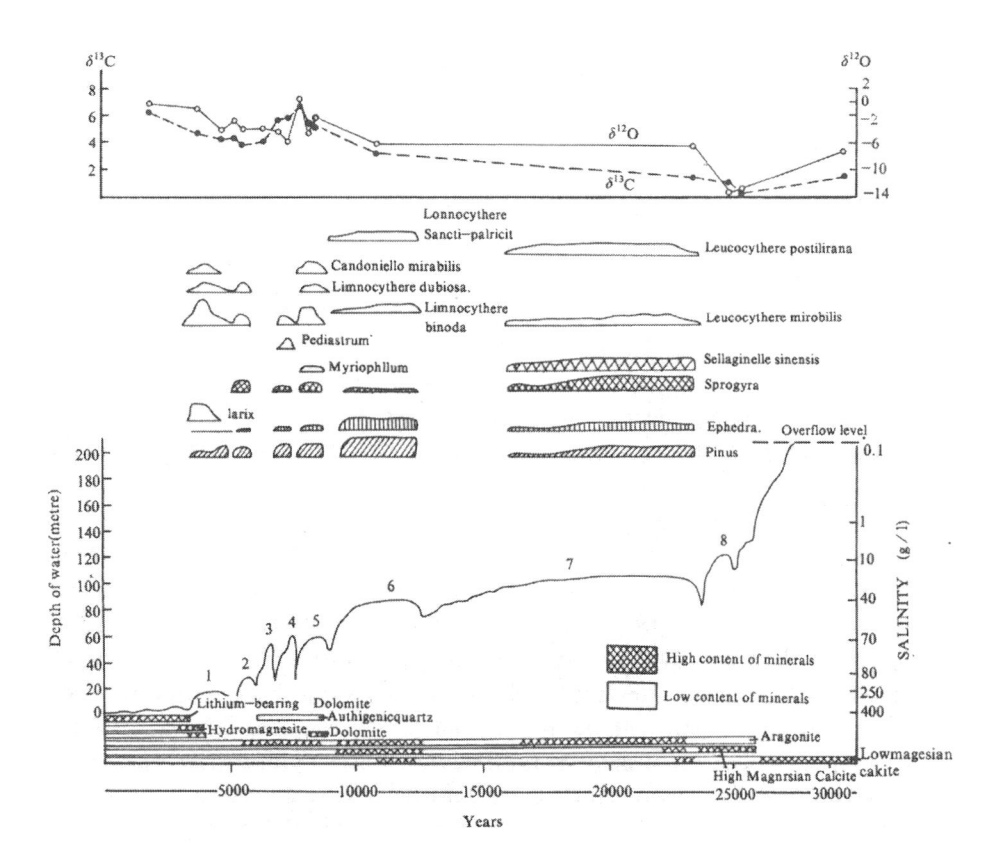

Figure 1.31. Variations of the water level, salinity and other features of the Zabuye—Mami Co pan—lake.

180—200 m higher than the modern saline lake surface①, and the ancient lake strandline and lake—eroded geomorphology are present throughout the area. The deposits corresponding to the high lake surface are dominated by low—Mg calcite bearing clay deposits of deep lake facies. The yielded ostracods such as *Limnocythere dubiosa*, which is common in fresh waters, weakly salt—tolerant and hypothermophilous *Leucocythere mirabilis*, fresh—water *Pediastrum simples* and

---

① According to the anthor's recent investigations, late Late Pleistocene lacustrine sands and gravels are still present at an elevation of 200 m above the present lake level.

charophytes that prefer clear and quiet waters (Fig. 1.30) (Zheng et al., 1989a; Liu, 1994).

Warm humid—warm slightly humid stage during 28—19.6 ka
As late as about 28 ka, the high lake surface in the southern part of the plateau fell to the medium—high lake surface, but the duration of this lowering was relatively long. For example, at about 28 ka the level of the Zabuye lake changed from a high lake surface to a medium lake surface, corresponding to sand levees 8 and 7, respectively (Fig. 1.31), At this time the lake surface was about 130—120 m above the modern lake surface. Deposits associated with this stage are dominated by silts and clays, with a relatively high concentration of organic matter containing abundant ostracods and charophytes. After about 26.7 ka thermophilic Ilyocypris gibba appeared amongst the ostracods, together with large tooth fossil lizards and the charophytes *Chara contraria—vulgaris* and *Tolypella* sp. Carbonate deposits yield $\delta^{13}C$ values of 0.6—0.9 and $\delta^{18}O$ values of 13.0—13.9 in the initial stages and $\delta^{13}C$ and $\delta^{18}O$ values of 2.7 and 6.6, respectively, in the middle and later stages. These sedimentary and micro—palaeonotological features indicate that the lake waters were initially relatively deep and alternately cold and warm and became progressively warmer and clearer. Generally, the climate was warm and humid throughout 8,000 years of this stage.

During this stage the Qarhan lake area in the northern part of the plateau was characterised by low lake surface and alternation between saline lakes and playas (Fig. 1.28). At about 28—24 ka evaporite—bearing clay was present in the sediments of these lakes (Chen et al., 1985). A halite bearing silt—layer has also been discovered at a depth of 52 m (^{14}C age 31.4± 1.8 ka) in a drill core in Qarhan lake (Shen et al., 1993). Sporo—pollen analyses indicate a relatively high proportion of thermophilic aquatic plants, pteridophytes and arbores between 30—26 ka; xeric shrubs and herbs were dominant from 26—25 ka, thermophilic and hygrophilic aquatic plants and ferns (pediastra etc.) increased between 25—24 ka; shrubs and herbs (*Ephedra* and *Artemisia*) predominated (Du, 1983; Fig. 1.32). Overall, this data indicates that during this stage the temperatures rose somewhat and that there were two slightly humid (30—26 and 25—24 ka) and slightly dry (26—25 and 24—20 ka) subcycles.

Slightly cold and dry stage 19.6—13 (13.5) ka
During this stage some of the lakes in the northern Tibet lake area and in the southern part of the plateau entered the saline lake stage. For example, at about 19.6 ka, micro—organisms such as ostracods and charophytes in Zabuye lake suddenly disappeared from this section; at about 15.5 ka the lake surface fell markedly; at about 13 ka the lake level reached its lowest point of this stage, corresponding to the shallow—medium lake surface (80—30 m above the modern lake level) (Figs. 1.28 and 1.31). In Bangkog lake some cold—facies mirabilite deposits were formed at about

TABLE 1.35. Division and Correlation of the Late Late Pleistocene–Holocene Climatic Stages on the Qinghai–Tibet Plateau

(绘图)

Age ka (B.P)	Europe (period)	Zabuye L. (Zheng et al.,1989)	Northern Tibet Lake Ban Kog Co (Zheng et al., 1989)	Siling Co (Gu et al.,1993)	Qinghai L. (Zhang ,1988)	(Du et al., 1992)	Qarhan L. (Chen.1980; Du et al., 1983)
1–	Subatiatic	similar to recent	warm dry / Cold dry	Cold dry slightly humid	cold dry	Cold dry V	
2–		6 Alternately cold dry and warm dry	waum dry slightly warm dry	cold dry		III	
3–			cold dry	sitghtly warm humid	Warm dry	warm dry IV	
4–	Subboreal		slightly warm humid	Cold dry			
5–		warm dry	cold dry	tending cold dry	cool humid		Cold dry slightly warm humid
6–	Atlantic	5 Cold dry–warm humid	slightly warm humid		warm humid	warm humid III	
7–		slightly warm dry		warm humid	II	humid	
8–	Boreal	cool dry	cold dry		Warm slightly dry	tending warm humid	
9–	Preboreal	slightly warm humid	slightly warm humid cold dry	slightly warm humid		II	
10–			Slightly warm bumid				
11–	Younger Dryas	cold dry	Cold dry	Cold dry	cold humid	sightly warm humid	I
12–	Allerod	Slightly warm humid	sligthly warm bumid	cool slightly dry	cold dry	slightly cold dry	cold dry
13–	Older Dryas Older Dryas	cold dry	Cold dry				
14–	The Second subinterglacial	warm slightly humid	warm slightly humid				
15–		cool–progressively –dry	cool slightly dry				
16–		Cold dry					
17–	Middle Weichselian glacial period (second subglacial) The coldest	3 cool slightly dry	cold slightly dry				Cold dry slightly warm humid
18–							
19–							
20–	(or 22)	warm slightly dry					
21–							
22–	Cool humid	warm humid	warm humid				dry
23–							
24–	The first subinterglacial Warm humid	2 slightly cold humid					
25–							
26–							
27–		waum humid					
28–							warm humid
29–							
30–	(or 30)	1 cold humid					
31–	cold						

* Bolling

18 ka (Zheng et al., 1983). In the northern part of the plateau, salt was continuously precipitated in Qarhan after 19.6 ka; halite beds were formed at about 19.5–19 ka; between 19–18 ka the sediments consist of evaporite–bearing silty clay containing high levels of arbor, aquatic plant and fern pollen grains; the deposits were again marked by halite beds between 18–17 ka; and the period after 16 ka was marked by continuous deposition of halite, together with some K–Mg salt deposits (Chen et al., 1990). At about 20 ka the Da Qaidam lake at the northern margin of Qaidam, became a saline lake and at about 16.8–14 ka salt deposition reached its culmination with the precipitation of large amounts of cold–facies mirabilite (Zheng et al., 1989a). This indicates that the climate became cold and dry in the northern part of the plateau after 16–13 ka.

Alternately warm (humid) and cold (dry) stage during 13 (13.5)–9 ka
This change witnessed frequent changes in the surface levels of the plateau lakes and in the deposition of saline minerals, reflecting the fact that the climate on the plateau was beginning to enter a new stage characterised by warm (humid) and alternately warm and cold periods. At about 13 (13.5) ka the climate first became warmer in the southern part of the plateau and its adjacent southern areas. Gradually this warming trend extended to the north. In sediment layers <13 ka from Bangkog lake, Chagcam Cake and Zabuye lake in the northern Tibet lake area there are 1 or 2 sections containing arborescent pollen, ostracod shells and increasing $CaCO_3$ contents (Fig. 1.32), together with an appreciable fluctuation in the ancient lake surface levels. During this stage there were at least three warm and slightly humid and two cold dry climatic intervals, as revealed in the sediments of Zabuye lake. At around 13.5 ka, large amounts of thermophilous *Ilyocypris gibba* pollen and high $CaCO_3$ levels were present (Fig. 1.30), together with an increase in the relative content of Li–bearing carbonates. The lake surface was about 80 m higher than the present day level. All together, these observations make the beginning of a warm stage on the plateau. At around 12.5 ka dry cold–facies microcrystalline borax deposits appeared and the ostracods *Ilyocypris gibba* disappeared, reflecting the fact that the climate had turned cold and dry. At around 12.5–11 ka, the relative contents of $CaCO_3$, colloidal Li–bearing carbonates and borax contents decreased abruptly, signifying a change to warmer, more humid conditions. At about 11–10.5 ka, the borax contents increased again, there was a further decrease in $CaCO_3$ contents and ostracods disappeared. The ancient lake level was now only 30–40 m above its present level, marking a cold–dry sub–stage. By about 10.5–9 ka, levels of warm–facies lithium dolomite increased dramatically together with NaCl contents. In contrast, there was a decrease in deposition of cold–facies borax and mirabilite. Ostracods were rare, reflecting relatively high salinities in lake water and a climate that had become warmer and slightly drier.

Geochemical studies from well CK1 from Bangkog lake in the eastern part of the northern Tibet lake area (Zheng et al., 1989a) suggest that three warm, slightly

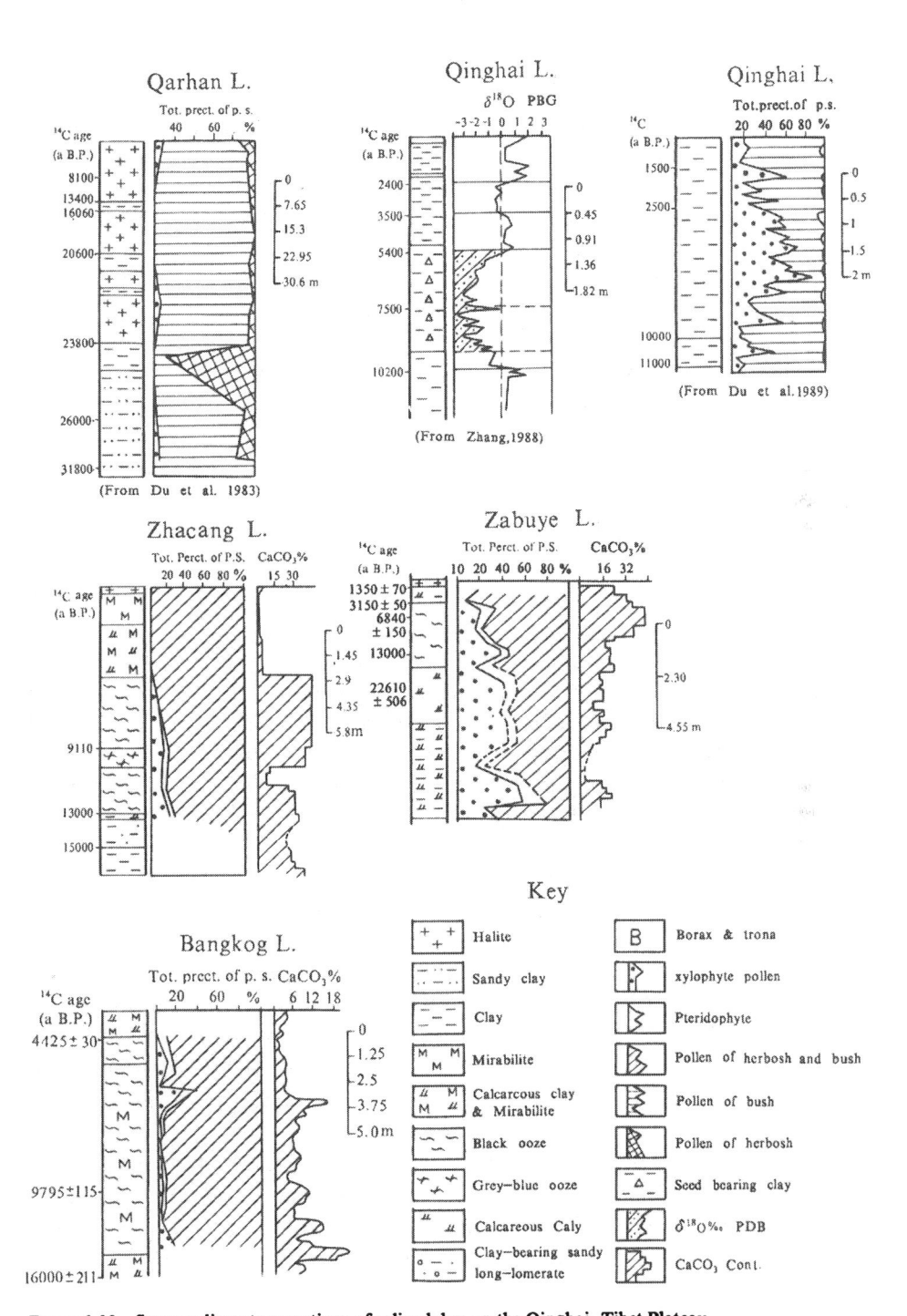

SHORELINE DEPOSITS AND PALEOCLIMATIC ANALYSIS 157

Figure 1.32. Some sedimentary sections of saline lakes on the Qinghai–Tibet Plateau.

humid–cold, dry cycles (Table 1.34) may be distinguished in the period from 13.5 to 10.5 ka. At around 13 ka the climate became warmer and more humid, beginning in the southern part of the plateau. This was not an isolated event. Glaciers in southern Tibet also started to recede at 13 ka and the temperature rose by about 0.4–0.8°C as compared to the Rongpusi glacial stadial phase. Accordingly, Li (1977) used 13 ka to define the lower boundary of the Holocene in Tibet. To the southwest, in the Himalayas and the Didwana of the Thar desert of India there were large halite deposits before 13 ka. These may be correlated with the last glacial maximum (LGM). At about 13 ka the halite deposits were replaced by clastic deposits and the salinity fluctuated strongly (Wasson et al., 1984). Studies of sporo–pollen from the Erhai and Dianchi lakes of western Yunnan on the southeastern margin of the plateau also indicate that the climate started to warm and slightly humid at 13.5 ka. In the vicinity of Eryuan, the period of 15–10.5 ka there were great changes in vegetation (Worker, 1985; Sun et al., 1987). Studies were also conducted from northern Tibet northward to the Karakorum Mountains lake area. At 12 ka the lake surface of Lungmu Co was medium–high (about 70 m above the modern lake level) (Fontes et al., 1993). Sporo–pollen studies of well CK 2020 in the Biele Playa in the western sector of Qarhan of the Qaidam basin (Du, 1983) indicate that after 11 ka the air temperature and humidity around the lake again increased significantly. Hence, Du (1983) suggested that 11 ka should be used as the lower boundary of the Holocene. The climate around Qinghai lake also turned warm at about 10 or 9.5 ka and sporo–pollen analyses (Du et al., 1989) indicate that during 10.6–9.1 ka there was a mixed coniferous and broad–leaved forest (Fig. 1.32) that suggests the climate had become warmer. Geochemical studies of ostracod shells also suggest that the climate became warmer and more humid at about 9.5 ka (Zhang, 1988; 1994; Fig. 1.32).

Alternately cold dry and slightly warm humid stage during 9.3–3.5 (4) ka
During this stage there were at least two cold dry or slightly warm and dry and one or two humid or slightly warm humid substages on the plateau. In the south the lake surfaces were mostly in a state of medium–low lake surfaces and the great majority had higher lake levels than their modern equivalents. In the warm and humid or slightly humid substages the lake levels were mainly at a medium levels and salt deposits were mostly in the sulphate and carbonate phase. In contrast, in the north the waters in Qaidam maintained a low lake surface and even dried up with the precipitation of chloride and some K–Mg salts.

In the zone around Zabuye lake–Bangkog lake–Siling lake in the northern Tibet lake area, the lake surfaces were 50–60 m above their modern lake levels at about 9–7 (7.5) ka, and their was a distinct low lake surface at about 8 ka. In this period there was significant deposition of cold–faices mirabilite and the climate was mainly cold and dry. Oxygen and carbon and isotope studies indicate that the period between 7(7.5)–5.1 ka was a slightly warm humid substage. During the early stages (7.1–6.4 ka) large amounts of fine–grained borax and mirabilite were deposited. Appreciable amount of thermophilic ostracods *Ilyocypris gibba* have been found in salt–bearing sludge (with an isotopic age of about 6.7 ka) reflecting the fact that there was a hot climatic oscillation within the slightly cold humid background. At about 6.4–5.1 ka the warm–facies lithium dolomite content reached a maximum and large amounts of ostracods were present, indicating the presence of a warm–slightly humid climate existed. During about 5.1–3.5 ka, large amounts of borax and mirabilite were deposited. This, and the $\delta^{18}O$ and $\delta^{13}C$ values, imply that there was a cold dry climate. In the northern part of the plateau, Qarhan had been in the late stage of saline lake, with the deposition of large amounts of K–Mg salt–halite beds. At about 6–5 ka the great majority of the saline lakes dried up and new lakes were only present on the southern margin. A geochemical study of ostracod shells (Zhang, 1988) revealed that the climate of Qinghai lake, that was located in a transition area, was hot–slightly dry and warm slightly dry at about 9.1–7.5 ka and warm dry in the early part of the 7.5–3.6 ka interval and warm humid in the late part. However, sporo–pollen analyses (Du et al., 1989) suggest that the climate there was slightly warm dry at about 9.3–8 ka, and mainly warm humid at 8–3.5 ka. Remnants of *Picea purpurea* have been found in lacustrine strata of Qinghai Lake 100 m or more above the levelof the modern lake, with a ^{14}C age of 6.2 ± 0.2. The seventh levee of the lake has a ^{14}C age of 4.5 ± 0.1 ka and is located 46 m above the modern lake level, indicating a medium level lake surface.

Cold dry (warm dry) stage from 3.5 (4) ka to the present
In this stage the lakes in the southern and northern parts of the plateau had a general trend of development in the direction of contraction and desiccation. In the southern part of the plateau, the lake surfaces were lowered from low–medium stages to low levels and locally playa surfaces. Salt–forming processes were prevalent and this stage witnessed the maximum amount of salt formation in the areas in the Quaternary. Some lakes entered the chloride phase. According to the cold– and warm–facies association of saline lake deposits and their clast content as well as the variation in oxygen and carbon isotope compositions of carbonates (Gu et al., 1993), at least two warm dry–cold dry climatic cycles may be distinguished in the region. These may correlate with the first glacier (3 ka) and third glacier advance (17th–19th century) of the neoglaciation of Tibet (Li et al., 1987). An obviously dry cold climate appeared between 3(3.1) ka and 2.5 ka and between 990–230 a. During this time thick–bedded borax–bearing mirabilite was deposited. In the last 100–200 years the climate has become drier. In the northern part of the plateau, the great majority of saline lakes in the Qaidam basin dried up and became sandy. Only in saline lakes on the basin margins, such as Xiao Qaidam lake, Dabsan lake and Gas Hure lake, are surface brines preserved locally. The data from Qinghai lake on the northeastern margin reflect the existence of three warm dry–cold dry cycles (Zhang, 1988).

8.1.2. SOME UNDERSTANDING OF THE FEATURES OF CLIMATIC EVOLUTION

There was a well established pan—lake stage during 40—28 ka, which is the youngest Pan lake stage on the plateau. During this stage pan—lakes were widespread and mostly contained freshwater. Rivers and lakes were connected and the water expanse was wide. According to the scale of the pan—lakes and micropaleontological and geochemical data, the climate at this time was mainly humid and cold. Precipitation was 1—2 times higher than that today, but the air temperature was similar. Until the Holocene "climatic optimum", the precipitation and temperature on the plateau were also higher than today. For example, the water expanse, depth and volume of Qinghai lake during the Holocene "climatic optimum" were, respectively, 0.7, 5 and 2 times those to today. An analysis of the growth conditions of Picea purpurea indicate that the mean annual water temperature was 3℃ higher than today and the precipitation was over 100 mm higher①. However, carbon and oxygen isotope analyses of ostracod shells indicate that the water temperature were 1.8℃ higher than today, and even in the Younger Dryas interval (from about 11—10.2 ka), the water temperature of Qinghai lake was also 0.3—0.5℃ higher than the modern temperatures (Zhang, 1988).

From 40 ka to the present, the temporal trend in climatic evolution was from cold—humid (1) → warm humid (2) → alternately warm humid and dry humid [(3)—(4)] → cold dry and slightly warm humid (5) → cold dry (warm dry) (6) (Table 1.34; Fig. 1.33). The duration of the freshwater—brackish lake stage of the lakes was long in the southern and southeastern parts of the plateau and short in the northern and northwestern parts, and accordingly the time of salt deposition was late in the southern and southeastern parts of the plateau (2—3 ka in southern Tibet and 11—18 ka in northern Tibet) and early in the northern and northwestern parts (30—25 ka). Overall, the general framework of the paleoclimate since 40 ka is that the climate was humid in the south, but dry in the north and that the humid climate lasted long in the south (east) but short in the north (west).

8.1.3. GENETIC MECHANISM

Many researchers consider that the uplift of the high Himalayas and the plateau in the Quaternary was the main cause of climatic changes on the plateau. The large amount of uplift of the Qinghai—Tibet Plateau since the Early Pleistocene led to the uplift of the Himalayas and a series of nearly E—W trending mountains to the

---

① The modern Qinghai lake has a mean annual temperature of 0.9—2.7℃, a mean annual precipitation of 407 mm and a mean annual evaporation loss of 954.5 mm.

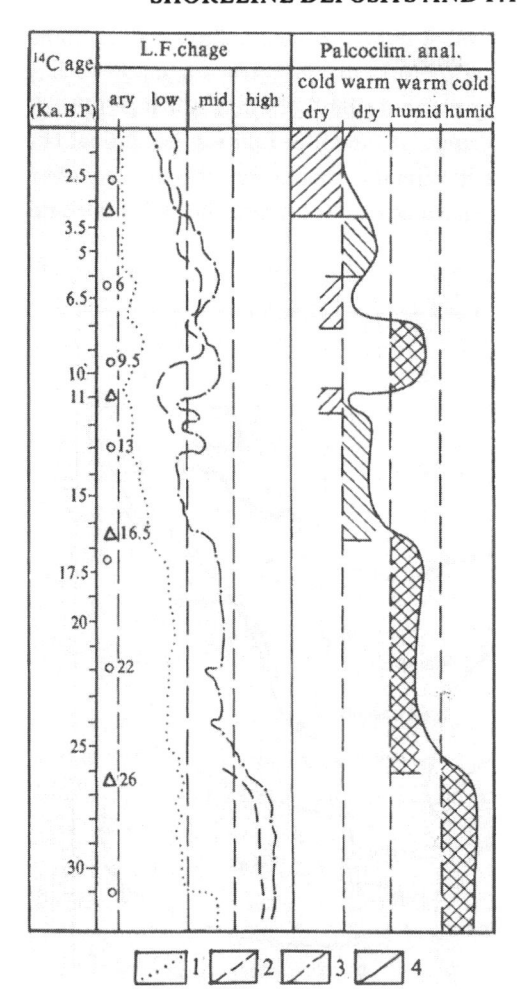

Figure 1.33. Changes of the lake surface and paleo–climate on the Qinghai–Tibet Plateau since 33 ka. 1. Qarhan L.; 2. Qinghai L.; 3. Zabuye L.; 4. Trend of paleoclimate change.

north, which impeded the warm moist air flow from the Indian Ocean. This caused the climate of the plateau to become progressively arid. Some Quaternary deposits with ages of 10^5–10^6 years on the plateau result from these changes. In particular, in the Late Pleistocene the height of the plateau above sea level had reached approximatedly the same height as that of the modern plateau, hence the climate also became arid during this time. The 10^5 year time scale reflects the fact that there was a nearly unidirectional change from humid cold (warm) climate to dry cold (warm) climate (Fig. 1.33; Table 1.35) and the important effect of the strong uplift of the plateau on the climate of the plateau. Over the time scale of 10^3–10^4 year there appear to be some reversals in this climate change. An example is provided by the existence of pan–lakes between 40–28 ka. Although the Himalayas towered aloft on the southern edge of the plateau, they could not completely stop the South Asian warm humid air masses from entering the lake areas of the plateau. Evidence for this ingress of warm air is provided by the global warm (humid) turning stage when voluminous warm humid air masses not only reached the south–central part of the plateau, but also Qaidam basin in the north, even though it was impeded by numerous mountains. This reversal in the general direction of climatic change is difficult to explain by the hypothesis that the continuous uplift of the plateau could only bring about cold and dry changes. During the Late Pleistocene the plateau was uplifting, with the mean height attaining 4000 m, and there was no subsidence phase. As shown in Table 1.35, the pan–lake stage of the plateau roughly corresponds with the cold stage (44–29 ka) of the global Late Pleistocene first interglacial stage (i.e.,

W_1–W_2 of the Würm glacial stage); the warm humid–warm slightly humid stage (28 (29)–19.6 (20) ka) of the plateau may correlate with the warm stage (28–23 ka) of the first interglacial substage and the warm cool slightly humid substage (23–20 ka) of the early part of the second glacial stage. It therefore follows that global climatic changes played an important role in climatic change on the plateau over timescales of 10^3–10^4 in the late Late Pleistocene. Not until the Middle–Late

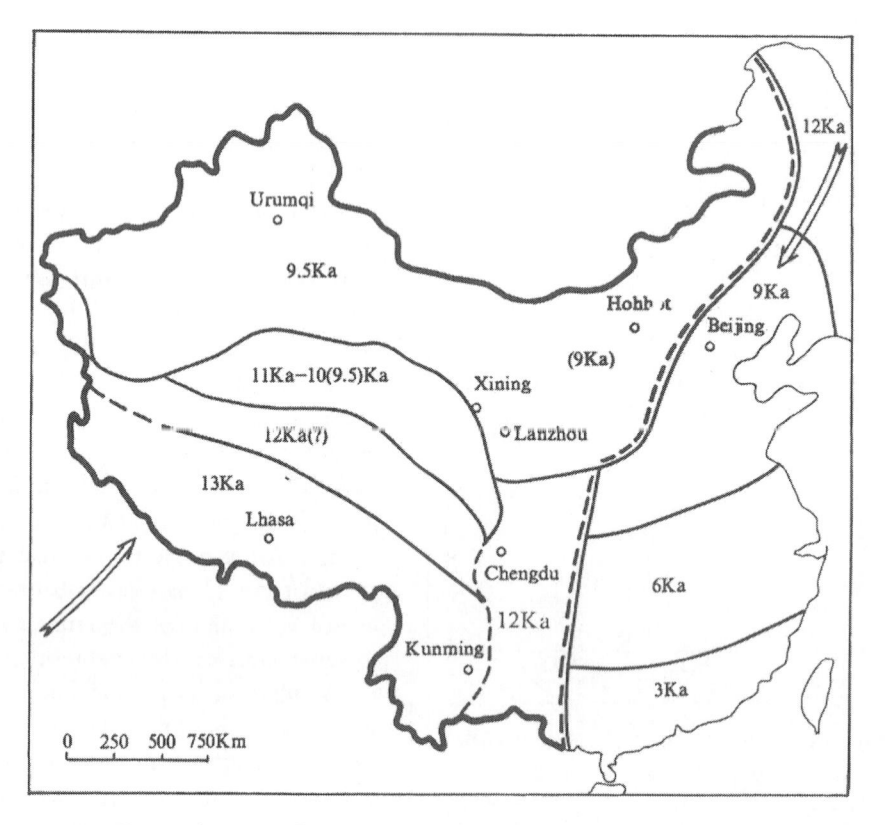

Figure 1.34. The starting ages of temperature and precipitation increments in the terminal Late Pleistocene–Holocene (summer) (the data of the eastern part from An Zangsheng, 1993; the western boundary modified).

Holocene was the trend of becoming cold and dry on the plateau intensified. In comparison with the global climate the Atlantic warm stage was shortened or it was less warm and humid. Hence, during this time salts continued to be deposited in many saline lakes of the plateau and they even dried up in some areas (e.g., in the northwestern part of Qaidam), suggesting that the influence of the plateau uplift was strengthened to some extent. The universal recession of the plateau lakes in recent time is due to the combined effects of the global climatic change and the continuous uplift of the plateau.

Diachronism of monsoons on the plateau in the Early Holocene

The Holocene warm-(humid-) transition stage occurred at 13 ka in the northern Tibet lake area in the southern part of the plateau and also at about 13 ka in the Erhai lake of Yunnan on the southeastern edge of the plateau and in the Didwana lake of the Thar desert of India to the south of the plateau (Figs. 1.32 and 33). This stage occurred at 11 ka in the area north of those mentioned above to Qarhan in the northern part of the plateau, 10 (9.5) ka in the Qinghai lake area, and 9.5 ka in the Lop Nur area of the west wind zone in the northwestern part of the plateau (Zheng et al., 1992). This indicates that the Holocene warm-(humid-) transition stage in western China had a tendency to occur early in the south and southeast and late in the north and northwest. This diachronism is the opposite to that observed in eastern China, which is characterized by south-younging of the Holocene warm humid increment stage (Fig. 1.34). The origin of this latter trend is considered to be "due to a decrease in the sun's radiation increment from high to low latitudes" (An et al., 1993), while the case in western China cannot be explained by this origin. The climatic change seen in the sediments from Didwana lake and the sporo-pollen analyses are consistent with the results of sporo-pollen, oxygen isotope and faunal analyses of cores from the Arabian Sea. In addition, it has been found that the sporo-pollen elements of the Somalian tropic zone began to appear in the ca. 13 ka deposits of Didwana Lake. This suggests that as early as 13 ka the South Asian monsson had moved across the equator. As a result, the humidity increased rapidly, so that Didwana salt lake was turned into a fresh water lake full of life (Wasson et al., 1984) and synchronous monsoon deposits appeared in the southern and southeastern parts of the Qinghai-Tibet Plateau. The South Asian monsoon was bound to be stronger than the late Early Holocene East Asian monsoon. According to Prell et al. (1987), the peak index (12 ka) of the South Asian monsoon appeared 3 ka earlier than the peak (9 ka) of the East Asian monsoon (Fig. 1.35). Under this climatic situation, the rising Tibet Plateau not only served as a barrier that impeded the northward flow of the East Asian monsoon, but is also appeared to be an importan factor responsible for the formation of the warm humid South Asian monsoon and its northward transition across the plateau. Modelling of the mountains indicates that the high-temperature air flowed southward over the Tibet Plateau onto the South Asian plain. This produced a low pressure zone, whose position was south of that produced by models that did not include the mountains. The mountains also produced a relatively strong meridional atmospheric pressure gradient and promoted the flow of humid-air northward over the ground northwards over Asia (Hahn et al., 1975).

The South Asian monsoon has a higher wind speed and influences more of the atmosphere (up to a height of 6 km) than the wind speed and influence (only 4 km) of the East Asian monsoon in recent time (Zhang, 1989). Hence, the Southeast

164   CHAPTER 8

Asian monsoon can pass through Himalayan mountain passes with elevations of 4500 to > 5000 m and into the plateau, in addition to entering the plateau through the valleys to the southeast. The evidence is provided by determination of the passage of moisture in the area of Zhongba. At 13 ka the elevation of the Qinghai–Tibet Plateau was about 300 m lower than in the recent times. When the intensity and speed of the monsoons increased, the warm humid air mass first reached the southern part of the plateau promoting earlier rejuvenation of the lakes in the southern part of the plateau.

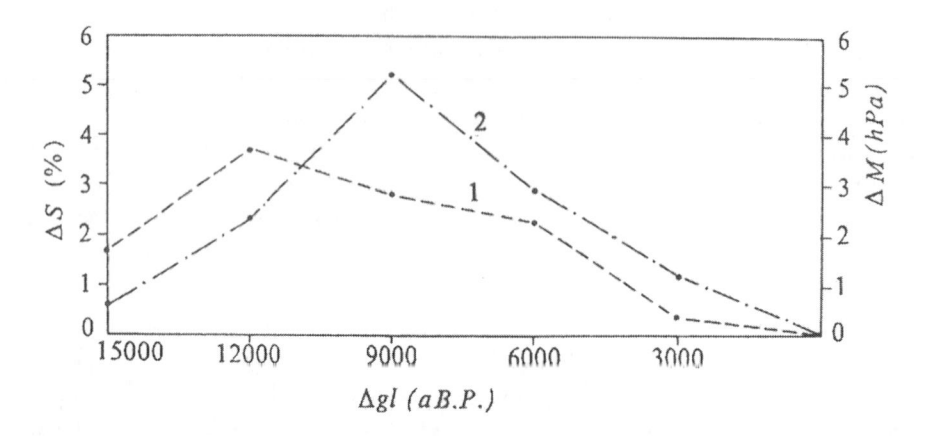

Figure 1.35. Evolution of indices of the East Asian and South Asian monsoons in summer (July) since 15 ka in the northern hemisphere. 1. $\triangle$MAA–South Asian monsoon index departure, which is expressed by the mean sea level atmospheric pressure difference between latitudes 43 ° N and 15 ° S in the 45 ° –120 ° E longitudinal zone; 2. $\triangle$MEA–East Asia monsoon index departure, which is expressed by the mean sea level atmospheric pressure difference between longitudes 160 ° E and 110 ° E in the 25 ° –50 ° N latitudinal zone (after Prell and Kutzback, 1987).

8.2. Shoreline Deposits of the Zabuye Lake Area and Its Paleo–climatic Analysis

As stated above, Quaternary lake basin shoreline geomorphology and deposits are widespread and well preserved on the Qinghai–Tibet Plateau. This section will use Zabuye lake in the interior of the plateau as an example to further analyse the paleoclimatic significance of shoreline deposits and geomorphological features.

8.2.1. DIFFERENCES BETWEEN THE PALEO–LAKE SHORELINE AND STRUCTURAL TERRACES

Both the palaeo–lake shoreline and terrace are well developed in the Zabuye lake area, and often lake strandlines with identical elevations, structural terraces and paleo—lake shoreline landforms were formed simultaneously in different tectonogeomorphological positions in the lake basin. However, they have different genetic implications, so that a clear distinction must be made between them.

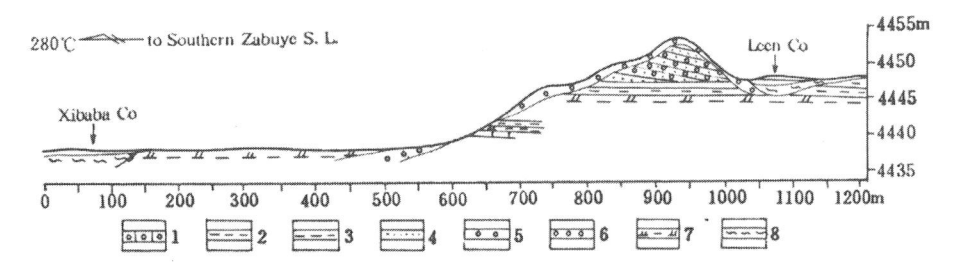

Figure 1.36. Section of spit 3 at Zhaduixiong. 1. Greyish—yellow calcareous sand and gravel; 2. Grayish—red—grayish brown clay; 3. Grayish—green carbonate—bearing clay; 4. Grayish—red—grayish—yellow sand; 5. Gray sandy gravel; 6. Gray loose sand and gravel; 7. Grayish—green clayey carbonate; 8. Black mirabilite—bearing marl.

Characteristic features of the palaeo—lake shoreline

The paleo—lake shoreline deposits in the area include the accumulations of sand barriers, sandspits, sandbanks, interbarrier lakes and lakeshore deltas; their eroded landforms comprise the lake cliff, lake—eroded bench, lake—eroded stack and a variety of unusual lake—eroded stones. All these features are products of lake water action in a high energy environment. Their morphological and accumulational features main reflect dynamic conditions of the lake water and paleoclimatic characteristics. Among the various types of the paleolake shoreline landforms in the area, sand barriers and spits are best developed. The sand spits and interbarrier lakes show a laterally alternating arrangement on the specific lake retreating background of the area. This demostrates that in different stages of lake retreat caused by the drying paleoclimate, knap ridge—like sand barriers and spits of undirectional extension were formed due to the action of lake waves. These resulted in intermittent transport of materials along the lake bottom when sediments entered the gentle lake bankslope in a direction perpendicular to the lake waves. Moving from the higher position to lower positions, the sand barriers are arranged in parallel lines that are slightly curved towards the center of the lake, with single barriers being approximately equally wide and tens of centimetres to 5 m high in most cases. The number of sand barriers varies with the relief of the lake shore where the occur. There may be up to 50 sand barriers at one location (Plate 1.27), but the barrier bodies of sand spits vary in width and often project sligthly towards the center of the lake. They exhibit feather like, or sickle shaped forms in plan view, and there are often small interbarrier lakes (lagoons) between the sandspits. The lateral surfaces of the sand barriers and sand spits shows a streamlined curved form. The lakeward face of the sand barriers are gently sloped or stepped with a general slope of 0.5–3 °. The slope of the back face varies in places, from nearly horizontal on the relatively steep lake shores and up to 2–5 ° on the gently sloped shores. The spits have a gently stepped lakeward faces, with up to 10 steps and average slopes of 2–5 °, but the back faces have steeper slopes (3–9 °) and no stepped undulation (Fig. 1.36). The height of the single sand barriers is generally in the range of 3–5 m, but may be up to 10 m in individual cases. The height of the spits is greater, reaching up to 10–25 m.

Sedimentary features and occurrence modes
The sediments in the upper part of the sand barriers and spits are most composed of loose sandy gravels or gritty granules and often contain carbonate–bearing clay or carbonate clay in varying amounts. The carbonate components change from low–Mg calcite in the upper barriers, towards high–Mg calcite, aragonite, dolomite, magnesite and even alkali–metal carbonates in the low barriers. In these sediments, oblique or tabular bedding is commonly observed, including major low–angle oblique bedding (plus bedding planes) (Plate 1.29). These beds are tens of centimetres to several metres thick. The sediments in the lower part of the sand barriers and spits consist of argillaceous (carbonate clay or carbonate–bearing clay) with horizontal bedding or bedded lamination, together with coarse–grained clastic sediments. The beds are up to several metres thick.

In vertical sections of the spits, the middle and upper parts on its higher side are composed mainly of sandy gravels and gritty granules. On its lower side, the spit consists of sandy gravels or gritty gravels in the lower part and usually of fine sand and an argillaceous bed in the upper part. The argillaceous bed occurs as a wedge–shaped body lying directly on the coarse–grained clastic layer of the spit, representing a trough in the latter stage of formation of the sand barrier or spit, and is the product of a tide channel low–energy environment (Plate 1.29).

Characteristic features of structural terraces
The lake structural terraces are stepped landforms extending along the edges of lake basins. They formed as a result of differential tectonic movement after lake water erosion or sedimentation. The structural terraces in the Zabuye lake area show the following characteristic features. (a) The appear mainly as erosional and basal terraces in morphology and are usually developed parallel to fault zones. They are distributed in nearly straight lines and several to scores of metres high. They are particularly well developed in the vicinity of WNW and nearly N–S fault zones and are often seen in association with triangular fault planes and cliffs (Plate 1.30) to give rise to erosional terraces, with wave eroded caves and lake eroded stacks often being observed on terrace slopes (Plate 1.31). (b) The contact relationships between the terraces usually take the form of internally stacked terraces. Terraces which are

comprised of younger lacustrine beds have their slope feet located on the underlying bed rocks, which often leads to dissection of the lakeward face of the strata in the upper part of the strata. (c) The terrace faces of the structural terraces are concordant to the nick points formed as a result of the rise of the lake edges and descent of the basal plane at the lake bottom. For example, the nick point observed to the east of the Langmenga River conforms with terrace II 4440 m above sea level.

TABLE 1.36. Spits and terraces in Zabuye s.c.

Stage	Area (km^2)	Volume (10^{8}m^3)	Time of shrinkage	Aver. depth used for vol. calcn (m)	Lake water vol. Cumu. shrinkage (%)	Shrinkage at one stage compared to that at the next (%)
Overflow	9780	11700	0	120	0	0
Spit 8	1381	665	17.6	48.2	94.32	94.32
Spit 7	1350	576	20.3	42.7	95.08	13.38
Spit 6	1150	451	25.9	39.2	96.15	21.7
Spit 5	810	267	43.8	33	97.72	40.80
Spit 4	720	199	58.8	27.6	98.30	25.47
Spit 3	550	101	115.8	18.3	99.14	49.25
Spit 2	480	68.6	170.6	14.3	99.41	32.08
Spit 1	350	24.5	477.6	7	99.79	64.29
Modern saline lake	242	1.7	6882.4	0.7	99.985	93.06

According to field measurements, the number of steps in the structural terraces in the area vary from 3–11 (Table 1.36). Among these, the 7 tectonic terrace steps may coexist with the 1–9 steps of the sand spits (sand barriers), but the former mostly occur on the sedimentary banks in the eastern, northern and northwestern parts. At the same time, the exposed height and distribution range of the sand barriers indicate an inflow lake sedimentary environment. Lacustrine coarse–grained clastic deposits and traces of lake erosion are frequently observed on the structural terraces of over 10 steps, suggesting that the highest lake water level in the area was higher than the height of the ninth step sand barrier. However, it is difficult to reconstruct the original height of the lake water level due to subsequent structural events. Hence, the spits (sand barriers) in the accumulation zones of the Zabuye lake area have been chosen as the main objects for paleoclimatic studies of lake basin in the area.

168   CHAPTER 8

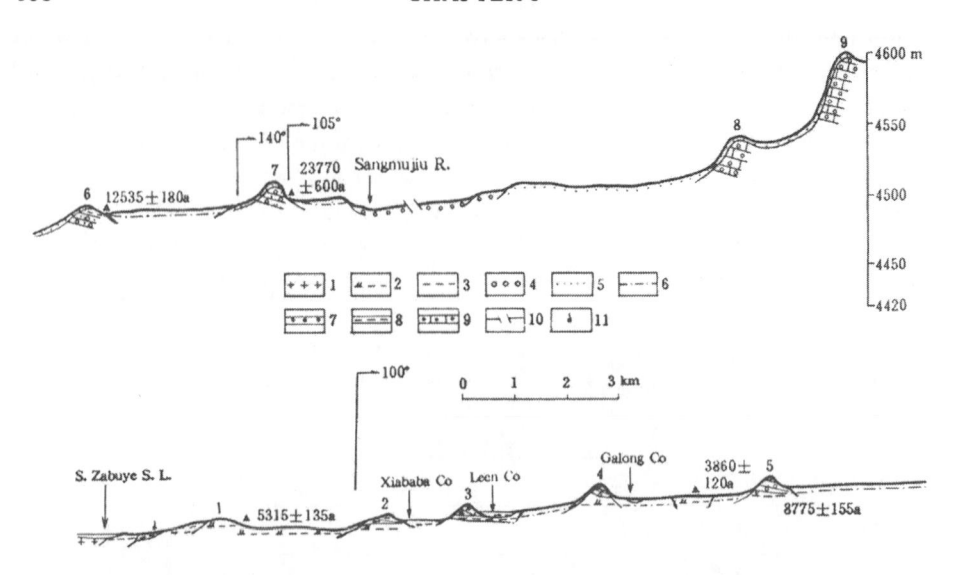

Figure 1.37. Measured geomorphological section of the Zhaduixiong spit swarm. 1–Halite; 2–Carbonate–bearing clay; 3–Marl; 4–Loose alluvial sand & gravel; 5–Loose lacustrine sand and gravel; 6–Sandy carbonate clay; 7–Semi–cemented sandy conglomerate; 8–Calcareous conglomerate; 9–Line of facies change; 10–Ellipsis of profile; 11–Spring–water overflow zone.

8.2.2. DEPOSITS OF THE ZHADUIXIONG SPIT SWARM AND THEIR AGES

The Zhaduixiong valley in this lake area extends from the lakeside of South Zabuye Caka eastward to Jiaobuqu pass. Over a distance of 31 km there is a spit swarm of 9 spits, which successively descend in elevation from spit 9 to spit 1. The spits are seperated from one another by depressions, which were ancient interbank lakes and are usually occupied by seasonal or perennial lakes, even today. The spits are separated by 2–8 km with the height difference between the spits and the neighbouring depressions generally being in the range 10–20 m (68 m in maximum and 6 m in minimum). On the lakeward slope of each spit there are 2–9 crest lines (benches), yielding a total of 52 crest lines on the nine spits. The height difference of the crest lines ranges from over 10 cm to 18 m (Fig. 1.37).

Sedimentary features

Structures and textures of sediments   From spit 9 to spit 1 and even down to the modern lake bed, the clastic sediments decrease in grain size. Spits 9–7 consist of calcareous conglomerate and calcareous sand–bearing conglomerate. The rocks show a gravely texture with well rounded to subangular sands or gravels. The sands and gravels are enclosed by cements, which consist mainly of calcareous material with minor clay. In the upper part of spit 6 there is a thin bed of carbonate clay which shows a sand–bearing pelitomorphic texture. In spits 5–2, large amounts of carbonate clay or clayey carbonate sediments begin to appear in the middle and upper parts. They show pelletoidal and pelitomorphic textures and pelletoids are commonly seen to have a concentric texture (Plate 1.32). Spit 1, however, is composed mainly of carbonate clay or clayey carbonate sediments of pelitomorphic texture, partly containing sandy gravels or gritty granules, whereas the sediments in the modern saline lakes are dominated by black marl and salts.

Carbonate and evaporitic minerals   The cement in the calcareous conglomerate of spit 9 is low–Mg calcite. The cement and micritic carbonate in the calcareous conglomerate of spits 8 to 6 are mostly high–Mg calcite and aragonite with minor low–Mg calcite. The calcite and cementing materials of spits 5 to 2 are dominated by aragonite and contain high–Mg calcite and minor ankerite (ferrodolomite). In the sediments of spits 5 to 3 trace amount of interstitial quartz are often observed (Plate 1.33), and the clasts in the carbonate often occur as pellets, oolites or nodules (Plate 1.34). Alternatively, the clay and carbonate may be separately assembled to form pellets, which are cemented by late stage carbonate. In spit 1 minor hydromagnesite is also observed① in addition to a relatively large amount of dolomite. At the same time mirabilite and borax deposits were also locally developed. From sand barrier 1 down to the modern saline lake there is a great abundance of Li-bearing magnesite, torona and even natural lithium carbonate (zabuyelite).

Variation of carbon and oxygen stable isotopes. In association with the above mentioned variations in the abundance and mineralogy of carbonate minerals the stable isotope compositions of C and O in the sediments show a general vairation trend of gradual increase from spit 9 to the modern lake bed. The $\delta^{13}C$ and $\delta^{18}O$ values change, respectively, from 0.6 to 1.4‰ and −7.9 to −13‰ in spits 9 and 8 to 6.3‰ and −0.5‰ in the modern saline lake sediments (Table 1.36), with three intervals of increase. The first increase of $\delta^{13}C$ and $\delta^{18}O$ values, which begin from spit 7, is concordant with the bulk appearance of high-Mg calcite; the second increase parallels the appearance of dolomite and authigenic quartz; the third increase appears in the modern lake sediments (Table 1.36).

Palynological variation trend. Sporo-pollen sediments are found in sediments from spit 7 to the modern lake bed, except in the coarse clasts of spits 9 and 8. The sediments of spit 7 yield arboreal pollen (up to 60% of the spore-pollen assemblage) dominated by *Pinus, Abies* and *Selaginella sinensis* (the latter making up 36%) and contain abundant Pteridophyta spores, of which Polypodiaceae amounts to 50%. Hence, the sporo-pollen assemblage is characteristic of the temperate coniferous-broadleaf mixed forest. Spit 6 has the highest concentration of arboreal pollen (up to 67%), which is composed mainly of *Pinus* with minor *Abies*. There is also a remarkable decrease in the concentration of Polypodiaceae spores. Among the herbaceous and shrub pollen the content of Ephedraceae also reaches the highest value (19%) seen in the spits, and is characteristic of the sporo-pollen assemblage of a coniferous forest in a warm and more or less dry temperate zone. In the sporo-pollen assemblage of spit 5, arboreous pollen accounts for 31-58% of the total assemblage. It consists mainly of *Pinus*, with minor *Abies* and a negligible amount of Betulaceae, *Alnus*, etc.; the herbaceous and shrub pollen is dominated by *Artemisia*, with *Epheadra* being of secondary importance; Pteridophyta spores contained in this spit consist of Polypodiaceae and Chenopodiaceae; *Myriaphyllum* is occasionally present. The basic aspect of the assemblage characteristic of a temperate coniferous forest in a more or less dry and cold climate. In spit 4 the content of arboreal pollen increases (29-78%), and is dominated by *Pinus* with minor *Abies, Tsuga, Alnus, Betula*, etc. The herbaceous and shrub pollen is mainly composed of

---

① Hydromagnesite deposited in the same period (ca. 3500 ka) is also seen in spit 5 (Plate 1.35).

TABLE 1.37. ^{14}C ages of spits, sand barries and lakebed sediments

Ser. No.	Locality	Position of spit	Altitude (m)	Relative height (m)	Sampling horizon	Age (a B.P.)	Analyst
1	CK 4 in northern S.Zabuye Caka	Lakebed	4421	0	Sample taken from upper part of a bed of grayish–black medium to coarse–grained sand bearing B, crystalline saltpetra and carbonate, with a bed of halite and mirabilite 20 cm thick above	1350 ± 70	Institute of Palaeovertebrates, Academia Sinica
2	As above	As above	4421	0	Sample taken from grayish–green nitrate–bearing carbonate clay in upper part, with carbonate clay in lower part	3150 ± 80	As above
3	CK 8 in southern N.Zabuye Caka	Lakebed, app. upper part of spit 1–1	4421	0	Sample taken from light gray carbonate clay in middle–upper part, with upper part being a salt–bearing bed	3530 ± 70	As above
4	West bank of S.Zabuye Caka	Higher part of barriers 1–2	4440	19	Grayish–green clayey carbonate. Sample taken from its upper part	5315 ± 135	Institute of Geology, State Seismlogical Bureau
5	S. Spit of N.Zabuye Caka	Barriers 1–2	4440	19	White clayey carbonate. Sample taken from middle part; with weakly cemented sand and gravel above and calcareous– argillaceous sand and gravel below	9510 ± 165	As above
6	Zhaduixiong in eastern S.Zabuye Caka	Southern part of spit 5	4470	49	Sample taken from middle–upper part of light gray carbonate clay, with weathered gravel–bearing carbonate clay at top	8725 ± 135	As above
7	Jiadonglongba	Spit 6	4485	74	Sample taken from upper part of a light gray carbonate clay bed; with lower part being a bed of fine sand	12535 ± 180	As above
8	Jiuer in eastern S.Zabuye Caka	Corresponding to lower part of spit 6	4480	69	Sample taken from lower part of a grayish–yellow calcareous clay bed, with sand–gravel and clay beds in middle and upper parts	22670 ± 380	Institute of Palaeovertebrates, Academia Sinica
9	W.bank of S.Zabuye Caka	Barrier 7	4510	89	Sample taken from upper part of alight gray Ca–carbonate bed	23770 ± 600	As above

Ephedra and *Artemisia*, containing minor amounts of elements of Chenopodiaceae and *Dianthus*, and hygrophilous *Podiastrum* is present in small amounts. The sporo—pollen assemblage still represents a temperate coniferous in a more or less warm and humid climate. The sporo—pollen assemblages in spits 3 and 2 are generally the same as those in spit 4, but they contain less arboreous pollen and are marked by the absence of *Podiastrum*. In the spore—pollen assemblage of spit 1, the arboreal pollen increases again (46—96%), and consist mainly of *Pinus* and *Larix*, while the herbaceous and shrub pollen is dominated by *Artemisia*, with minor amounts of *Epheadra* and Chenopodiaceae, but without Polypodiaceae. This assemblage represents a vigorous coniferous forest in a more or less warm—dry climate. In the recent deposits on the modern lake bed, herbaceous and shrub pollen predominates and arboreal pollen is largely absent.

Features of Ostracoda assemblages   Analyses indicate that, with the exception of spits 3, 8 and 9, fossil ostracods are present in all other spits. Spit 7 yields the *Canadona—Leucocythere—Limnocytherellina* assemblage. Among these, *Canadona* has been discovered in the present day Yamzho Yumco, Bangong Co and Yagedong Co whose salinity values are 7.8, 19.6 and 39.4 g / l, respectively. Spit 6 yields *Limnocytherellina binoda* and *Limnocythere sancti—patricci*. The latter has been seen in the brackish water of the modern Bangong Co. *Leucocythere mirabilis* has been found in spits 5, 4, 2 and 1 and is particularly abundant in spits 5 and 1. This element is seen in modern lakes of Tibet, such as Co Nag Lake, Zhari Namco, Yagedong Co and Cam Co, of which Cam Co is a saline lake with a salinity of 174 g / l, and the others are brackish lakes. Spits 5 and 1 also yield *Candoniella mirabilis* and *Limnocythere dubiosa*. The latter species, which has also been found in spit 2, is common in brackish and saline lakes of Tibet.

Basis for Geochronological Division of Sediments
In the last few years, paleolake shorelines have been described in individual reports, but the age of their formation can only be inferred as absolute age data are largely lacking. Some authors consider that "the highest ancient lake shorelines of the lakes were formed in the Pleistocene, particularly the Middle Pleistocene" (Yang et al., 1983), while others assume that they were formed in the Mid—Holocene (Han, 1984).

On the basis of the above mentioned stratigraphic studies of the Quaternary in the area, the author of this monograph made a classification and chronological study of the sediment of the spits and barriers of various orders in the area, through a detailed survey of three sections of sand barriers and spits in the lake area. With a total of 24 ^{14}C samples collected from the area, relatively systematic age data have been obtained for the first time for some of the sand barriers (spits), and are listed in Table 1.37.

TABLE 1.38. Recession time of the Zhaduixiong Spit Swarm

Order of spits	Time of beginning and ending of recession (a B.P.)	Duration of residence and recession (a)
9–8	27460–25000±	2460
8–7	25000–23770±	1230
7–6	23770–12535±	11235
6–5	12535–8775±	3760
5–4	8775–7642±	1133
4–3	7642–6855±	787
3–2	6855–6085±	770
2–1	6085–5315±	770
1–0	5315–3530±	1785

Based on Table 1.37, the ages of the Zhaduixiong spits can be placed in the following order; about 23.8 ka for spit 7, 12.5 ka for spit 6, 8.7 ka for spit 5, 5.3 ka for spit I–2, 3.53 ka for spit I–2. The studies of the Malan loess section and sea–level changes of the Yellow Sea (Liu et al., 1979; Xie and Xu, 1979) indicate that China witnessed a period of humid climate (interglaciation age) at about 36–25 ka, and a period of dry–frigid climate (glaciation age) after 25 ka. The study of borehole sections at Qarhan playa in Qaidam Basin has led to a similar conclusion that the playa experienced a humid period before 24 ka (Chen and Bowler, 1985). The presence of a relatively large drop in the water level between the formation spits 9 and 8, together with the sedimentary evidence of an increase in salinity levels, suggests that the formation time of spit 8 to be in the range of 23.7–25.0 ka. The intensity of the vertical retreat of spit 8 has been extrapolated to suggest that spit 9 formed at about 27.5 ka, or possibly as early as 30 ka. This approximately corresponds with the latter half of the interglacial epoch. Even above the highest spit there are still 100 m high lake erosional terraces of 2–5 orders. Although these terraces are tectonic in nature, they indicate a height difference of over 179 m between the highest level of the ancient lake and that of the modern lake. This suggests that their formation age should be considered earlier than 27.5 ka. Hole ZK 91–2 from Zabuye lake (Fig. 1.33) has revealed that there are still flood plain deposits at a depth of 20 m, which correspond to an age of 38 ka as estimated from the deposition rate. In addition, ^{14}C age dating in the adjacent Karakorum–West Kunlun mountains lake area suggest that the flooded lake confined by the highest lake shoreline appeared at 40 ka. Hence, it is inferred that the flooded lakes in this district were initiated at about 40 ka.

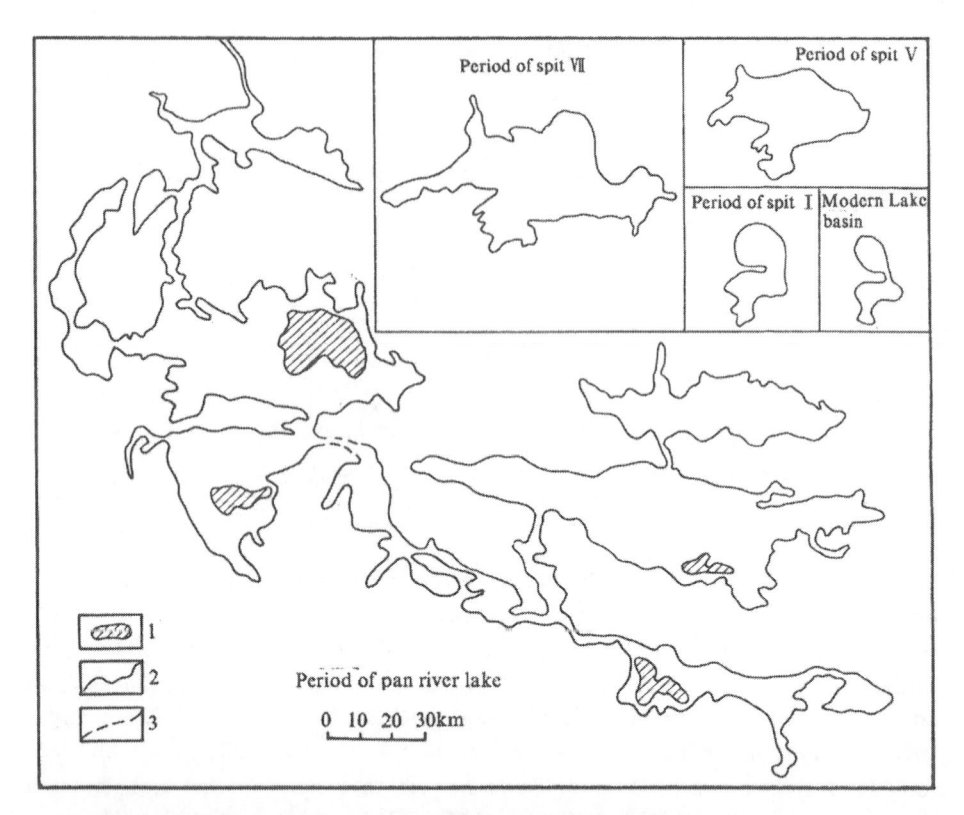

Figure 1.38. Distribution of the lake water of the Zabuye Caka basin since 30000 a B.P. 1—Islet; 2—Boundary of flooded river—lake; 3—Inferred boundary.

8.2.3. EVOLUTION CHARACTERISTICS OF THE ZHADUIXIONG

The Zhaduixiong spit swarm is one of the localities with the most distinct traces of water ressession of Zabuye Caka. The swarm indicates that the lake water in this area experienced 9 major ressessional or stagnation events. The duration of these events ranged from 0.77 ka to 3.76 ka in most cases, although the duration of stagnation for spits 7—6 was as long as 11 ka or longer (Table 1.38). The duration of the stagnation intervals was longer in the early and middle stages and shorter in the latter stage. In each of the spits there are 2—9 ressessional sand ridges (benches), and the horizontal and vertical distances of the ressession were usually smaller at the beginning and greater duringthe late stage (Table 1.38), reflecting 2—9 transcient stagnations in each major ressession event and a tendency towards increasing intensity of ressession in the late stage.

TABLE 1.39. Estimates of volumetric shrinkage of lake waters in the Zabuye Caka since 30000 a

Stage	Area (km²)	Volume (10⁸m³)	Multiple of shrinkage volume	Average depth used in volumetric calculation (m)	Volume of lake water Cumulative reduction (%)	Volume of lake water Relative reduction (%)
Overflow surface	9780	11700	0	120	0	0
Spit 8	1381	665	17.6	48.2	94.32	94.32
Spit 7	1350	576	20.3	42.7	95.08	13.38
Spit 6	1150	451	25.9	39.2	96.15	21.7
Spit 5	810	267	43.8	33	97.72	40.80
Spit 4	720	199	58.8	27.6	98.30	25.47
Spit 3	550	101	115.8	18.3	99.14	49.25
Spit 2	480	68.6	170.6	14.3	99.41	32.08
Spit 1	350	24.5	477.6	7	99.79	64.29
Modern salt lake	242	1.7	6882.4	0.7	99.985	93.06

8.2.4. PALEOCLIMATIC ANALYSIS

Overflow surface and high lake level
As noted by the authors①in 1960s, there were several lake flooding stages with highlake levels on the Tibetan Plateau during the Pleistocene. However, the high lake levels in the Early Quaternary had already been influenced by large neotectonic movements. Only the last trace of the last flooding of the lake is well preserved, and therefore suitable for tracing its general paleoclimatic aspect. The highest spit in the area has an elevation of about 4600 m above sea level, and is 179 m above the present lake level. This elevation is 34 m higher than that of Taro Co, which is still replenishing Zabuye Caka. It therefore reached and exceeded the overflow surface of the ancient Zabuye Caka. The carbonate minerals in spit 9 are dominated by low–Mg calcite and its $\delta^{13}C$ $\delta^{18}O$ values are comparatively low. The sediments also

---

① Zheng Mianping et al. (1974) Report on the study of boron deposits in saline lakes of Tibet. In: Report of scientific and technological achievements, Institute of Scientific and Technological Information of China, p.43.

contain brackish water, low temperature ostracods, such as *Leucocythere mirabillis* and fresh water *Pediastrum simptes*. Ostracods and charophytes are not developed in hole ZK 91–2 section, equivalent to the deposits of the flooding stage (ca. 38–28.5 ka), but a few thermophilous *Ilyocypris* forms are found. This suggests that the climate was cool and humid, with occasional warm climatic fluctuations. The evaporation loss was small and the water area was vast during the overflow period of the lake. The water area outlined by the water lines is estimated to be as much as 9780 km^2 or more (for the main part of the lake only) and the volume of the lake water was 1.17×10^{12} m^3 (Table 1.39). The ancient lake had a lower salinity than that of the present day Taro Co (0.8 g / l) and should be considered as a fresh water lake (Fig. 1.38).

Paleoclimate after the river–lake flooding period

Spit 8 (28.5–25 ka)   A water level drop of 68 m is recorded from spit 9 to spit 8, but the sediments of this period are dominated by coarse clasts. According to the water lines the water area of the lake at the time reached about 1380 km^2, and the volume of the water body was as high as 6.65×10^{10} m^3. This indicates that there was a drastic volume decrease of 94% during this period. At this time there were large amounts of high–Mg calcite and aragonite appeared for the first time in the carbonate cements and thermophilous forms such as *Ilyocypris gibba* occurred. This demonstrates that there was a remarkable increase in the salnity of the ancient lake water. A comparative study of the depositional conditions of carbonates indicates that the salinity of the ancient lake water was no less than 5‰. The $\delta^{18}O$ and $\delta^{13}C$ values of the sediments of spit 8 are the lowest (Fig. 1.37). During that period (28.5–25 ka) (equivalent to the Jiabula interglaciation) the air temperature increased compared to that in the previous period. Although the increase in evaporation loss caused the lake level to drop to some extent the environment was more suited for the reproduction of organisms. The annual average air temperature in southern Tibet at that time was 5–8°C higher than of today (Li et al., 1983). This last Quaternary lake flooding period in the Qinghai–Tibet region caused the thawing of glaciers and a rise in the lake water level. This resulted in the deposition of boulders in a high energy environment. In recent years, large amounts of fine and old stone implements of this period have been discovered along the lake shores in northern Tibet (An et al., 1979). This suggests that the physical environment at that time was suitable for human habitation.

Spit 7 (ca. 28.5–23.8 ka)   In this period the water level dropped by 22 m compared to that in the period of spit 8, and the volume of the lake water decreased by 8.7×10^9 m^3 (Table 1.39; Fig. 1.37), but this water level was only maintained for a short time interval. The carbonate minerals in the spit consist mainly of aragonite, with high–Mg calcite being of secondary importance and low–Mg calcite present in minor amounts. Compared to spit 8, this spit shows an apparent increase in $\delta^{13}C$ and $\delta^{18}O$ values (Table 1.36; Fig. 1.41), indicating a significant increase in the salinity of the lake water. The spit contains euryhaline ostracods *Leucocythere postilirata*. According to investigations of modern lakes in Tibet this species of ostracod lives in lakes with salt contents of 1.8–39.4‰ (Yang et al., 1982). The $\delta^{18}O$–salinity correlation suggests that the salinity of the lake water of this period was 8–10 g/l. The sediments of this period are still dominated by coarse clasts, implying the presence of a vast water area and a high energy depositional environment. Based on the water line of the spit, the volume of water in the ancient Zabuye Caka was about 5.76×10^{10} m^3. Palynological analyses indicate a sporo–pollen assemblage of a temperate zone coniferous broadleaf forest, suggesting a more or less dry climate, but a colder air temperature marked by the increased proportion of *Abies* in the arboreal pollen. This implies that in the interval of about 25–23.8 ka (approximately at the initial stage of the last (Rongbushi) glaciation) the air temperature dropped abruptly and the climate was drier and colder than before.

Spit 6 (about 12.5–13.5 ka)   During this period, the water level dropped by about 14 m and the lake had an area of 1150 km^2 and a water volume of about 4.5×10^{10} m^3 (Table 1.39). The spit sediments of this period became finer and they contained intercalations of argillaceous material, which suggests there was a slight decrease in the energy of the lake water. However, the association of carbonate minerals in the spit is similar to that in spit 7 and the $\delta^{13}C$ and $\delta^{18}O$ values are slightly higher (Fig. 1.31; Table 1.36). *Limnocythere sancti–patricii* that lives in modern lakes with a salinity of about 20 g/l is seen in the assemblage of fossil ostracods, just in concordance with the position of the $\delta^{18}O$–salinity correlation of carbonates at this time. The sporo–pollen assemblage in this period is characteristic of a temperate zone evergreen coniferous forest and has a peak content of arboreous pollen, but a decreased content of *Abie* and Polypodiaceae, etc. It is important to note that there was an apparent lake recession after about 16 ka, resulting in the lowest lake level of this period at about 13.5 ka. But hole ZK 91–2 section (Fig. 1.33) has revealed that there was a climatic fluctuation turning to warm humid conditions in this area at about 13.5–13 ka. The lake recession at about 13.5 ka was coincident with the uplift and emergence of the continental shelf of the Yellow Sea at 12.4 ± 0.2 ka (Xie and Xu, 1979), which implies that this lake recession was related to global events.

Spit 5 (ca. 13.5–8.9 ka)   During this period the water level dropped by 25 m and the water area was remarkably reduced to about 810 km^2, and the volume of the lake water decreased to about 2.67×10^{10} m^3 (Table 1.39; Fig. 1.37). In the spit of this period the contents of fine clasts and carbonate minerals greatly increased, and for the first time a large amount of dolomite appeared, together with a trace amount of authigenic quartz and a minor amount of soluble salts (halite). However, the carbonates often occur as pellets, indicating that a turbulent depositional environment still prevailed. The $\delta^{13}C$ and $\delta^{18}O$ values of the spit sediments showed a dramatic increase, but varied over a great range, averaging 2.6‰ (5 samples) and 5.5‰ (5 samples), respectively. These correspond to the $\delta^{13}C$ and $\delta^{18}O$ values of the salt-bearing carbonate clay bed lying below the halite bed in the modern Zabuye Caka (Fig. 1.31; Table 1.36). Moreover, the sediments yield halophilic ostracods (*Cunadimella mirabilis* and *Limnocythere dubiosa*), of which the latter form is seen in modern brackish lakes (with a salinity of 19.6–29.8 g/l) and in the brine of saline lakes (35.4–39.6 g/l) in Tibet. The sporo-pollen assemblage shows a basic aspect of a temperate zone coniferous forest. In the herbaceous and shrub elements the kerophilous plants *Artemisia* and *Ephedra* occupy a dominant position. Hygrophilous *Myriaphylla* appears occasionally. This shows that the water in the ancient lake was remarkably salinised during this period. According to the position of the carbonates of this period on the $\delta^{18}O$-salinity correlation diagram, the salinity of the lake water may have reached 70 g/l or more, and sometimes as high as 100 g/l. However, the appearance of *Myriaphylla* implies that the local marginal water body was surrounded by desalinised marshes. From an analysis of this data, together with the data in the section in Fig. I-30 (Fonte et al., 1993; Qi and Zheng, 1995), at about 13–10 ka the lake surface in the area rose and fell frequently and the temperatures rose and dropped abruptly. Hence, this period approximately corresponds to that stage in the late to recession stage of the last glaciation.

In short, the air temperature during that period rose again, but the climate as a whole was dry and cold.

Spit 4 (ca. 7.7–6.8 ka)   During this period the water level of the lake fluctuated widely, with coarse and fine clastic sediments occurring in alternating interbeds. At the beginning of this period the water rose slightly again as compared to that in the period of spit 5 and then dropped to approximately 10 m below spit 5. At that time the lake had a water area of about 720 km^2 and a water volume of about 1.99 × 10^{10} m^3, showing a reduction of 25% compared to that in the previous period (Table 1.34). At the same time, the assemblages of carbonate minerals deposited in this period showed a predominance of aragonite and a decreased proportion of dolomite that reflects the increase salinity, but there was a minor amount of authigenic quartz. Euryhaline *leucocythere mirabilis* is the only form of ostracods seen in the sediments of this spit. This species is still living in modern lakes with a salinity range of 0.5–175 g/l in Tibet. In the sporo-pollen assemblage there is a remarkable increase of arboreous pollen, in which there appears a small amount of *Tsuga* that has not been found in sediments of previous periods. However, *Pinus* remains the dominant species. Among the shrub and herbaceous species, the proportion of *Ephedra* is higher than that of *Artemisia*, and *Cyperaceae* and *Dianthus* appeared, as well as hygrophilous *Myriaphylla*. As far as the $\delta^{13}C$ and $\delta^{18}O$ values are concerned, in comparison to those in the previous period, $\delta^{13}C$ shows an increase of 0.4‰, while $\delta^{18}O$ drcreased by 1.9‰ (averaging −4.4‰ for two samples). This indicates that the salinity of the lake water was slightly lower than that in the period of spit 5 (Table 1.36; Fig. 1.38).

It can be seen from the above that the air temperature in this period was higher than that in the period of spit 5, showing identical features to the Middle Holocene, as revealed by lake bed drilling. This period corresponds to the early Atlantic climate stage. Thanks to the melting of glaciers, the climate in this early stage was slightly warm and humid and then became arid. Consequently, the lake water should have changed from a more or less fresh to a brackish state, suggesting a subalkaline salt water lake depositional environment.

Spit 3 (ca. 7.7−6.8 ka)   This spit is similar to spit 4 in its basic features and is characterized by the presence of interbeds of coarse and fine clastic sediments. There are identical association of carbonate minerals to spit 4 and there is also authigenic quartz present. The $\delta^{13}C$ and $\delta^{18}O$ values are 4.2‰ and 4.3‰ respectively, which are close to those of spit 4. According to Fig. I−38, the lake water should have a corresponding salinity of about 65 g / l. As in the period of spit 5, the water level of the lake also showed wide fluctuations. There was a slight rise in the early stage, but there was an approximate 10 m drop in the late stage, relative to the water level in the late period of spit 4. The volume of water in the lake in this period is estimated to have been about 1.01×10^{10} m^3 (Table 1.39). Few samples have been collected from this spit and no sporo−pollen and ostracod fossils have been found in sediments. This period corresponds to the "relatively warm−humid" middle Atlantic stage in North Europe. It is inferred that in this period the climate was similar to that in the period of spit 4. The air temperature and humidity changed from slightly high to slightly low and there was a subalkaline salt water lake depositional environment.

Spit 2 (ca. 6.8−6. ka)   In this period the water level continued to drop (down to 5−6 m below that in the period of spit 3), and the lake had a water area of about 480 km^2 and a water volume of 6.86×10^9 m^3 (Table 1.39). The sediments consist mainly of dolomite and high−Mg calcite, reflecting a decrease in the energy of the depositional environment and increase in the salinity of the water. The sediments of this spit show a slight increase in $\delta^{13}C$ and $\delta^{18}O$ values, which average 4.2‰ (for two samples) and −3.55‰ (for two samples), respectively. The fossil ostracod assemblage contains *Limnocythere dubiosa* and *Limnocytherellina binoda*. The former species occurs in both present day brackish and salt water lakes of Tibet (Yang et al., 1982). According to its position in Fig. 1.39, the salinity of the lake water is estimated to have been about 80 g / l. The sporo−pollen assemblage consists largely of *Pinus* with minor *Abies*, with *Pinus* being especially abundant in the latter stage. The herbaceous and shrub sporo−pollen are dominated by *Artemisia* and *Ephedra*, containing abundant Polypodiaceae (50%) in the late stage. This reflects a change from a more or less warm–dry climate to a more or less warm–humid one. This period approximately corresponds to the late Atlantic stage. However, the subalkaline salt water lake remained as the depositional environment of this period.

Spit 1 (ca. 6.0–5.3 ka)   In this period the level of the lake water kept falling drastically (down to 7–8 m below that in the period of spit 2). This resulted in a water area of about 350 km^2 and lake water volume of about 2.45×10^9 m^3 (Table 1.39 and 1.37). The sediments consist mainly of carbonate clay, with aragonite, high–Mg calcite and even hydromagnesite being of secondary importance. Mirabilite deposits with planar borax layers are also seen locally in the upper part of the horizon corresponding to the lake bay, indicating a drastic decrease in the energy of depositional environment and a great increase in salinity. The $\delta^{18}O$ of the carbonate minerals suggest that the salinity of the lake water may have reached 10–20% (Fig. 1.37). This is supported by the presence of halophilic blue and green algae at this level. In modern brines these algae grow most vigorously at salinity of 12–18%. The presence of hydromagnesite in the sediments also indicates a remarkable increase in the Mg/Ca of the brine to values above 7. The early stage sporo–pollen assemblage of arboresous pollens consist mainly of *Pinus*, with minor *Abies* and up to 4% Pteridophyta spores. In the late stage assemblage the arboreous pollen increases to 92–98% and contains abundant *Larix* pollen in addition to *Pinus*, *Betula* and *Alnus*. Other spores and pollen includes *Artemisia*, *Ephedra* and *Chenopodiaceae*.

All the above indicates that the climate was more or less dry and cold in the early stage, but dry and warm in the late stage. This period roughly corresponds to the "relatively warm and dry" North stage in North Europe. Its climatic trend is also consistent with the sea level fall in the Yellow Sea after 5.0 ka (Xie and Xu, 1979).

It should be emphasised that spits 3–1 (ca. 7.7–5.3 ka), which reflects a relatively high air temperature and a dry and slightly warm climate in the lake area during that period, are coincident with the transgression climax of the Yellow Sea at 7.7–5.0 ka. In adjacent areas e.g., Didwana Lake (27° 20'N, 74° 35'E), Thar desert, India) the climate became humid, from being dry before 6.0 ka (Wirsson et al., 1984). The air temperature in this area also went up to some extent, and the humidity was slightly higher than in the previous period, suggesting the influence of worldwide climatic changes on the plateau. Since ca. 5.3 ka, the lake water dropped abruptly from the surface of levee 1 to the present lake level. In the modern saline lake stage, the lake water is highly concentrated and the lake water has a water volume of about 1.7×10^8 m^3. This represents a 99.985% reduction in the water body as compared to that in the overflow level period and a 93% reduction as compared to that in the period of spit 1 (Table 1.39).

PART 2 STUDY OF HALOPHILIC ORGANISMS

CHAPTER 9 DISCOVERY AND STUDY OF HALOBACTERIA AND HALOPHILIC ALGAE IN THE ZABUYE CAKA

9.1. Introduction

It is a long time ago when there appeared reports on the existence of microorganisms in hypersaline environments (e.g. Artari, 1913; Baas–Becking, 1928; Zobell, 1946). However, since the 70s such microorganisms have demonstrated their major significance in fundamental theories and important prospects in practical application as a result of the discovery of the quantasomic function of the rhodopsin membrane of the halophilic bacillus and the abundance in β–carotin and glycerol of *Dunaliella salina*, thus attracting great attention of biologists, biophysicists, biochemists and geologists as well as people from the food industry and astronautic sector. So far, halobacteria and halophilic algae have been found in some coastal saline lakes (e.g. Hatt Lagoon) in Australia, the Dead Sea, the Great Salt Lake in the United States, certain soda lakes in Egypt, etc. A great deal of research work has been carried out and relevant reports published abroad on the biophysics and biochemistry of halobacteria and halophilic algae as well as the extraction and application of β–carotin and glycerol from halophilic algae (Ami Ben–Amotz, 1982; Bayley et al., 1964; Borowitzka et al., 1977, 1981, 1983; Brock, 1978; Brown et al., 1979; Khorana, 1979; Kudhner, 1978; Mastui et al., 1979; Nissenbaum, 1980; Stoeckenius et al., 1979)①. In the field of earth sciences, a study on the ecology of hypersaline environments was made by some scientists in 1980 (Nissenbaum, 1980). In China, this also aroused the interest of the scientific com—

---

① Haevey Steve, 1983. Pink algae future glouling. The Sunday Times (Australia).

munities (Tan and Tan, 1979)①② and a proposal on the development of the geoecology of hypersaline environments was put forward③. The authors have begun to pay attention to the investigations and studies in this realm since 1980. Following the discovery of *Artemia salina* in the Bangkog—Ⅲ Co and abundant *Artemia salina* and *Daphniopsis tibetana Sars* in the Siling Co in 1980, the authors and others found, for the first time, natural halobacteria and halophilic algae over large areas in the Zabuye Caka in the summer of 1982. Afterward, they continued substantial laboratory and field studies and successively discovered *Artemia salina* in the Jibu Caka, Lagor Co and Bangkog Co and various types of halophilic algae in the Da Qaidam Lake. The present chapter mainly deals with the formation of the halobacteria and halophilic algae in the Zabuye Caka, field observations of their geoecological conditions and laboratory experiments on the conditions of their culture.

9.2. Physiogeographic Conditions

The Zabuye Caka is a N—S trending tectonic lake (Zheng et al., 1985). With an ENE—trending concealed fault passing across the middle part of the lake, there has formed a presque isle— Zabuye Island. The Zabuye Caka is separated into two lakes (the South Zabuye Caka and the North Zabuye Caka) by a sand barrier accumulated along the domal part of the fault (Fig. 2.1). The two lakes were still connected by a shallow water channel more than 100 m wide in 1982, but they were completely separated in the summer of 1984 as a result of the change in climate. The North Zabuye Caka is replenished by abundant glacial and snow melt water through rivers and deep—circulating ascending springs, while the South Zabuye Caka is recharged by lake—shore springs which mostly have a relatively small flow rate except for the ascending spring on Zabuye Island in its northeastern part, which has an outflow rate up to 53000 t / h and thus offers a comparatively substantial replenishment. As a result, there appears some difference in the geochemical environment between the two lakes. The North Zabuye Caka is almost completely a brine lake with a surface area of brine up to 97 km^2 and a water depth of scores of centimetres to 2 m. The suitable water depth plus the absence of massive suspended material have created favourable conditions for the incidence of sun light into the lower part of the lake water. When viewed from a distant high point,

---

① Wang Dazheng et al., 1981. Advances in ecology of microorganisms. To be published.

② Zheng Mianping, 1983. Peculiar biotic resources to be developed. Geological Newspaper of China, July 25, page 4.

③ Chen Shi, 1983. First discovery of halophilic algae in soline lakes of Tibet, China. Geological Newspaper of China, July 11, page 1.

the lake exhibits a grayish–red span of surface water, indicating the occurrence of abundant halobacteria and halophilic algae (Plate 2.1). Red halobacteria and halophilic algae are also seen to occur in discontinuous zones in the salt deposits on the south and north banks of the lake. On the salt flat of the south bank, for instance the zone of red halobacteria and halophilic algae is 10–20 m wide and discontinuously extends for about 7 km (Plate 2.2). The South Zabuye Caka represents itself a semi–playa with an area of 145 km^2. Its surface brine is mainly distributed in the northeastern part with a water depth of a few centimetres to 1 m. Its

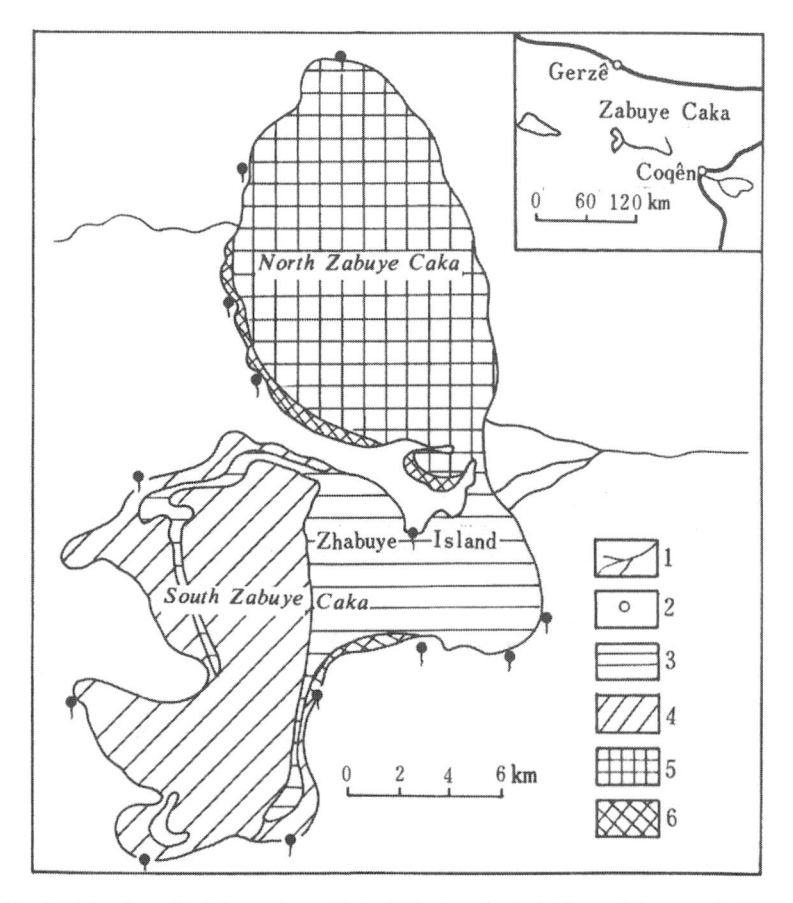

Figure 2.1. Distribution of halobacteria and halophilic algae in the Zabuye Caka area. 1–River; 2–Spring; 3–Surface brine; 4–Semi–dry salt flat; 5–Halobacteria and halophilic algae grown in brine; 6–Those growing in halite deposits.

southwestern part has become a dry salt flat (sabkha). In the marginal sabkha in the eastern part of the South Zabuye Caka discontinuous zones of halobacteria and halophilic algae are also seen at 3 localities. One of the zones which occurs in

CHAPTER 9

the middle part of the east bank of the lake is about 1 m wide and hundreds of metres long (Plate 2.3), with a 1—2 cm salt crust bearing red organisms on the underlying black sandy clay.

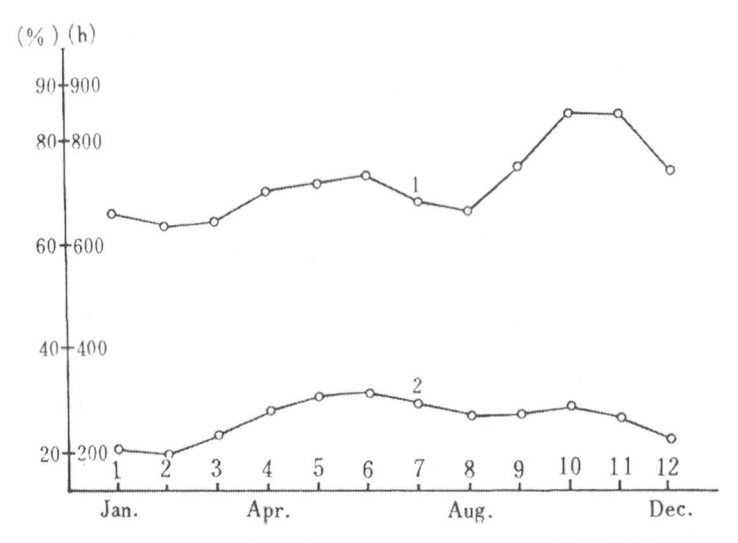

Figure 2.2. Sunshine percentage and hours from January to December in Gêrzê County. 1—Sunshine percentage; 2—sunshine hour

The Zabuye Caka lies in a semi-arid climatic region of the plateau temperate zone (Chen and Bowler, 1985). According to the observation records of the Gêrzê Meteorological Station in the north for the period of 1973—1981, the average annual precipitation is 192.6 mm and the average annual evaporation 2269.1 mm, the evaporation being 11.8 times as much as the precipitation; the annual atmospheric temperature averages $-0.4°C$, with the highest temperature in June—August ($9.3-11.7°C$ in monthly average) and a relatively low temperature in December—February ($-9.8- -13.4°C$ in monthly average). The air temperature measured on the spot by the author in June—July, 1982 ranges from $6°C$ to $20°C$ in the daytime and is $-3°C$ in the lowest at night. The highest temperature measured at the lake in July 1984 is $29.5°C$. As indicated by the above data, the air temperature in the lake area is relatively low and it is only in individual summer periods that it approaches the optimal temperatures for the accelerated metabolism of the algae and bacteria ($30-40°C$ and $40-50°C$ respectively) (Nissenbaum, 1980; Borowitzka et al., 1983). It is of interest that the water temperature of the Zabuye Caka is relatively higher than that of other saline lakes in neighbouring areas where no red-coloured algae and bacteria have been observed. According to field measurements, the temperature of the brine in this lake often exceeds the air temperature at the time of measurements (Tables 2.1 and 2.2), which proves that the

TABLE 2.1. Ecological observations of halophilic algae from North Zabuye Caka

Time (h)	Weather	Air temp. (°C)	Water temp. (°C)	Illumi-nance (lx)	State of lake water	Specific gravity of lake water	pH value	Amount of algae in visual area under microscope (× 80)	Algal activity	Water samp. No.	Remark
10:00	Clear	13.8	16.0	118000	Slightly undulating waves	1.273	9.18	Up. part:1–2 Low. part:0–1		T229 T230	
11:00	Clear	16.4	16.6	—	As above	1.271	—	—			
12:00	Clear	15.0	17.8	122000	As above	1.271	9.16	Up. part:3–5 Low. part:3–5	Monomer	T231 T232	A "chilli oil"-like layer of algal aggregates floats on water surface
13:00	Clear	17.5	19.0	125000	As above	1.270	—	—			
14:00	Clear	21.3	20.0	125000	As above	1.270	9.07	Up. part:20–30 to 40–50 Low. part:2–3	Zygotes seen	T233 T234	
15:00	Clear	26.0	21.2	123000	As above	1.270	—	—			
16:00	Clear	20.0	22.0	120000	10 cm-high small waves	1.270	9.11	Up. part:2–3 Low. part:2–3	Mostly monomers	T236 T237	In sample T239 collected around the island hundreds of algal elements are observed under microscope, with abundant zygotes present
17:00	Clear to overcast	17.8	21.0	98000	0.5 cm–high waves	1.270	—	—			
18:00	Overcast	17.0	21.0	78000	Small waves	1.270	9.18	Up. part:2–3 Low. part:1–2	Monomers	T238 T239	
19:00	Clear	18.5	21.0	97000	Small waves	1.270	—	—			
20:00	Clear	16.0	21.0	295000	Small waves	1.270	—	Up. part:1–2 Low. part:1–2	Monomers	T240 T241	

(Date of observation : July 19, 1984)

TABLE 2.2. Chemical compositions of brines from the Zabuye Caka of Tibet and Wadi Natron of Egypt

	Position	Date of samp.	Air temp. (℃)	Water temp. (℃)	Spec. grav.	Water colour	Smell	pH
Zabuye Caka, Tibet	Surface brine from the N. Zabuya Caka area ①	July 8, 1982	18	18.5	–	Slightly reddish	Slightly aromatic	9.5
	Surface brine from the eastern South Zabuye Caka area ②	July 2, 1982	14	14	1.279	Slightly reddish	Aromatic	8.9
	Interstitial brine from the southwestern salt beach of the S. Zabuye Caka①	June 30, 1982	16	11.5	1.304	Colourless	Smellless	9.1
	Desalted surface brine to the northwest of the S. Zabuya Caka ②	July 7, 1982	18.5	20	1.265	Slightly gray	Smellless	9.3
Wadi Natron, Egypt	Zugm L. ③	–	–	–	–	–	–	11.9
	Gcar L. ③	–	–	–	–	–	–	10.9

Content of principal chemical components (g/l)											Total salinity (g/l)	Remarks
Na^+	K^+	Mg^{2+}	Ca^{2+}	Cl^-	SO_4^{2-}	HCO_3^-	CO_3^{2-}	NO_3^-	PO_4^{3-}	B_2O_3		
107.09	24.70	0.026	0	144.55	30.54	6.98	15.86	0.001	0.37	7.21	338	With red halophilic bacteria and algae
135.34	27.60	0.001	0	162.14	56.20	0.0	25.42	–	0.15	8.58	418	As above
133.74	39.06	0.005	0	154.88	38.29	0	51.61	0.01	0.78	12.24	435.66	Without halophilic bacteria and algae
46.02	10.00	0.002	0	52.36	21.20	0.30	8.86	–	0.13	3.18	143	As above
142.0	2.3	0	0	154.6	22.6		67.2	–	–	–	388.7	With red halophilic bacteria and algae
137.0	1.4	0	0	173.7	48.0		6.6	–	–	–	366.7	As above

① Analysed by Chen Daxian; ② analysed by Xiao Huixiang; ③ after Larsen (1980)

red pigments of the halobacteria and halophilic algae are not only able to protect their cells from the strong sunlight but also help absorb it to increase the temperature of the brine they live in. Meanwhile, the lake area has a long duration of sunshine. At the Gêrzê Metaorological Station, for instance, according to the statistics for the period of 1974–1979, the average annual sunshine hours total 3 179, the annual sunshine percentage averaging 72.1% (Fig. 2.2). Particularly, the lake lies in a high-value area of the total solar radiation in the southwestern part of the plateau, where the intensity of direct radiation of the sun is high and the highest value of total radiation is as much as 8.37×10^5 J / cm^2 · a, and therefore is one of the areas with the highest value of total radian quantity on the earch (Research Group on the Radiation of the Qinghai–Tibet Plateau, 1982). The illuminance intensity correspondingly is also high. According to the authors' field observations made in July 1984, the illuminance from 8:00 h to 20:00 h in fine clear days is generally in the range of 10 000–130 000 lx (Table 2.1) and is as much as 145 000 lx or more in maximum. Practical observations in the brine area of the North Zabuye Caka in the above time interval indicate that the relatively high illuminance favours not only the accumulation of β–carotin in the algal bodies but also the propagation of such halophilic algae. According to the observations on Julu 17, 1984, for instance, the density of the halophilic algae in the upper and lower layers of the brine is relatively low in the early morning and in the evening, while at noon time these algae show not only an increased density but also a frequent reproduction and massive homogamy to give rise to zygotes (Table 2.1). This also demonstrates that the plateau's high radiation and strong light illumination are favourable for the development of these β–carotene–rich halophilic algae. Moreover, as the surface brine of the salt lake absorbs a great deal of heat energy due to the strong sunshine and the brine is possessed of the heat–storaging property, there has been created a favourable environment for the halobacteria and halophilic algae to multiply and resist low–temperature variations. In addition, the plateau is a region of winds, with the annual wind velocity averaging 4.1 m / s according to the data from the Gêrzê Meteorological Station. The periodical blowing leads to the constant reciprocation of the water in the brine lake that yields halobacteria and halophilic algae, which in turn not only provides hydrodynamic conditions for the convection of the brine but also may accelerate the gas exchange in favour of respective absroption of carbon dioxide and oxygen gas by the bacteria and algae.

9.3. Composition of Halobacteria and Halophilic Algae

Halobacteria and halophilic algae occur extensively in the brine layer of the North Zabuye Caka, making the lake water dark red in colour. The salt samples containing such bacteria and algae collected from the lake side salt flats originally also show a red colour but gradually turn to be orange yellow, and they are finger–staining and smell aromatic. A preliminary experiment on the culture of halobacteria in the samples was carried out by Wang Dazhen and Ma Guihong by inoculating the test specimens into culture media with 0%, 5%, 10%, 12%, 15%, 20% and 30% NaCl respectively. After 7 days' oscillating culture, microscopic determinations indicated that halophilic bacteria, including rod–like and ball–shaped ones, grew in the culture medium with a salinity of 25% whereas they were not observed to grow in the media containing less than 15% NaCl.

The halophilic algae associated with the halobacteria have been determined to be *Dunaliella salina* belonging to *Chlamydomonas*①, which is similar to *Chlamydomonas proboseigera Karschik ff* and *Chlamydomonas debaryana Gor* occurring in the former USSR, but shows some different features. However, while specimens of *Chlamydomonas* were collected from the northern part of the South Zabuye Caka, *Dunaliella* prevails in the North Zabuye Caka. Under the optical microscope the algal body is observed to be 21–26.5 μm long and 13–23 μm wide, orange–red in colour and pear–shaped. It has no cell wall, and is liable to deform. It has a slightly sharp top with two equally long flagella stretching out from the cell end and an obtuse rounded base with a cup–like chromatoblast, which has near its base a large proteosome surrounded by a starch sheath. Its core is in the middle of the cell. The above morphological features indicate that the alga is characteristic of *Dunaliella salina*.

TABLE 2.3. Compositions of amino acid in assemblages of halophilic algae

Amino acid	Content (%)	Amino acid	Content (%)
ASP	4.51	TYR	0.89
THR	1.88	PHE	1.97
SER	1.53	LYS	1.28
GLU	5.19	NH_3	0.53
GLY	1.93	HIS	0.15
ALA	2.59	ARG	2.05
VAL	2.53	PRO	1.78
MET	0.82	TRP	0.14
ILE	1.62	CYS	0.35
LEU	3.33	Total	35.07

Analysed by: Xie Yaqin et al. of Beijing Institute of Nutriology (1985).

The algal bodies of *Dunaliella salina* are abundant in β–carotin and other components. They have formed not only large patches of halite containing red–coloured bacteria and algae in saltshore zones but also large amounts of natural dark–red disseminated (Plates 2.2 and 2.3) and rope–shaped (Plate 2.4) algal ag-

---

① The determinations of *Dunalialla salina* have also been made by Zhu Haoran of the Nanking University and Chen Shufen of the Qingdao Institute of Oceanology, Academia Sinica.

gregates distributed along the lakeshore against the principal wind direction. These aggregates of the red-coloured algae seem to have formed by a process as follows: the algae multiply substantially at the noon time of the day, leading to an increased density of the algal bodies on the lake surface, then pushed by the wind and wave, they converge to form a patch of an oily microlayer of algal bodies on the surface of the brine (Plate 2.5), and finally they gradually assemble in the shallow water part of the saltshore zone and get reciprocally winded there. Under the microscope, numerous algal individuals are seen enveloped by thin transparent films (Plate 2.6). The algal sample collected is soft and elastic and sends out a strong aromatic smell. Chemical analyses indicate that it contains 22828 mg / kg of β-carotin, 5670 Iu / g of vitamin A, 35.04% of crude protein, 18.88% of crude fats and 1.29% of coarse fibre, and having an in vivo salt content of 2.26% and a total water content of 37%, including 4.29% of in vivo moisture. The contents of amino acids of various types it contains are given in Table 2.3.

The aromatic matter contained in the algal aggregates totals about 2%. An analysis with PE8320 fine-tube gas chromatograph[①] indicates that their products at the low boiling point (100℃ -200℃) contain large amounts of $C_6H_{14}O_2, C_7H_{12}O_2$, $C_{17}H_{34}, C_9H_{16}O, C_8H_{14}O, C_9H_{14}O, C_{13}H_{20}O$, etc.

9.4. Geological Conditions and Their Implications

Both the South Zabuye Caka and North Zabuye Caka have a K-rich alkaline high-salinity environment with a carbonate type of water (Zheng and Zhang et al., 1982). The South Zabuye Caka is a semi-dry saline lake, with the distribution of halobacteria and halophilic algae varying with the salinity of water in different parts of the lake. Halobacteria and halophilic algae have not been observed in the interstitial brine with a speicfic gravity of 1.300-1.305 and a salinity of 418-512 g / l at the dry salt flat of the South Zabuye Caka; they are developed in small patches at the lakeshore with a relatively lower salinity (with a salinity of about 404 g / l and a specific gravity of about 1.279) in the northeastern surface brine part of the South Zabuye Caka; in the surface brine part with infiltration of spring water, e.g. near Zabuye Island at the northern shore of the South Zabuye Caka, where the brine has a specific gravity of 1.265 and a salinity of 125.8 g / l, euryhaline green algae rather than red-coloured halophilic ones grow due to the even lower salinity of the water there. Filamental cyanophytes are also observed in the desalinized areas at the edges of some travertine isles in the North Zabuye Caka, where the specific gravity of the brine is 1.259. The North Zabuye Caka is almost completely a brine lake, where the surface brine has a specific gravity generally in the range of

---

[①] Analysed by Lin Zhumin etc. from the Department of Chemistry, Peking University.

1.268–1.275 and a salinity roughly in the range of 300–360 g / l, and red–coloured halobacteria and halophilic algae are extensively developed in the brine. The Zabuye Caka is also equivalent to Wadi Natron in Egypt in terms of salt content (Table 2.2). Therefore, according to the classification scheme proposed by H. Larsen in 1962, these bacteria should be assigned to the extremely halophilic bacteria; in accordance with the classification of halophilic algae by F. Hwstedt (1953), the algae should belong to the *halophilie formen* (Fordi, 1971). In the light of Ami Ben–Amotz's data of 1982 (Горбов, 1976), phosphates and nitrates are necessary for halophilic algae, but relatively low content of nitrates is favourable for the accumulation of β–carotin and glycerol of *Dunaliella salina*. The brine in the North Zabuye Caka is relatively high in PO_4^{3-} but comparatively low in the content of NO_3^-, a fact which conforms to the abundance in β–carotin of the algae.

The brine in the lake area that yields halobacteria and halophilic algae is characterically poor in Mg and Ca but rich in K, B, Li, etc. Based on his investigations, Larsen (1967) considers that the need of Mg ions (1–5% $MgCl_2$) by halophilic bacillus and coccus is a common feature of these organisms (Mastui et al., 1979). However, in the Wadi Natron area of Egypt and the Zabuye Caka area where the lake water is of the carbonate type, the halobacteria all occur in brines poor in Ca and Mg, thus suggesting that the demands of halobacteria and halophilic algae for Mg and Mg may vary in lake waters of different types. As for the biochemical demand of such halophilic microorganisms for special elements such as B and Li, this represents a new subject, which will be dealt with in Chapter 10.

Another important feature of the brine of the Zabuye Caka which yields red–coloured bacteria and algae is the relatively high content of potassium, which averages 26 g / l for the brine of the North Zabuye Caka and amounts to 25.38 g / l for the surface brine in the South Zabuye Caka. Moreover, determinations indicate that the halite samples that yield the red halophilic bacteria and algae often contain a potash mineral—glaserite. This very much conforms to the fact that halobacteria are found only in the culture medium mixed with K during the experiments on the culture of bacteria. According to some studies, the salt in the interio of the cells of halobacteria is essentially KCl (Nissenbaum, 1980). Halobacteria are capable of concentrating K^+ and erepelling Na^+. According to the studies of the enzyme of *Halobacterium salinarium* in 1968, for instance, L. I. Hochislein et al. hold that K^+ has a stronger activization effect on the enzyme as compared to Na^+. Bayley et al. discovered in 1964 that the ribosomes of *H. cutiroubrum* seem to have a particular need of highly concentrated KCl to gain its activity, which is attributed to the presence of acidic protein in the ribosomes (Bayley et al., 1964). The authors have also noted that some saline lake yielding extremely halophilic bacteria and algae also have a relatively high K content, e.g. up to 13.92 g / l KCl in the Dead Sea. Such an interesting feature of certain halophilic bacteria and algae indicates that it can, perhaps, serve as one of the prospecting criteria for K–bearing saline lakes. And therefrom derives a subject which is worth further studying, i.e. whether the fact that the potash ores mined in ancient times were usually red, orange—yellow and yellow in colour is also related to the activities of such halophilic bacteria and algae in addition to the presence of goethite in them.

It is of interest to note that the author has found numerous rarely—seen swarms of travertine isles in the brine area of the North Zabuye Caka which teems with halophilic bacteria and algae. They are comprised of as many as 500 or more white travertine cones (Fig. 2.7). According to the observations till 1984, most of the travertine cones stand in the brine only with the conical tops being visible scores of cm to about 1 m above the water surface or still remaining below the surface of the brine. With each of the isle swarms consisting of several to ten or twenty travertine cones, the travertine isle swarms spread all over the lake in an ENE—NE direction and extend to join the NNE—striking Cenozoic volcanic terrain to the east. The travertine cones are probably lake calc—sinters formed by the thermal springs. According to a few ^{14}C data available (dated by Liu Guangnian et al.), they were formed at 6000—odd a B. P. in the Early—Middle Holocene. Although the activities of thermal springs terminated long ago, small streams of cold spring water with a constant temperature are universally observed to surge upwards along the fissure of the sinters and, as the spring water contains HCO_3^-, CO_3^{2-} and NO_3^-, the deposition of minor calcium carbonates goes on and is accompanied time and again by the gas release. This indicates that at present these travertine cones are still supplying abundant CO_2 and fresh water to the lake, thus providing nutrients such as CO_2 and NO_2 necessary for the halophilic algae and playing a crucial role in regulating the salinity of the brine to the effect that the lake is not dried up. It is just the coincidence of the physiogeographic and geological conditions that have enabled the Zabuye Caka to become a locality of abundant halophilic bacteria and algae which is known to be the highest in elevation in the world.

In the Zabuye Caka, traces of extensive activities of bacteria and algae can be observed almost throughout the entire course of salt—forming process. The complicated exchanges between the organic and inorganic materials in the salt—forming geological bodies have left enormous records in the sediments, and these materials have usually become the important components of some sedimentary beds in the saline lake. Abundant sporopollen and well—preserved *Pediastrum boyanum* and *Zygnemo* (determined by Kong Zhaochen and Du Naiqiu) have been discovered in the lake—bed sludge deposits representing the freshwater to brackish lake stage of this saline lake; substantial amounts of reddish—brown to black, sludge rich in organic matter and dark red to light yellow salt deposits bearing bacteria and algae have been observed in the sediments of the saline lake stage. Euryhaline filamental cyanophytes marking the salt—water lake stage have been found by the author in the upper part of the ancient travertine cones of the Zabuye Island over 20 m above the modern lake level and they are developed, too, in the freshwater percolating—mixing zone of the travertine isles in the North Zabuye Caka. Large amounts of bacterial and algal aggregates still have been discovered in the water–soluble chlorides, sulfates and carbonates in the upper part of the modern salt deposits. Algae are a plant on the earth which makes use of the sunlight energy to produce organic carbon, and in this respect they contribute far more than terrestrial plants. According to relevant reports, it is estimated that marine algae of one square kilometre can produce 375 t of organic carbon per year (Shandong Institute of Oceanography, 1962). Hence it can be seen that bacteria and algae played an important part in the circulation of carbon and oxygen in various evolutionary stages of saline lakes. It seems that such a role is still far from being duly evaluated by salt geologists. Algae absorb a great deal of carbon dioxide from both the lake water and air and release oxygen to synthesize organic carbon. While bacteria may destruct organic matter (algae, etc.) by absorbing oxygen and releasing carbon dioxide, some pigment–producing halobacteria are able to perform autophototrophism to reduce and fix carbon dioxide, thus forming organic carbon. Such a function will inevitably result in the great increase of the carbonate content in the saline lake. Therefore, the "fixing of carbon from the air" by bacteria and algae represents another important source of materials in the saline lake that must be taken into consideration.

Besides, the above–mentioned lake–bed deposits of this lake are abundant in black organic matter–bearing sludge commonly with a stinking H_2S odor, which is clear evidence showing that sulfate reducing bacteria are much active. Surely, these bacteria have contributed to some extent to the consumption of sulfates in the lake and the increase in alkalinity of the lake water. Unfortunately, there has been little knowledge on the sulfate reducing bacteria and even the phototrophic sulfur bacteria in hypersaline environments so far (Nissenbaum, 1980). However, a comprehensive and correct understanding of the composition, evolution, material source and organic–inorganic mineralization of the modern saline lake and ancient saline deposits would be impossible unless the gigantic role of microorganisms in saline lakes is ascertained. For instance, as compared to the modern alkali lakes, the Tertiary soda deposits discovered in the United States and China all lack the sulfate mineral phase but are found to abund in dolomitic oil shale with large amounts of algae (Matter et al., 1979) and pyrite. Such a fact is difficult to explain purely by the deposition of inorganic salts and probably is the result of the decomposition of sodium sulfate under the participation of bacteria and algae into H_2S and sulfide and the accumulation of sodium carbonate and sodium hydrogen carbonate.

9.5. Preliminary Study on the Practical Application and Culture Conditions of Halophilic Algae

The study of halophilic algae and bacteria in hypersaline environments is a new realm of contemporary scientific and technological research. Particularly, the study and development of halophilic algae have increasingly attracted great attention of the Chinese and foreign scientific and technological circles as they have demonstrated major economic benefits and a wide perspective as a supplement to the agriculture. Since the late 70s, Australia, the United States, Israel, etc. have successively put in large amounts of funds and research forces in this realm. At present, they have succeeded in artificial culture of halophilic algae in saline lakes for the factory—scale production of β—carotene, glycerol and protein and are preparing to produce astaxanthin, phycoeyanin, tetroterpene oil, chlorophyll, fatty acid, EPA (d), etc.①②. At the same time, Australia, Japan and other countries are also taking advantage of the biological function of halophilic algae to clean the common salts in salt fields to raise their purity and have thus gained marked economic results. Some halophilic algae can be used to extract β—carotene, glycerol, protein, astaxanthin, phycocyanin and aromatic matter (tetroterpene oil, etc.). These products are of wide use in foodstuffs, medicines, paints, feeding stuffs, etc. and are high in price. For instance, β—carotene, which is the source of vitamin A, has a nursing effect on the human body and, as a new—type raw materal for foodstuffs, it is more and more extensively used in artificial butter and various soft drinks. Currently, β—carotene of various grades are sold at a price of US \$ 300—500 on the international market. In China, the development study of halophilic algae has gradually

TABLE 2.4. Synthetic culture fluid

Nutrient salt	Composition (g / l)	Nutrient salt	Composition (g / l)
$NaCO_3$	0.5	$MnCl_2 \cdot 4H_2O$	1.8
KH_2PO_4	0.05	$ZnSO_4$	0.11
$MgSO_4$	0.3	$CuCl_2 \cdot 2H_2O$	0.05
Na_2EDTA	5 mg / l	$CoCl_6 \cdot 6H_2O$	0.05
H_3BO_4	3 mg / l	KBr	1
$NaHCO_3$	4	NaCl	180 g / l

aroused greater interest. Since 1985, the Ministry of Light Industry has conducted experiments on the culture of halotolerant algae by using the algal strains brought back by the author from the Zabuye Caka and Australia and provided by A. W. Nonomura, Chief Advisor from the United States, respectively, in the salt field at Tanggu, which indicate that, while other halophilic algae decrease in amount or completely halt to grow at 5℃, the algal strain from the Zabuye Caka is able to

---

① Business Review, Nov. 8, 1985.

② Western Biotechnology Lt. Dec. 19, 1986 News Release.

grow at −4℃, thus representing a major extension of biological productivity in the cryophilic range (A. W. Nonomurs, personal communication of 15 Sept. 1986). With a view to gaining economic benefits out of the halophilic algae in China as soon as possible, the author has successfully conducted cultures of the halophilic algae from the Zabuye Caka under indoor conditions and have accomplished, in cooperation with the Group of Algal Culture, Academia Sinica, preliminary experiments of indoor culture. The experimental cultures are highlighted below 3①.

9.5.1. EXPERIMENTAL CULTURE OF ALGAE IN SYNTHETIC CULTURE FLUID

(1) Experimental conditions: All experiments were carried out under natural and artificial illumination in indoor conditions. For the experiments 25 ml or 125 ml triangular flasks were used and 10 ml or 50 ml of culture fluid were added into each of the flasks. Through preparatory experiments, the make−up of the synthetic culture fluid provided by Van Auken McNulty (1973) was selected, with only some modifications in the composition or amount of nutrient salts (Table 2.4).

TABLE 2.5. Influence of various mixtures of lake water with culture fluid in different rates on algal growth

Group	Lake water / culture fluid	Cell density after 10 days' culture × 10^4 (ps / ml)
1	10:0	1.4
2	9:1	2.0
3	8:2	3.3
4	7:3	5.2
5	5:5	7.5
6	4:6	12.1
7	2:8	15.8
8	0:10	20.7

(2) Experimental results: In order to determine whether the above synthetic culture fluid is suitable for the growth of the halophilic algae, the original lake water was boiled, filtered and added to with 0.5 g/l $NaNO_3$ and 0.05 g/l KH_2PO_4 and then was mixed up with the above culture fluid in different propor-

---

① The experimental work was led by Chen Jiaofen of the Institute of Oceanography, Academia Sinica and assisted by the author.

tions to make up 8 groups of cultural fluids. Each group was inoculated with 1 ml alga-bearing lake water for a 10 day culture at 26.5-30℃ under natural illumination (with a light intensity of about 840-5800 lx). The results of the experiments (Table 2.5) indicate that the scarcity of algae in the originar lake water is attributed to the lack of N and P nutrients. The algae increased in quantity remarkably when a certain amount of N and P nutrients was added. However, the growth rate of the algae in the pure lake water without the synthetic culture fluid added remained the lowest. Along with the decrease in amount of the lake water and the increase of the synthetic culture fluid, the growth speed of the algae drastically accelerates and the highest density of cells was achieved in the pure synthetic culture fluid, being 14.8 times that in the pure lake water. Besides, the algal fronds cultured in various groups show remarkably different tincts. The fronds exhibited an orange red tinct, in the pure lake water group but they showed a gradually increasing proportion of green tinct along with the increase of the synthetic culture fluid, and became green when the ratio of the lake water to the synthetic fluid was 5:5. This implies that the β-carotene content in the fronds is related to the salt content of the lake water.

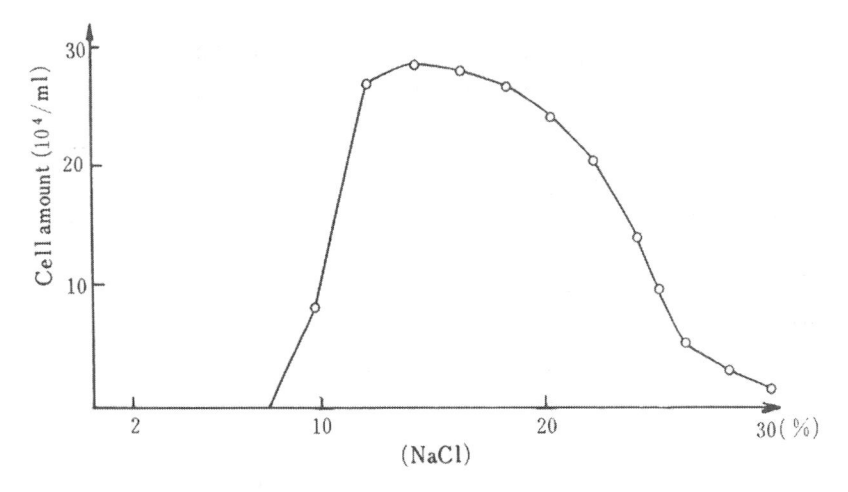

Figure 2.3. Growth curve of halophilic algae cultured in sythetic fluids with various NaCl concentrations.

9.5.2. EXPERIMENTAL CULTURE OF ALGAE IN CULTURE FLUIDS WITH DIFFERENT NaCl CONCENTRATIONS

Comparative experimental cultures were carried out, at 29°-30.5℃ under artificial illumination of 5200-5700 lx, in 16 groups of culture fluids made up of the above synthetic culture fluid and NaCl in various amounts (0, 2%, 4%, 6%, 8%, 10%, 12% 14%, 16%, 18%, 20%, 22%, 24%, 26%, 28%, and 30% respectively). The cell density at the beginning of the experiments was 2×10^4 ps/ml. The growth of the algae in 10 days is shown in Fig. 2.3.

As indicated by the experiments, the cytomembranes all ruptured and chromatoblasts escaped out in 4 hours after the fronds had been inoculated in the culture fluid without NaCl added; the cells of the fronds swelled into round balls, the chromatoblasts turned to be granular, and the fronds became inactive and then faded and died out in 24 hours in the culture fluid with 2% NaCl due to the low permeability of the solution; the fronds also died out successively on the second or third days in culture fluids with 4% NaCl and 6% NaCl respectively. Although the cells in the group of culture fluid with 8% NaCl did not increase in quantity within 10 days, a minority of them survived and were still slightly rotating. The amount of cells began to increase in culture fluids with the NaCl concentration from 8% up, and the growth of the cells was most flourishing in culture fluids with NaCl concentrations in the range of 12–18%. The growth rate of the cells gradually slowed down in the case with 20% NaCl or more. The algae could tolerate a NaCl concentration up to 30% but in this case the cell amount no longer increased. Besides, in terms of the colour tint of the fronds, the cells showed a green colour in the case with the NaCl concentration below 20%, gradually exhibited an orange tinct in the case with 22% NaCl and became basically orange–yellow in the case with 30% NaCl. The fronds cultured in culture fluids with the same NaCl concentrations under the natural illumination appeared more orange–red than those under the artificial illumination.

9.5.3. EXPERIMENTAL CULTURE OF ALGAE WITH N– AND P–NUTRIENTS

Different amounts of $NaNO_3$ and KH_2PO_4 in the ratio of 10:1 and 5:1 were added to the above synthetic culture fluid (with 14% NaCl only) to make up 8 groups of cultural fluids for comparative experimental cultures. The experimental results of the seven–day cultures are given in Table 2.6.

TABLE 2.6. Influence of various $NaNO_3$ and KH_2PO_4 concentrations upon algal growth

Concentration of nutrient salts (g / l)		Cell density after 7 day's culture $\times 10^4$ (ps / ml)
$NaNO_3$	KH_2PO_4	
0.5	0.05	26.0
	0.1	26.0
1.0	0.1	47.0
	0.2	46.5
1.5	0.15	48.3
	0.3	48.8
2.0	0.2	50.6
	0.4	46.0

TABLE 2.7. Influence of various K_2SO_4 and Na_2CO_3 concentrations on algal growth

	Na_2CO_3		Cell density after 8 days' culture $\times 10^4$ (ps / ml)
	Concentration M	Amount (g./ l)	
K_2SO_4 0.05 M (8.7 g / l)	0.2	21.0	28.2
	0.4	42.4	21.9
	0.6	63.6	17.8
	0.8	84.8	14.9
	1.0	106.0	12.5
K_2SO_4 0.5 M (8.7 g / l)	0.2	21.2	23.8
	0.4	42.4	17.4
	0.6	63.6	11.2
	0.8	84.8	7.2
	1.0	106.0	5.7

As indicated by the experiments, the nitrogenous fertilizer is an important factor confining the growth of the algae, while the concentration of sodium nitrate has a remarkable influence on the cell density of the algae, and at a low content of $NaCO_3$ the cell density can not be increased even by increasing the amount of KH_2PO_4. The experiments have proved that 1.5 g / l $NaNO_3$ and 0.2 g / l KH_2PO_4 are the appropriate concentrations for the growth of the algae and their excess concentrations do not help accelerate the propagation of the fronds. Besides, it can be seen from the tints of algal cells that the algal cells in two groups with a low nitrogen concentration show a yellowish tint whereas in other 6 groups they are all green in colour, thus suggesting that the formation of β—carotin is also related to the nitrogen insufficiency.

9.5.4. EXPERIMENTAL CULTURE IN FLUIDS WITH DIFFERENT pH

The experimental culture of the halophilic algae was carried out in 5 groups of culture fluids with pH values of 6.5, 7.0, 7.5, 8.0 and 8.5, which were made up of 0.1 NHCl and 0.2 MTris solutions in appropriate proportions prepared with the synthetic culture fluid (with $NaHCO_3$ added separately). The experimental results are given in Table 2.7 and Table 2.8.

It can be seen from the experimental culture that the halophilic algae far more flourish in the alkaline solution than in the acid one, with the optimal pH values ranging from 8.0 to 9.0. Different concentrations of $NaHCO_3$, K_2SO_4 and

Na$_2$CO$_3$ seem to have no significant influence on the growth of the algae.

9.5.5. EXPERIMENTAL CULTURE IN THE CONCENTRATED SEA WATER

The algal species from the Zabuye Caka (*Dunaliella Salina*) can grow substantially in the sea water processed in various ways, thus implying a step foward for them to be rebreeded in coastal areas.

TABLE 2.8. Influence of pH on algal growth

Experiment	pH At the beginning of experiment	pH At the end of experiment	Cell density after 8 days' culture × 10^4 (ps / ml)
With NaHCO$_3$ added	6.4	6.7	14.5
	7.0	7.3	16.1
	7.4	7.6	24.4
	8.0	8.1	33.0
	8.6	8.4	46.9
Without NaHCO$_3$ added	7.1	8.8	31.1
	7.3	8.7	43.8
	7.7	8.7	44.0
	8.0	8.8	43.7
	8.7	9.0	43.0

9.6. Some Ideas of the Halobacteria and Halophilic Algae in Saline Lakes

1. It is reported, for the first time, in this monograph that the halophilic algae from the saline lakes in Tibet are dominated by *Dunaliella salina*, with *Chlamyadomonas* being of secondary importance. The halobacteria from this area, including the rod—like and ball—shaped ones belong to the extremely halotolerant bacteria.

2. The halophilic alga from the Zabuye Caka (*Dunaliella salina*) is a valuable algal species which is strongly cryophilic and adaptive as well as rich in useful component such as β—carotin and protein. Such a strongly cryophilic algal species is rare abroad and its cryophilic ability has been verified by the field and laboratory cultures. In the meanwhile, it is, too, an algal spicies which has a strong ability to

adapt to the environment. It is able not only to multiply in highly frigid alkaline brine which is high in salinity, poor in Ca and Mg, and rich in K, B and Li but also to propagate substantially in the synthetic culture fluids and processed seawater. Its growth can not be checked by the large amounts of Ca and Mg in the seawater.

3. The North Zabuye Caka abounds with natural halophilic algae, particularly in summer, when massive accumulations of halophilic algae rich in β-carotin and protein are formed. However, the halophilic algae in this lake vary greatly in quantity in different seasons of a year. As no multiannual data on the natural growth variation of the halophilic algae are available, the establishment of semi-fixed long-term obstravation stations is necessary to collect systematically data on the variations of the halophilic algae and solid-liquid phase salt mineral deposits. At the same time, a study on the small-scale utilization of natural dry fronds, first of all, should be carried out in adaptation to local conditions. By doing so, an initial success may be achieved with less capital investment within a short period of time. Then, an experimental study on the culture of the algae in man-built canals can be further planned.

4. It is pointed out that, despite the relatively low temperature on the plateau, the long sunshine duration and periodical wind blowing and the appropriate hydrochemical conditions provided by the geologicl environment are all in favour of an areal development of the halobacteria and halophilic algae. The experimental studies have proved that the strong illumination, suitable salinity and PO_4-NO_3 containing weakly alkaline brine, *inter alia*, play a leading role in the development of the carotin-rich algae.

5. The study and development of halophilic algae is a new realm in modern science and technology as well as a newly-emerging industry of far and wide prospects, which is an important supplement to the agriculture.

6. The activities of halobacteria and halophilic algae have occurred almost throughout the whole process of sedimentation of the lake basin and have played an unnegligible role in the mineralization process of saline lakes, and to some extent they are facies-marking, too.

As stated above, the author suggests that the geoecology of the salt-forming environment be studied from an integrated point of view. It will be a science specialized in the study of the laws governing the interaction of the minerogenic environment, biotic activity and geological environment and their geological and minerogenic implications. That is to say, it not only studies the general physicoecological conditions of the halophilic organisms but also probes into the geological origin and significance of their physico-ecological conditions, the influence of the activities and biochemical evolution of these organisms on the material composition, deposition, diagenesis and minerogenesis of saline lakes, and the facies implication of the palaeoecological environment of the halophilic organisms. As a new borderline science, however, it has still many subjects worth further probing as well as problems to be settled through deep-going studies in the future.

CHAPTER 10 BIOMINERALIZATION OF BORON AND OTHER HALOTOLERANT ORGANISMS[①]

Biomineralization has long been known well, yet the biominerolization in a superhaline environment remains a new realm. Based on the field investigations and studies in the areas of the Zabuye Caka, Da Qaidam Lake and Bangkog Co, a preliminary discussion on the subject is made by the author in the present chapter.

10.1. Diversity of Boron Mineralization

Borate deposits have been invariably considered to be formed by an abiogenic process. However, author's, observations in the modern saline lakes on the Qinghai—Tibet Plateau have indicated that their abiogenic processes consist of diversified mechanisms of boron mineralization, i.e. evaporation—concentration, dilution and condensation—cooling and that a biochemical process may also result in substantial mineralization of boron or play an important part in promoting the formation of borates. The borates formed by the abiogenic process differ to some extent from those formed predominantly by the biochemical process in mode of occurrence and type, structure and texture of ores. The former occur as thick beds of pure fine—grained borate minerals (e.g. the magnesian borate deposit on the Chagcam Caka terraces) or bedded deposits of powdery borates (e.g. the thin beds of borax interlayered with those of mirabilite and soda in the Dujiali Lake), whereas the latter very often appear as nest—like bodies of platy coarse—crystalline borax containing organic matter (e.g. in the Zabuye Caka and Bangkog Co) or concretional or lamelliform bodies of pinnoite (e.g. in the Da Qaidam Lake).

10.2. Basic Facts

1. Close relationship between borate minerals and organic matter: Enormous amounts of large coarse—grained platy boraxes can be found in both the Bangkog Co and the Zabuye Caka. Occurring in the black "sludge" (marl) bearing organic

---

[①] A project supported by the National Natural Science Foundation of China.

matter, the platy boraxes are mainly enriched in the salt marsh zone of the salt–mud flat. It has been also found that in the salt marsh zone of the northern South Zabuye Caka the contents of both organic matter and platy borax are relatively high in modern sediments in which halophilic algae and bacteria are developed. In the meanwhile, the platy borax crystals are observed to show gray to grayish–black ring–shaped bands containing organic matter. Sometimes, halophilic algae are found to be enclosed in boron crystals. The analyses of samples from salt–bearing sediments of various types (Table 2.9) further indicate that the contents of platy borax and organic matter show a positive correlation. The relative high contents of kerogen and moderate pyrolytic peak temperatures (generally in the range of 300–430°C) suggest a relatively low maturity of the organic matter. Moreover, the comparatively high H / C values (up to 758) imply that the organic matter in the sediments is mainly derived from inferior organisms, halophilic algae in particular. All this is very concordant with the massive multiplication of halophilic algae in the said lake.

TABLE 2.9. Comparison between borax and organic matter contents in the Zabuye caka

Sam. No.	Type of sediment	Gaseous compon. (%)	Source compon. (%)	Kerogen (%)	Peak temp. (°C)	Total organic carbon (%)	H / C	Content of platy borax (%)
1	Halite crust	0.00	0.00	0.18	428	0.01	*	0
2	Halite crust	0.00	0.00	0.08	423	0.00	0	0
3	Mirabilite	0.00	0.00	0.31	432	0.02	*	0
4	Mirabilite	0.00	0.00	0.15	384	0.01	*	0
5	Borate–bearing silty clay	0.00	0.00	0.23	429	0.01	*	0
6	Slightly borax–bearing clay	0.00	0.00	1.03	434	0.08	0	5
7	Borax–bearing "sludge"	0.00	0.02	2.35	279	0.31	758	30±
8	Black borax–bearing "sludge"	0.00	0.03	2.14	368	0.37	578	40

* Trace; analyst: Cheng Peng.

There is a similar case in the Da Qaidam Lake. At the lake bottom of the shallow brine area in the eastern part of the lake there occurs a most recently deposited calcareous lamina 0.5 cm in thickness. This lamina is a cryptocrystalline pinnoitic calc–crust, which is composed of micro–granular calcite and ankerite and contains organic matter. Observed on this thin crust is a brownish–red microlamina, which is determined by Zeng Zaoqi to consist of *Microcoleus* sp., *Auabaena* sp. and

Chroococcus sp. of Cyanophyta. This lamina discovered here is analogous with the pinnoite—cemented and encrusted sand and gravel layer in the lake—beach rock of the Xiao Qaidam Lake. According to the studies by Xia Wenji et al., the sand—gravel encrustation is made up of primary protodolomite. The deposition of the lamina is related mainly to the action of halophilic bacteria and algae, and the formation of some dolomitic algal fragments and nodules in them is obviously related to Cyanophyta (Xia and Li, 1984). The encrusted sand particles are basally cemented by pinnoite, thus proving that the encrustation of the sand grains, i.e. micritic protodolomite, was formed at the early--middle syngenetic stage while the formation of the pinnoite should be no later than the syngenetic stage and so is, likewise, related closely to the process of activity of Cyanophyta.

2. Special borophylicity of halophilic algae: The borophylicity of organisms has been universally acknowledged by many scientists. Biophysiologists have ascertained that boron is necessary for organisms in that it promotes the oxidation of enzyme in them (Dobrovolisky, 1983), but excess absorption of boron would cause poisoning of the organisms. It has been identified through investigations that the organisms growing in brine and brackish water and lower organisms, particularly algae in lower plants, have a higher content of boron. The author has made a comparative analysis of the halophilic algae (*Dunaliella Salina*) from the Zabuye Caka. Chemical analyses of the dry algal fronds washed with distilled water and the unwashed natural saliferous dry fronds indicate that the former contain 2.47% B_2O_3 and the latter 7.01% B_2O_3 (analyzed by Ding Fan). This suggests that the halophilic algae show a special boronphylicity and are important "collectors" for the enrichment of boric components, thus becoming the significant source of boric components for the formation of borates in the strata.

10.3. Preliminary Experimental Studies on Biomineralization of Boron

In order to ascertain the exact mechanism of biomineralization of boron, it is still necessary to carry out experiments on the solubility of boron, decomposition of humic acid by algae and redecomposition of humic acid. Now, an introduction to the experiments on the solubility of boron is given below.

10.3.1. EXPERIMENTS ON THE SOLUBILITY OF BORON (BASED ON THE EXPERIMENTS CARRIED OUT BY CHENG PENG)

With a view to excluding the impact of the difference in solubility of various borate minerals on experimental results, the borates were first transformed into B_2O_3 and then the solubility of B_2O_3 under different pH conditions was determined. B_2O_3 was obtained by treating borax with hydrochloric acid:

$$Na_2B_4O_7 + 2HCl + 5H_2O \rightarrow 2NaCl + 4H_3BO_3$$

$$H_3BO_3 \xrightarrow{Heating} HBO_2 + H_2O$$

$$4HBO_2 \xrightarrow{Heating} H_2B_4O_7 + H_2O$$

$$H_2B_4O_7 \rightleftharpoons 2B_2O_3 + H_2O$$

To ensure that no remarkable change in pH and no chemical reaction would take place during the experiment, a buffer solution was used in the experiment which was carried out at a temperature of 20℃. The experimental results are listed in Table 2.10.

TABLE 2.10. Solubilities of B_2O_3 under different pH values

\multicolumn{7}{c	}{Amount of chemical reagent in 100 ml of water (g)}	pH value	Solubility of B_2O_3 (g)
NaAc3H₂O	HAc	NH₄Cl	NH₄Ac	NH₃·H₂O	NaCl	Na₂SO₄		
4.000	36.662	/	/	/	10.00	2.00	4.0	14.320
10.000	9.302	/	/	/	10.00	2.00	5.0	13.711
/	/	/	15.400	/	10.00	2.00	7.0	2.200
/	/	10.000	/	0.368	10.00	2.00	8.0	1.509
/	/	7.000	/	2.520	10.00	2.00	9.0	1.144
/	/	5.400	/	20.685	10.00	2.00	10.0	1.022

It can be seen from Table 2.10 that the solubility of B_2O_3 under acidic conditions is greater than that under alkaline ones. For instance, the solubility of B_2O_3 in the medium with the pH value of 4.0 is as much as 14 times that in the medium with the pH value of 10.0. A comparison between the acid—intermediate interval of media with pH values in the range of 4—7 and the alkaline interval of media with pH values in the range of 8—10 indicates that the concentration of B ions in the acid solution is markedly greater than that in the alkaline solution. The dead lower organisms such as halophilic algae often produce large amounts of humic acid to make their media acidic, thus facilitating the increase of the media in the solubility of boron and creating a local environment for the enrichment of boric components to form and accumulate borate minerals.

10.3.2. PRELIMINARY DISCUSSION ON THE MECHANISM OF BIOMINERALIZATION OF BORON

Nest-like bodies of platy borax: Halophilic algae may become "collectors" of boron as they are able to absorb boric components in high concentration. Moreover, when dead and accumulated, they produce plenty of humic acid at the same time when boric components are released. The humic acid plays a dual role in the enrichment of boric ions. Firstly, as an acidic substance, the humic acid causes the surrounding medium to be acidic where it is present in a great abundance. Secondly, humic acids occur mainly as a negative gel in the media and such a gel can adsorb a large amount of boric ions to turn them into chelates, thus raising the concentration of boric ions in the media.

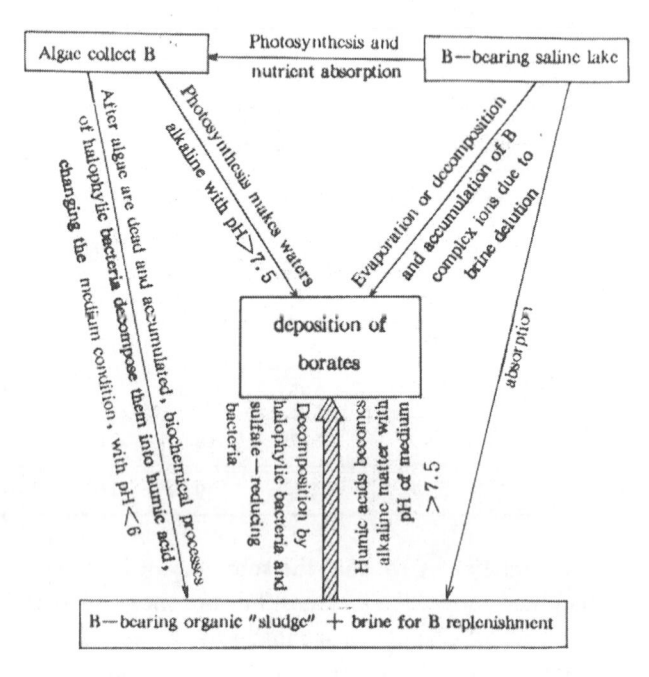

Figure 2.4. Diagram of biogenic mineralization of boron in the B-bearing saline lake.

After the humic matter is formed from the accumulated halophilic algae, it is liable to the decomposition by abundant halophylic bacteria and sulfate-reduing bacteria into an alkaline substance and thus favorable conditions for the deposition of substantial platy boraxes are created in the following three ways: 1) following the decomposition of the humic matter, B^{3+} is released from the chelates, thus providing a prerequisite for the combination of B^{3+} with Na^+ to form borax; 2) owing to the decomposition of the humic matter into an alkaline substance, the surrounding, originally acidic medium is transformed into an alkaline one, resulting in the substantial decrease in the solubility of boron in the medium and creating an optimal condition for the deposition of borax; 3) the saliferous "sludge" containing abundant humic and other organic acids not only has a good plasticity and porosity but also contains "sludge brine" for slow replenishment of boric components, which provides a good space and favorable condition of slow crystallization for the formation of large platy borax crystals. This is the basic reason why the nest—like bodies of platy borax are almost unexceptionally concentrated in the organic matter containing "sludge" (marl) of the salt marsh zone.

Flaky and tubular pinnoite: Flaky and tubular pinnoites in the Da Qaidam Lake are mostly associated with dolomite (calcite). According to synthesization experiments and field investigations in other areas, the deposition of primary protodolomite should have the following prerequisites: 1) the salinity is 5—8 times higher than that of the normal sea water; 2) the Mg/Ca ratio is high; 3) the pH value is greater than 9 in most cases. In the Da Qaidam Lake the salinity and Mg/Ca ratio of the water conform with the above conditions except for the slightly lower pH value. However, the pH value of the water may be raised by the photosynthesis of the halophilic algae, thus leading to the formation of protodolomite (Xia and Li, 1984). According to laboratory experiments (Dobrovolisky, 1983), low—Mg alkaline Na_2SO_4—NaCl solutions containing boron and carbonates favour the synthesization of pinnoite. Hence it can be seen that the halophilic algae in the Da Qaidam Lake have played a dual role in the mineralization of boron, namely, they have become the "collectors" for the enrichment of boric components on the one hand and have created a medium favourable for the precipitation of pinnoite on the other, thus leading to the formation of the alga—bearing calc materials of flaky and tubular pinnoites.

In summary, the biomineralization of boron in the boron—bearing saline lake is generalized as shown in Fig. 2.4.

10.4. Halotolerant Organisms in Other Lakes

10.4.1. *DAPHNIOPSIS TIBETANA* SARS

On August 19—30, 1980, investigations were carried out by the author in the second largest lake on the plateau— the Siling Co and its eastern marginal lake— the Jiangdongrurui Co. Abundant planktons, which are brownish—black in color, long elliptical in form and 1.5—2.5 mm in size, were observed in both these two lakes. They were found along the lakeshores and river mouths and in the superficial water layers from the Jiangdongrurui Co to the Siling Co and even at the water depth of 8 m in the Siling Co. The plankton of this kind has been determined by Dai Ailian et al. from the Institute of Zoology, Academia Sinica, as *Daphniopsis tibetana* Sars, which is a cryophilic species, and its characteristic features are shown in Fig. 2.5 (Dong et al., 1982). *Daphniopsis tibetana* Sars was first discovered in the Yurba Co

TABLE 2.11. Hydrochemical composition and type of water in the Siling Co and Jiangdonggrurui Co

Locality	Salinity (g/l)	pH	Na⁺	K⁺	Ca²⁺	Mg²⁺	Li⁺	Rb³⁺	CO₃²⁻	SO₄²⁻	Cl⁻	B₂O₃	Br⁻	F⁻	HCO₄⁻	Classification	Organisms developed
Surficial water in the Jiangdongrurui Co	17.10 ①	9	5400	584	0	255	18.8	0.48	359	6845	3320	191	3.4	3.0		Weak carbonate subtype	*Daphniopsis tibetana Sars*
Eastern Siling Co at a water depth of 8 m	1750 ①	9	5500	584	1	263	16.5	0.44	417	6990	3390	184.18	3.8	4.07		As above	*Daphniopsis tibetana sars*, Amphipoda. etc.
Eastern Siling Co at a water depth of 18 m	17.71 ②	9	5500	590	0	261	16.8	0.44	419	7130	3450	194.17	3.5	1.3	1.02	As above	As above

Dates of salinity sampling: ① Auegust 19, 1980; ② August 20, 1980.

TABLE 2.12. Distribution of artemia in saline lakes of Qinghai and Tibet

Prov.	Name of lake	Longi-tude (E)	Latitude (N)	Salinity (g/l)	pH	Na	K	Ca	Mg	Cl	SO₄	CO₃	HCO₃
Tibet	Chagcam Caka	82°23'	33°47'	250	8.0	77.6	11.4	0.4	8.5	123.9	43.1	–	2.1
	Nau Co	82°1'	32°8'	215.5	9.5	71.0	5.0	0.004	0.26	48.7	71.8	13.1	3.1
	Oma Co	83°2'	32°3'	7.0	9.6	1.5	0.2	0	0.47	0.98	2.8	3.9	0.54
	Jibu Caka	84°10'	32°02'	82.7	–	23.4	4.8	0.03	1.4	33.8	13.2	0.8	2.0
	Dong Co	84°7'	32°1'	114.3	8.7	22.3	9.0	0.056	4.2	31.1	38.5	0.8	6.0
	Bangkog Co	89°5'	31°7'	49.0	9	14.48	1.07	0.006	0.03	115.87	22.7	1.24	3.5
	Co Nyi L.	87°15'	34°34'	62.73	8.5	18.83	0.84	0.71	2.13	30.95	8.4	0.17	0.2
	Cam Co	83°42'	32°58'39"	174.88	8.9	18.97	3.63	0.15	2.20	36.3	6.5	0.33	0
	Cêdo Caka	88°59	33°10'	168.35	8.3	58.19	1.85	0.58	2.59	86.91	17.7	0.01	0.17
	Shiquanhe L.	80°30'	32°00'09"	–	–	–	–	–	–	–	–	–	–
	Laggor Co	84°13'	32°03'	59.79	8.8	16.67	2.62	0.008	0.79	13.15	19.4	1.2	2.9
Qing-hai	Suhan L.	94°0'	38°9'	32.3	8.9	8.1	0.4	0.07	1.9	9.8	10.7	0.2	0.94
	Da Qaidam L.	95°3'	37°8'	340.6	8.0	102.3	5.1	0.5	15.7	186.2	27.7	–	–
	Xiao Qaidam L.	95°6	37°4'	239.5	7.8	81.2	1.1	0.4	4.4	108.5	41.0	0.28	1.1
	Toson L.	96°9'	37°1'	15.9	8.9	–	–	–	–	–	–	–	–
	Gahai L.	97°47'	37°2'	90.6	8.3	27.3	0.3	0.3	4.2	45.3	13.1	0.14	0.034

in northern Tibet (collected on August 4, 1978). Abundant amphipods were also observed on the southern lakeshore of the Siling Co (the determinations were made by Dai Aiyun from the Institute of Zoology, Academia Sinica), and they had been washed to by lake waves caused by the north wind and thus accumulated along the lakeshore zone to form a layer of the latest accumulations which is several centimetres thick and 10-odd centimetres wide. These amphipods look like shrimps and may be used as fish feed.

The above two organisms both belong to halophilic organisms of the halotolerent type. The salt contents of lake water in the Siling Co and Jiangdongrurui Co are 17.76 g/l and 17.37 g/l respectively (average values for 8 and 10 water samples respectively) and therefore these two lakes are of the alkaline, weakly carbonate subtype (Table 2.11). However, the lakewater of the Yurba Co has a salt content as much as 314 g / l (according to the Regional Geological Survey Party of Tibet) and therefore is a supersaline brine (of Mg–sulphate subtype). As this salt content of the lakewater is far greater than those in the above–mentioned two lakes and is far beyond the halotolerance range of the known elements of Daphniopsis, the writer doubts if the specimen was probably collected from the desalinized lakewater area at the lakeshore of the Yurba Co.

TABLE 2.13. Climatic data of saline lake areas

Name of lake	Climatic type	Hydro–chemical type	Elev. (m)	Aver. ann. air temp. (℃)	Max. temp. (℃)	Mini. temp. (℃)
Chagcam Caka	Semi–arid subfrigid plateau zone	Mg–sulfate	4326	−0.3±	24.9±	−33.0±
Nau Co	"		–	−1.5	23.0±	−39.0±
Oma Co	"	Na–sulfate	4436	−0.35±	24.5±	−34.6±
Jibu Caka	"	"	–	−0.4±	24.0±	−34.0±
Dong Co	"	"	4394	−0.48±	24.0±	−36.0±
Bangkog Co	"	Carbonate	4380	−1.4	21.7	−35.8±
Co Nyi Co	"	Mg–sulfate	4902	−1.0±	23.0±	−36±
Cam Co	"	"	4450±	1.0±	23.5±	−30±
Cêdo Caka	"	Na–sulfate	4822	−1.0±	23.0±	−36±
Laggor Co	Arid temperate plateau zone	Carbonate	4490	1.6±	24±	−26±
Shiquanhe L.	"	/	4240	0.01	27.6	−34.6
Da Suban L.	"	Na–sulfate	2795	2.60	34.2	−34.3
Da Qaidam L.	"	Mg–sulfate	3148	1.10	29.9	−33.6
Xiao Qaidam L.	"	Mg–sulfate	3172	1.10	29.9	−33.6
Toson L.	Semi–arid temperate plateau zone	Mg–sulfate	2813	3.70	32.7	−37.2
Gahai L.	"	Na–sulfate	2851	3.70±	32.7±	−37.2±

10.4.2. *Artemia salina Linneans*

So far, *Artemia salina Linneans* has been found to occur in 16 saline lakes on the Qinghai–Tibet Plateau (Table 2.12). The lakewater salinities of these saline lakes differ greatly from one another, ranging from 7 g / l to 258 g / l, and are, in most cases, 10–15% higher or lower than the salinity range normally required by

Artemia salina; the lakewaters are weakly alkaline or alkaline with pH values in the range of 7.5–9.6 and mostly belong to the sulfate type according to Kypnakov–Valyshko's hydrochemical classification of saline lakes. As compared with the lakes elsewhere in China, these saline lakes show higher contents of trace elements such as B, Li, Cs, Rb and As in the brine (Table 2.12). The 16 saline lakes are located in areas 2813–4902 m above sea level. Among them, eleven lakes lie in the semi–arid region of the plateau temperate zone—semi–arid region of plateau subfrigid zone at elevations of 4240–4902 m with annual air temperatures in the range of 1.6℃ to −1.5℃ (Table 2.13).

The peculiar physio–geographic and hydrochemical conditions of the occurrence of *Artemia* in the plateau saline lakes are of major significance for the further study of the strain, coenology and ecology of Artemia.

Based on the studies by Jiang Xizhi et al., the diagnostic features of *Artemia salina Linneans* are as follows (Fig. 2.5, 1 a–b):

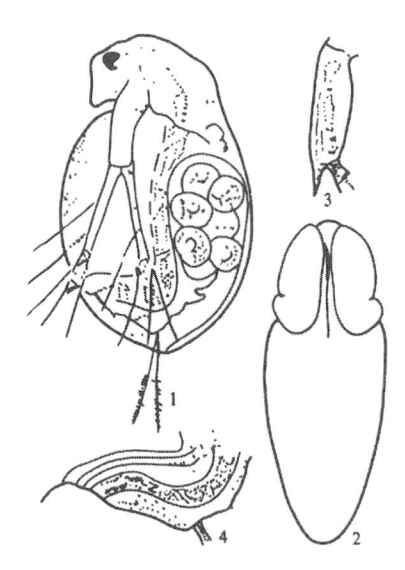

Figure 2.5. Daphniopsis tibetana Sars. 1–Whole of the female; 2–back view; 3–male, first antenna; 4–rear abdomen

Diagnosis of the male: Body thin and long (8.5–1.21 mm in length); abdomen longer than head plus chest and consisting of 8 segments, with the terminal segment being longest; first antenna thin and long; second antenna having a slightly bound coxa with inner margin as a small spheroidal haired excrescent prominance, and a flatter terminal segment with a sharp top; costal organ having a thick and strong coxa with a circular process on inner side, and a flat triangular terminal segment with a curved end; major jaw thick and strong with fine denticles on the masticating side; second minor jaw having clustered setae; chest and abdomen consisting 11 and 9 segments respectively, without abdominal appendage; caudal fork moveable and combined with last metamere, with each artus having 3–10 setae; body type slightly variable depending on salinity difference of environment of existence.

Diagnosis of the female: Body 7–10 mm long; second antenna unsegmented, showing a form of plough with a sharp–pointed end and having a half–round setous process on central inner margin and another two flatter processes near basal part; oecyst reversed pear–shaped, extending to third or fourth abdominal segment; adult carrying 0–40 eggs (17 eggs on average); caudal fork greatly variable, commonly with just a few, or even without, feather–like setae. The specimen incubated under the conditions of experimental culture has a more developed caudal fork but a relatively shorter abdomen than that in the field.

PART 3 DIVISION OF MINEROGENIC ZONES, PREDICTION OF RESOURCE POTENTIAL AND PROSPECTS OF DEVELOPMENT OF SALINE LAKES ON THE QINGHAI–TIBET PLATEAU

CHAPTER 11 DIVISION OF MINEROGENIC ZONES AND PREDICTION OF RESOURCE POTENTIAL[1][2][3]

Based on the afore-mentioned tectonic and geochemical analysis of the evolution of the lakes and their multi-order salt formation model in combination with the studies of their hydrochemical compositions, regionalization and minerogenic specialization of the lakes (Table 3.1)[4], and the estimation of geological reserves of mineral resources such as boron, lithium and potassium[5], a division of minerogenic zones of the saline lakes on the plateau has been made and the direction for the search of mineral resources of K, B, Li, etc preliminarily predicted (Fig. 3.1).

---

[1] Zheng Mianping et al., 1963. Preliminary opinions on the assessment, development and utilization of salt resources in northern Tibet. Material printed by Academia Sinica for internal reference.

[2] Zheng Mianping et al., 1983. Report on the researches on subjects related to saline lakes on the Qinghai–Tibet Plateau. Material for internal reference.

[3] First Geological Party of Qinghai Bureau of Geology, 1983. An overview of saline mineral resources in the Qaidam Basin. Material for internal reference.

[4] Table 3.1 is appended to the book.

[5] Zheng Mianping, Xiang Jun, et al., 1986. Mineral resources potentials and reserve estimation table of the saline lakes on the Qinghai–Tibet Plateau (Material for internal reference).

11.1. First—Class Minerogenic Zone of B, Li, K, Cs (Rb, Halite, Soda and Mirabilite)—bearing Carbonate—type Saline Lakes (A)

This minerogenic zone is located in the region of dispersion centres of such elements as B, Li and Cs. With the Gangdisê zone of igneous rocks (including the high—B and —Li andesite series) lying just in the central and northern parts, it embraces the entire Yarlung Zangbo and Bangong Co—Nujiang River Li, B, K, Cs and Rb—rich geothermal zones, and therefore has abundant material sources of special elements. Since it is possessed of numerous tectonic lake basins with closed and semi—arid or arid salt—forming conditions, which are hydrochemically of the carbonate type and have abundant sources of B, Li, K, Cs, Rb, etc., it has become the principal

TABLE 3.1. LiCl contents in the lake water of various minerogenic zones on the Qinghai—Tibet Plateau

Zone No.	A		B		C		D		E		Sum	
Salinity (g/l)	①	②	①	②	①	②	①	②	①	②	①	②
1–100	52	9	22	6	12	3	5	1	2	1	93	20
100–200	3	3	9	9	2	2	1	/	/	/	15	14
200–300	4	4	3	3	2	1	5	3	/	/	14	11
300–400	6	6	9	9	1	1	14	11	/	/	30	27
>400	2	2	/	/	/	/	1	1	/	/	3	3
Total	67	24	43	27	17	7	26	16	2	1	155	75
%	43	32	28	36	11	9	17	21	1	1	100	100

① Number of Li—bearing lakes; ② Number of lakes with LiCl ≥ 150 mg/l.

minerogenic zone of B, Li, K, Cs (Rb, Br, halite, soda and mirabilite)—bearing brine—borax (and K salt). Based on the statistics of 64 lakes in the zone, there are 24 lakes with LiCl contents equal to or greater than 150 mg/l, which account for 32% of all lakes with the same LiCl contents on the entire Qinghai—Tibet Plateau and of which 9 lakes have lakewater salinities of less than 100 g/l; according to an analysis of 81 lakes, there are 40 lakes with B_2O_3 contents equal to or greater than 400 mg/l, which make up 49% of such lakes on the entire plateau and of which 25 lakes have lakewater salinities of less than 100 g/l; in the light of an analysis of 83 lakes, there are 19 lakes with KCl ≥ 2500 mg/l, occupying 29% of all similar lakes on the whole plateau, and 10 of them have salinities less than 100 g/l in the lake water (Tables 3.2, 3.3 and 3.4). All the above indicates that the water in the lakes of this minerogenic zone is the highest in relative abundances of Li, B and K and the zone has more lakes with commerical grades of such elements.

TABLE 3.2. B_2O_3 contents in the lake water of various minerogenic zones on the Qinghai–Tibet Plateau

Zone No.	A ①	A ②	B ①	B ②	C ①	C ②	D ①	D ②	E ①	E ②	Sum ①	Sum ②
Salinity (g/l)												
0–100	66	25	27	13	8	2	2	/	3	1	106	41
100–200	2	2	9	6	2	1	/	/	/	/	13	9
200–300	4	4	3	2	/	/	3	1	/	/	10	7
300–400	6	6	13	9	/	/	14	6	/	/	33	21
>400	3	3	/	/	/	/	1	1	/	/	4	4
Total	81	40	52	30	10	3	20	8	3	1	166	82
%	49	49	31	37	6	4	12	10	2	1	100	100

① Number of B–bearing lakes; ② Number of lakes with $B_2O_3 > 400$ mg/l.

TABLE 3.3. KCl contents in the lake water of various minerogenic zones on the Qinghai–Tibet Plateau

Zone No.	A ①	A ②	B ①	B ②	C ①	C ②	D ①	D ②	E ①	E ②	Sum ①	Sum ②
Salinity (g/l)												
0–100	69	5	28	4	15	1	4		8		124	10
100–200	2	2	9	8	3	2					14	12
200–300	4	4	3	2	2	2	2	1			11	9
300–400	5	5	10	10	1	1	15	15			31	31
>400	3	3					1	1			4	4
Total	83	19	50	24	21	6	22	17	8		184	66
%	45	29	27	36	11	9	12	26	4		100	100

① Number of K–bearing lakes; ② Number of lakes with KCl ≥ 2500 mg/l.

The present minerogenic zone is further divided into two subzones which are bounded by the Gangdisê Mountains.

11.1.1. GANGBEI (NORTH OF THE GANGDISE) PRINCIPAL MINEROGENIC SUBZONE OF B, K, Li, Cs Br AND Rb—BEARING BRINE— BORAX (HALITE, SODA AND MIRABILITE) (A_1)

This subzone has large lake basins, which are dominated by saline lakes and brackish lakes. The saline lakes are the dispersion centres of special elements such as B and Li on the Qinghai—Tibet Plateau and have extremely abundant material sources; the subzone has comparatively good conditions of salt differentiation and the low—terrace lakes are enriched in salts bearing B, Li, K, etc, constituting saline lake areas with B, Li and K—bearing brines and Li—bearing minerals, borax and potash salt. There are two kinds of mineral resources: liquid and solid. The liquid mineral resources are brines rich in Li, B, K, Cs, (Rb and Br); the solid mineral resources include borax, soda, mirabilite, hydromagnesite and halite, and potential resources— Li—bearing magnesite, Li—bearing dolomite, zabuyelite, etc. In the light of mineralization conditions and mineral potentialities, the subzone is subdivided into the eastern Gangbei subordinate prospective subarea (A_{1-1}) and the western Gangbei most prospective subarea (A_{1-2}) (see Fig. 3.1).

A_{1-1} eastern Gangbei subordinate prospective subarea: In spite of the presence of more or less abundant material sources of B, Li, etc., this subarea is situated in a semi—humid to semi—arid region; with the contemporary aridity in the range of 5—8, it shows a lower degree of lakewater concentration except in its northwestern sector where the lakewater is more concentrated, thus giving rise to B, Li and K—bearing saline lakes of medium to small sizes such as the Pongyin Co and Pêli Co. However, these lakes have lower contents of Cs, Rb, Br and other elements in the lake water than those in the A_{1-2} subarea.

A_{1-2} western Gangbei most prospective subarea: This subarea is the most promising area of saline—lake mineral resources of B, Li, Cs, Rb and even Br, (W, U), etc. on the Qinghai—Tibet Plateau. This subarea lies just in the dispersion centre of special elements such as B, Li, Cs, U and Br at the northern foot of the Gangdisê—Nyainqêntanglha Mountains. In the subarea, the lake basins are large in size and have a longer history of evolution; a more or less arid climate prevailed at the late stage of the Late Pleistocene and the aridity of the contemporary climate is mostly in the range of 10—20; the low—terrace lakes are enriched with soluble salts bearing B, K, Li, etc. All this has created exceptionally advantageous conditions for the formation of special saline lakes and consequently speical saline lakes essentially of carbonate type such as the Chalaka Lake, Zabuye Caka, Dujiali Lake and Bangkog Co have been formed. Among these lakes, the Dujiali Lake and Bangkog Co are well—known localities of Borax occurrence and Li, B and K—bearing brine, while the Chalaka Lake is characterized by the high concentration of solid borax. Particularly, the Zabuye Caka is marked by high contents and complete varieties of useful elements and large size of mineral deposits, and therefore is highlighted below.

In 1982 and 1984, the author and their colleagues made two separate field investigations of the Zabuye Caka. During these investigations, an expanded drilling exploration in the South Zabuye Caka was carried out on the basis of the work done by previous scientists and a survey of the North Zabuye Caka initiated; hydrochemical and bacterial—algal geoecological observations as well as systematic lakewater sampling on a regular grid were made for the first time on rubber boats in the North Zabuye Caka, and drilling and pitting were undertaken at its southern lake shore. As a result, some new advances were achieved in the resource assessment of the lake through laboratory determinations by various means and an integrated analysis as follows:

1) It was for the first time identified that there occur large quantities of lithium salts and potash salt (sylvite and glaserite) in the South Zabuye Caka. The lithium salts are composed of zabuyelite (natural lithium carbonate) and Li—bearing dolomite. Because these two minerals are water—soluble lithium salts, they are more easily extractable than conventional lithium silicate ore, and a preliminary experimental technological test has yielded a fairly good result. For instance, the leaching rate of lithium may reach 74.84% when halite—type Li—bearing ore is immerged in a weakly acidic medium; a leaching rate of 95.48% is obtained by ore milling and cold water immersion[1]. According to the systematic analytical data from 7 drill holes in the South Zabuye Caka and 1 drill hole in the North Zabuye Caka, the Li_2CO_3 contents in the sediments from the sludge at the bottom of the saline lake upward mostly exceed the commerical grade for sedimentary Li—bearing mineral deposits[2]. Taking the CK3 section for example, the average contents of Li_2CO_3, from below upward, are 0.948% for the saliferous sludge layer, 1.624% for the carbonate clay, 0.43% for the mirabilaite lamina, and 0.90% for the halite layer. A detailed estimation indicates a considerable quantity of Li_2CO_3 reserves in the South and North Zabuye Caka. The KCl contents in the potash deposits above the saliferous sludge layer in the South Zabuye Caka reach the standards required for comprehensive utilization, i.e. 3—6% generally and 20% or more in maximum, and locally are of economic grades for separate exploitation. The total reserves of solid KCl in the South and North Zabuye Caka are estimated at tens of million tons.

---

[1] Sichuan Central Laboratory and Saline Lake Research Group of the Institute of Mineral Deposits, 1985. Report on the experimental study on comprehensive utilization of the intercrystalline brine and salts in the Zabuye Caka, Tibet.

[2] According to the Reference Handbook of Industrial Standards of Minerals (1972) published by the Geological Publishing House, the marginal grade of Li_2O is 0.05% (equivalent to 0.1236% Li_2CO_3 and the block average grade (commercial grade) of Li_2O is 0.08% (equivalent to 0.1978% Li_2CO_3) for a certain sedimentary type Li and Ga—bearing bauxite deposit in southwestern China.

Fig. 3.1

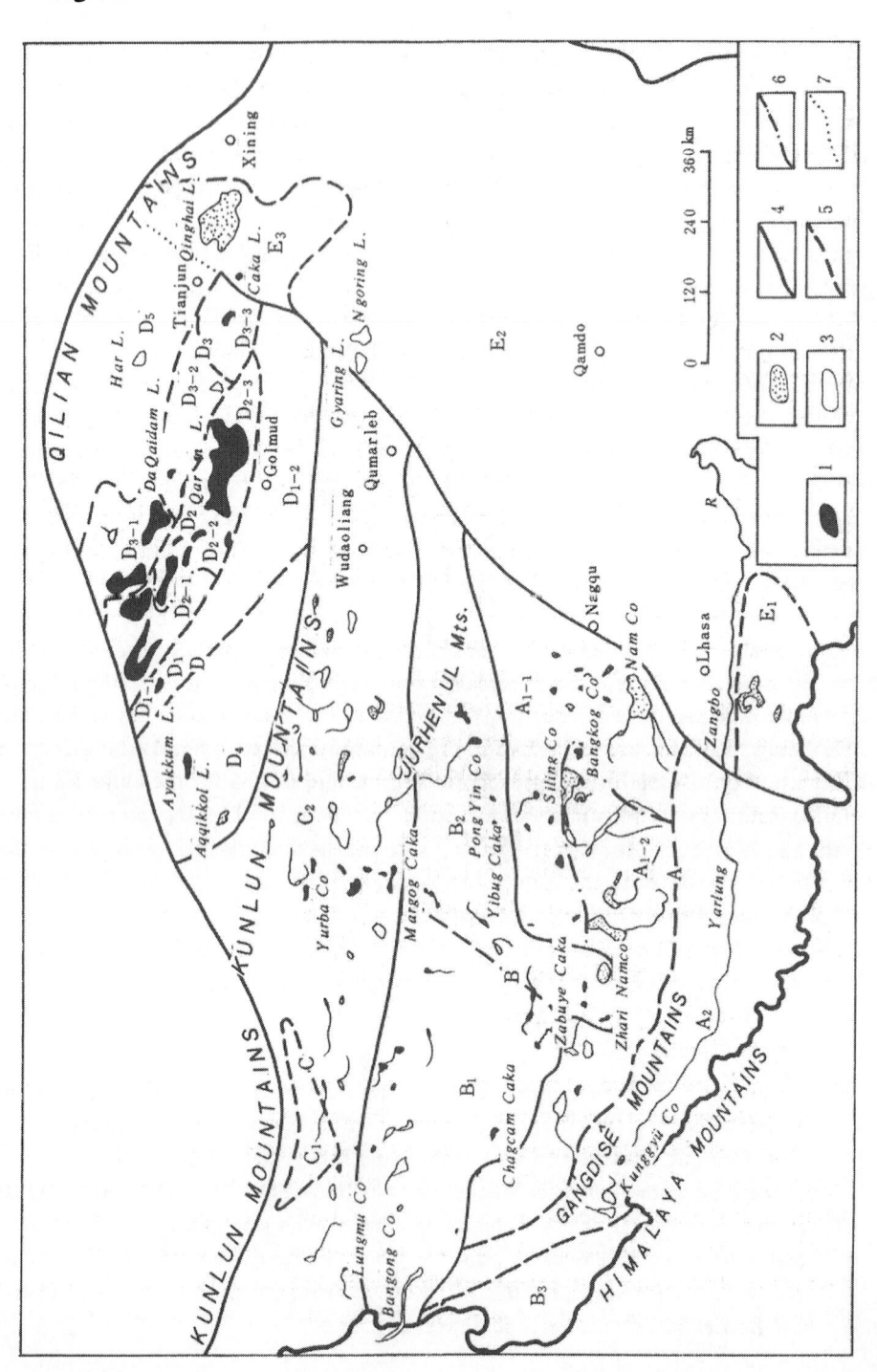

Figure 3.1. Prognostic sketch of minerogenic zones of B, Li and K-bearing saline lakes on the Qinghai–Tibet Plateau. 1–Saline lake; 2–Brackish lake; 3–Fresh-water lake; 4–Boundary of minerogenic zone; 5–Boundary of minerogenic subzone; 6–Boundary of minerogenic area; 7–Inferred boundary; A–First-class minerogenic zone of B, Li, K, Cs, (Rb, soda, halite and mirabilite)–bearing saline lakes of carbonate type: A_1–Gangbei (North of the Gangdise Mountains) subzone; A_{1-1}–eastern Gangbei subordinate prospective subarea; A_{1-2}–western Gangbei most prospective subarea; A_2–Gangnan (South of the Gangdisê Mountains) subzone; B–first-class minerogenic zone of B, Li, K, (Cs, Rb, halite and mirabilite)–bearing saline lakes of sodium (magnesium) sulfate type: B_1–West Qiangtang subzone; B_2–Shuanghu subzone; C–second-class minerogenic zone of Li, B, K, halite and mirabilite–bearing saline lakes of magnesium sulfate subtype: C_1–Shuoer Lake–Dulishi Lake subzone; C_2–Hoh Xil subzone; D–first-class minerogenic zone of K, Mg (B and Li)–bearing saline lakes of sulfate-chloride type: D_1–South Qaidam subzone; D_{1-1}–Gas Hure prospective subarea; D_{1-2}–marginal South Qaidam subordinate prospective subarea; D_2–Central Qaidam subzone; D_{2-1}–Dalangtang prospective subarea; D_{2-2}–Chahansilatu prospective subarea D_{2-3}–Yiliping prospective subarea; D_{2-4}–Qarhan prospective subarea; D_3–North Qaidam subzone; D_{3-1}–Kunteyi–Mahai prospective subarea; D_{3-2}–Da Qaidam–Xiao Qaidam prospective subarea; D_{3-3}–Delingha–Hoh prospective subarea; D_4–Kumkol inferred minerogenic subzone; D_5–Qilian area poor in saline lakes; E–third-class minerogenic area of halite and mirabilite–bearing saline lakes of sodium sulfate subtype: E_1–South Tibet brackish lake subarea; E_2–Sanjiang–Huanghe outflow lake subarea; E_3–Qinghai Lake–Gonghe halite and mirabilite prospective subarea.

2) It was for the first time discovered that the brine in the South Zabuye Caka has high contents of Cs and Rb and that in the North Zabuye Caka is rich in Li, Cs and Rb. Particularly, the average LiCl content in the North Zabuye Caka is as high as 11284 mg / l, which is higher than the Li contents in all other known saline lakes in China and only lower than that in Salar de Atacama of Chili which is the highest in Li content in the world. The lake water of the Zabuye Caka is low in Ca and Mg and therefore is technology in favour of Li extraction, as its Li / Mg ratios range from 57.87 to 326.57, which are much higher than those of Salar de Atacama (0.1627), Silver Peak (1.0), the Da Qaidam Lake (0.0896), the Yiliping Playa (0.023), etc. (Table 3.4). The water in this lake has a high CO_3^{2-} content, which hampers the further concentration of Li and this poses a subject to be tackled in the course of its future exploitation and utilization. In the meanwhile, the Cs content of the intercrystalline brine in the Zabuye Caka is as high as 47 mg / l, which is the highest among all saline lakes in the world. According to some estimates, both the Cs and Rb reserves in the brine of the Zabuye Caka are quite considerable.

3) A systematic analytical study of the solid and liquid phase components in the South and North Zabuye Caka and a comprehensive estimation of the reserves of useful constituents in them indicate that the Zabuye Caka is the largest of all known borax deposits in China. Moreover, the Zabuye Caka is also abundant in bromine. The Br content of the intercrystalline brine in the South Zabuye Caka exceeds the commerical grade and that of the surface brine at the South and North

Zabuye Caka is greater than the prospective grade (see Table 3.4). Besides, in this lake area there still occur large amounts of halite, trona, mirabilite and other common salt minerals, and the brine has a high content of tungstic acid, which can be exploited as a by-product. At the same time, as a natural producer of halophilic algae, the North Zabuye Caka is perspective for the extraction of food resources such as β-carotin.

Hence it can be seen that the Zabuye Caka of the A_{1-2} subzone represents a special saline lake with a multiple variety of minerals and high grades of useful components as well as a treasure house of peculiar biotic resources and therefore is of very high economic value.

11.1.2. GANGNAN (SOUTH OF THE GANGDISE) SUBORDINATE MINEROGENIC SUBZONE OF BORAX-B AND LI BEARING BRINE(A_2)

The lake basins here are large in area, covering scores to hundrends of square kilometres, with 3 / 5 of them occupied by lakes. The lake basins are poorly closed. A part of the ancient lakes have disappeared as a result of outflowing; some others have turned from exorheic into endorheic ones only recently, with short duration of accumulation of saline components. Therefore, the lake basins are dominated by fresh-water and brackish lakes, with saline lakes developed only in minor amounts. The Malashan Lake on the northern slope of the Xixiabangma Peak, for instance, has a content of dissolved solids of 116.68 g / l. The present subzone is also located in the dispersion centre of special elements such as B and Li and is passed through by the Yarlung Zangbo high-B, Li and Cs geothermal zone (I). Hence, it has an abundance of B, Li, etc. The lake water has high relative contents of B and Li, which may reach the economic grade even in brackish lakes. The Kunggyu Co in the western part has a salinity of 5.4 g / l, but its B_2O_3 and LiCl contents are as high as 1010 mg / l and 229 mg / l respectively. Consequently, deposits of borax, ulexite, etc. are liable to form in marginal mud-flat environments of such lakes or their derived small lakes and oxbow lakes, at the time of evaporation of these B-bearing water bodies. This region is prospective for small or medium-sized borax (ulexite) and B-Li bearing brine deposits.

11.2. First-Class Minerogenic Zone of B, Li, K (Cs, Rb, Halite and Mirabilite)-Yielding Saline Lakes of Sodium (Magnesium) Sulfate Type (B)

This zone lies on the north side of a B, Li and Cs dispersion centre as well as in the west sector of the Bangkog-Nujiang geothermal zone. It has an abundant source of special elements too. As the zone is possessed of numerous tectonic basins with favourable closed and arid conditions for salt generation, multi-step saline-lake subbasins have been formed. The present minerogenic zone is roughly coincident with the sodium sulfate sub-type transitional zone (Fig. 1.14). The saline lakes in the zone, which are dominantly of sodium sulfate subtype and occasionally of magnesium sulfate subtype and carbonate type, are mostly saline and brackish ones, which have relatively abundant sources of materials such as B, Li and K and good conditions for salt formation. According to the statistics of 43 lakes, 27 lakes show LiCl contents equal to or greater than the marginal grade, making up 36% of the total of such lakes on the entire plateau, and among them 6 lakes have lakewater saliniteis less than 106 g / l; according to the statistics of 52 lakes, 30 lakes have B_2O_3 contents equal to and greater than the marginal grade, making up 37% of the total of such lakes on the entire plateau, and 13 of them have lakewater salinities less than 100 g / l; an analysis of 50 lakes indicates that 24 lakes have KCl contents equal to or greater than 2500 mg / l, accounting for 36% of the total of such lakes on the entire plateau, and 4 of them show lakewater salinities less than 100 g / l (see Tables 3.1, 3.2 and 3.3). All the above shows that the present zone has the highest proportion of lakes with the marginal grades of Li and K in lake water as well as more or less high B, Li and K abundances of lake water. The zone can be divided into two subzones (B_1 and B_2). The B_1 subzone lies close to the West Qiangtang area in the northern part of the A_1 subzone and is abundant in source materials. The lake basins in the subzone are large in size and have a relatively long history of evolution and duration of closed flow as well as good conditions for the differentiation of salts. Consequently, the B_1 subzone has become the principal subzone of the present minerogenic zone. The two subzones are described respectively as below.

West Qiangtang Principal Minerogenic Subzone of B, Li, K, (Cs and Rb)-bearing Brine and Sodium, Calcium and Magnesium Borates (Halite and Mirabilite) (B_1)
The borates consist mainly of sodium borate, calcium borate and magnesium borate, and the boron ores are made up chiefly of kurnakovite and pinnoite and subordinately of ulexite, inderite, etc. High-grade ores are common. Other ores include those of halite and mirabilite and brine rich in B, Li, K, (Cs and Rb). Lakes of this subzone may be represented by the Chagcam Caka, which has relatively high contents of K, B and Li (Table 3.4). What is worth noticing is that the Rb content in the water of that lake reaches 3.1 mg / l①. The present subzone is a prospective saline-lake area most promising for sodium-calcium and magnesium borates and B, Li, K (Cs and Rb)-rich brine. This is preliminarily verified by the successive discoveries of the Chagcam Caka and Jibu Caka deposits of magnesium borates which were mistaken by Sven Hedin for gypseous deposits.

---

① From the viewpoint of saline-lake hydrogeochemical prospecting, due attention should be given to the possible presence of Ta-Nb mineral deposits in the lake area.

TABLE 3.4. Brine components of saline lakes (Zabuye Caka, etc.) on the Qinghai–Tibet Plateau and abroad (unit: mg/l)

Saline lake	Compon.	B_2O_3	*	KCl	*	LiCl	*	Cs	*	Rb
Zabuye	S.Z.	8438.98[17]①	1.0	25376[17]	1.0	6627[17]	1.7	15.26[12]	0.8	34.3[12]
		11569.14[32]	0.7	38559[32]	0.7	10528[32]	1.1	21.24[28]	0.6	50.9[28]
	N.Z.	8136.10[50]		26081[50]		11284[50]		12.15[49]		32.82[49]
Da Qaidam		2580	3.2	9743.24	2.7	916.38	12.3	0.22	55.2	0.51
Yiliping		1040	7.8	22365.59	1.2	3836.58	2.9	0.13	93.5	8.7
Qarhan		223.5	36.4	23261	1.1	473.46	23.8			14.7
Dalangtan		254.3	32.0	8770.82	3.0	397.1	28.4			
Hoh		167.6	48.5	8408.55	3.1	290.19	38.9			9.1
Ayakkum		800	10.2	1785.62	14.6	122.18	92.4	3.20	3.8	1.70
Aqqikkol		852.8	9.5	7515.89	3.5	439.86	25.7	1.92	6.3	1.60
Yurba				10258.01	2.5	1016.57	11.1			
Yupan		1009	1.9	10735	7.3	1985	29.6			
Yaggai		2759.9	2.9	24501.1	1.1	640.06	17.6			
Kongkong		948.71	8.6	10677.52	2.4	2020.13	5.6			
Chagnag		1935	4.2	65176.73	0.4	1832.76	6.2			
Chagcam (II)		2088	3.9	26369.66	1.0	3525	3.2			
Dujiali		6052.8	1.3	21528.3	1.2	3288.9	3.4			
Pongyin		3054.7	2.7	44797.92	0.6	1004.35	11.2			
Jianhaizi		1174.49	6.9	8325.6	3.1	1460.1	7.7			
Searles L.		3500	2.3	43854	0.6	489–200	23.1–56.4			
Silver Peak				15253.6	1.7	2442	4.6			
Dead Sea				12012.21–13919	2.2–1.9	110	102.6			60
Salar de Atacama		2478	3.3	4499.12	5.8	11500–13000	0.98–0.868			
Uyuni				3432.1–45761	7.6–0.6	489–9220	23.1–1.2			

* $\dfrac{\text{Water of N. Zabuye Caka}}{\text{Water of other lake}}$; The figures in brackets show average of the samples.

(Table 3.4. continued)

Saline lake	Compon.	*	Br	*	WO$_3$	*	U	Salinity (g/l)	Li/Mg	Rem.
Zabuye	S.Z.	1.0	242.03[17]	1.0	0.426[2]	1.9	2.952[5]	368.99[17]	1085.4 / 20.53	
		0.6	512.45[32]	0.5	1.713[5]	0.5	6.88[4]	439.82[32]	1724.3 / 5.28	①
	N.Z.		246.81[50]		0.792[3]		2.11[2]	393.50[50]	1848.1 / 23.8	
Da Qaidam		64.4	90.31	2.7				340.57	150 / 15670	
Yiliping		3.8						333.97	628 / 24440	①
Qarhan		2.2	12.5	19.8				323.104	77.5 / 36650	①
Dalangtan			44.8	5.5				382.43	65 / 84227.4	①
Hoh		3.6						335.66	47.5 / 27230	
Ayakkum		19.3						145.90	20 / 7781	
Aqqikkol		20.5						78.193	70 / 3210	
Yurba								314.03	166.4 / 4925.7	
Yupan								143.72	62.4 / 7232.6	
Yaggai								324.76	104.8 / 14030	①
Kongkong								350.61	330.7 / 37119	
Chagnag								349.87	300 / 20095	
Chagcam (II)								305.12	577 / 10190	①
Dujiali								220.50	538.4 / 198	
Pongyin								340.54	164.4 / 36.5	
Jianhaizi								98.35	239 / 34.38	
Searles L.			860 (U. bed)	0.3	40 / 70		0.02 / 0.01	342	80	
Silver Peak									400 / 400	
Dead Sea		0.5	5290		0.04				18 / 34000	
Salar de Atacama									1880 / 9650	
Uyuni									80 − 1510 / 1500 − 24000	

* $\dfrac{\text{Water of N. Zabuye Caka}}{\text{Water of other lake}}$; ① = Intercryst. brine.

Shuanghu Subordinate Minerogenic Subzone of K, B *and* Li—*bearing Brine and* Na, Ca *and* Mg *Borates* (B_2)   This minerogenic subzone, which is bounded by the Jangngai Zangbo fault on the west, lies in the Shuanghu—Zhari Namco zone in the neiborhood of the B_1 subzone. The most part of the subzone belongs to a semi—arid region, with a comparatively long duration of closed flow. The saline lakes make up a greater percentage of the lakes of all types in the subzone than in the A zone. According to the statistics of 36 lakes investigated, there are 20 saline lakes (55.5%), 14 brackish lakes (39%) and 2 fresh—water lakes (5.5%). The lake basins in the subzone are smaller than those in the A_1 subzone, and only individual lakes may reach around 500 km^2 in area. However, this subzone is also close to the dispersion centre of special elements (B, Li, etc.) and is passed through by B—Li—bearing geothermal zone III, and so the lake water is higher in B and Li and the subzone is particularly characterised by the higher relative content of Li. The B, Li and K contents in the Kongkong Caka and Yaggai Co, for instance, all reach the commercial grades (see Table 3.4); in some saline lakes, e.g. the Yibug Caka, it is only the Li content of the lake water that reaches the commerical grade (LiCl 263 mg/l, B_2O_3 219 mg/l, and KCl 2219 mg/l). In brackish lakes such as the Zhari Namco, both the Li and B contents reach the commerical indices. Hydrochemically, the saline lakes in this subzone are mostly of the sodium sulfate subtype, subordinately of the magnesium sulfate subtype, and locally of the carbonate type. It is inferred from the chemical characteristics of the lakes that it is worth while to look for Chagcam Caka type mineral deposits of Na—Ca and Mg borates on terraces I and in the lakes of the subzone and that the subzone may be locally prospective for borax deposits.

11.3. Second—Class Minerogenic Zone of Li, B, K, Halite and Mirabilite—Bearing Saline Lakes of Magnesium Sulfate Subtype (C)

This minerogenic zone is comparatively poorly studies. It roughly coincides in extent with the afore—mentioned Hoh Xil inflow lake district. The principal part of the zone lies in an arid region but is recharged by the snow—melt water from the Kunlun and Hoh Xil mountain systems. The lakes there have had a long duration of closing except for those at the eastern end of the zone, and so the lake water shows a relatively high salinity and brackish and saline lakes are widespread. According to statistics, saline lakes and brackish lakes make up about 32% and 57% of all lakes in the zone, respectively (see Table 3.1). The lakes are generally similar in size with those in the B zone, and the lakes ranging from scores to hundreds of square kilometres in dimension account for less than one third of the total amount of lakes in the zone. As the zone lies in an arid region and is possessed of numerous tectonic basins, it is generally arid and closed except in local areas which are recharged by Cenozoic volcanic and geothermal B and Li-bearing springs. However, in this zone subbasins are not developed, salts are poorly differentiated, and material sources of special elements such as B, Li, K, Cs and Rb are less abundant than in the A and B zones. Solid minerals consist of halite, mirabilite, epsomite, etc. The B, Li and K contents in the brine remain to be relatively high, with the B and Li or K and Li contents reaching the commercial grades in most cases and the B, Li and K contents reaching simultaneously the marginal grades in fewer cases. According to the statistics of 17 lakes investigated in the zone, 7 lakes show LiCl contents greater than 150 mg / 1, making up 9% of the lakes whose LiCl contents reach the marginal grade on the entire plateau, and 3 of them have lakewater salinities less than 100 g / l; according to the statistics of 10 lakes, 3 lakes show B_2O_3 contents greater than the marginal grade, making up 4% of all such lakes on the entire plateau, and 2 of them show lakewater salinities less than 100 g / l; according to the statistics of 21 lakes, 6 lakes have KCl contents greater than 2500 mg / l, making up 9% of all such lakes on the entire plateau, and one of them has a lakewater salinity less than 100 g / l (see Tables 3.1, 3.2 and 3.3). The above statistics indicate that this zone ranks the fourth on the plateau in terms of the lakes with marginal grades of B, Li and K, while it occupies the third position in their relative abundances. The Yupan Lake and Margai Caka, for instance, are both saline lakes with commercial or prospective grades of B and Li; the Yurba Co and Lungmu Co are saline lakes with commercial or marginal grades of Li and K (see Table 3.4). In the western part of the zone, local carbonate saline lakes have been formed probably duu to the southward compression of the Tarim Basin, which caused the outward escape, along faults, of B, Li and K-bearing materials and CO_2 in the low-land area between the northern edge of the Kunlun Mountains and the Karatax Mountains. In this area B and K-rich saline lakes have been discovered. In the Shuoer Lake, for instance, the brine contains 0.5–3.0% KCl and sediments of borax, mirabilite and halite have also been observed. Hence, this area is classified tentatively as the Shuoer Lake-Dulishi Lake subordinate minerogenic subzone of B, Li, K and borax (C_1). Other areas are temporarily referred to as the Hoh Xil subordinate minerogenic subzone of Li, B, K and mirabilite (C_2). It should be also pointed out that in the western part of the present minerogenic zone there occur playas of relatively large sizes, such as the playa of the Kushui Lake which covers an area of about 125 km², with a salt bed of mirabilite and halite more than 3 m in thickness. According to Norin's (1946) data, the Kushui Lake should be assigned to the sulfate type. A minor amount of pinnoite has been found in the mirabilite-halite bed. It is presumed that there is the possibility of finding Chagcam Caka type Na-Ca and Mg borate deposits and B, K and Li-bearing brines of commercial value in the playa.

11.4. First-Class Minerogenic Zone of K, Mg, (B and Li)-Bearing Saline Lakes of Sulfate-Chloride Type

This minerogenic zone consists of intermountain fault–block basins controlled by major faults, including the Qaidam and Kumkol Basins. It is a zone of large–scale and relatively strong subsidence resulting from tectonic movements of the Qinghai–Tibet Plateau, with structural–geomorphological conditions of deep basins surrounded by high mountains and sedimentary environments of multi–step subbasins. Barred by the mountain systems on the east and south, this zone has been of difficult access for warm and damp atmospheric currents. Since the Oligocene, the climate in this zone has time and again become dry. The zone is the earliest closed Cenozoic lake district at a comparatively lower altitude on the Qinghai–Tibet Plateau. As a result of the accumulation of salt components in lake basins over a long period of time, the district has become the largest saline lake zone with the longest duration of differentiation and accumulation of salts on the plateau. With the Oligocene–Pliocene Epoch excluded, the Quaternary Period witnessed, in general terms, two salt–forming stages: Q_1–Q_3^1 and Q_3^3–Q_4 (31 000 a to the present). According to the statistics of 26 lakes in the zone, there are 16 lakes with LiCl contents greater than the marginal grade (including 1 lake with the lakewater salinity less than 100 g / l), making up 21% of the total of such lakes on the plateau; according to the statistics of 20 lakes, there are 8 lakes with B_2O_3 contents greater than the marginal grade, making up 10% of all such lakes, and all of them show salinities greater than 100 g / l; according to the statistics of 22 lakes analysed, there are 17 lakes with KCl$\geq$2500 mg / l, accounting for about 26% of all such lakes on the plateau, and all these lakes have lake water salinities exceeding 100 g / l (see Tables 3.1, 3.2 and 3.3). Based on the above statistics, this zone ranks the third on the plateau in terms of the percentages of lakes with marginal grades of B, Li and K but occupies a middle position on the plateau in terms of the ralative abundances of these elements. In the light of the differences in mineralization conditions, this zone is preliminarily divided into 5 subzones (or prospective areas) and 1 area poor in saline lakes.

South Qaidam Subordinate Minerogenic Subzone of K, Li, (B), Halite and Soda (D_1) This subzone extends along the "Kunbei terrace" structural belt on the southern margin of the Qaidam Basin①. The basement of this zone is essentially a rigid terrain, which generally appears on the surface, as an inclined slope gently dipping toward the centre of the Qaidam Basin. The zone, bounded by the nearly E–W No. XIII fault in the eastern part of the Gas Hure Lake, is subdivided into the Gas Hure prospective subarea of K, Li, (B) and halite (D_{1-1}) and South Qaidam marginal mud–flat subordinate prospective subarea of soda (and B) (D_{1-2}). The

---

① Yang Shaoqing, 1985. Basic characteristics of the oil–bearing geological structures. Qinghai Petroleum Geology.

D_{1-1} subarea is located in the area of the over 100 km^2 Gas Hure Lake hydrochemically of magnesium sulfate subtype, where enormous halite deposits and minor carnalite deposits have been found in addition to brines with commercial grades of K and Li and the marginal grade of B. The Gas Hure Lake is a fault–depressional basin with a deeply block–faulted basement. Petroleum boreholes have revealed that there occur a number of Pleistocene–Tertiary salt beds in this area and the lake has become a semi–playa, thus suggesting that this subarea is one of the areas with the longest salt–forming stage in the Qaidam Basin and is prospective for K and Li–bearing brines. The D_{1-2} subordinate prospective subarea lies to the southeast of the D_{1-1} subarea. Extending along the Narin Gol and Urt Moron Rivers at the foot of the Kunlun Mountains, this subarea receives the B–Li bearing freshwater recharge from the Cenozoic volcanic–hot water zone of the Kunlun Mountains, resulting in the formation of the Xarag borax–bearing salt–pit zone along these rivers, which is prospective, to some extent, for boron. In this subarea and in the middle reaches of the Qaidam River in the east, as a result of the recharge of leaching water from the Paleozoic volcanic rock series, trona and soil soda deposits have been found to occur at Xiaoqiao, Zongjia, Barun, etc. along both banks of the Qaidam River. These deposits are products formed in the marginal salt–mud flat subenvironment of saline lakes of carbonate type but they are small in size and have no significant economic value.

Central Qaidam Principal Minerogenic Subzone of K, Mg, Halite, (Li and B) (D_2)
This subzone is located in the central part of the Qaidam Basin. It is the main part of the Neogene–Quaternary palaeo–salt basin and tectonically belongs to the central depressional zone. Hence, it has a long history of salt formation and evolution. According to the differences in the mineralization characteristics, it is subdivided into 4 subareas (see Fig. 3.1), namely the Dalangtan prospective subarea of K, Mg, halite and mirabilite (D_{2-1}), Chahansilatu prospective subarea of halite and mirabilite (D_{2-2}), Yiliping prospective subarea of K, Mg, Li and B (D_{2-3}) and Qarhan most prospective subarea of K, Mg and halite (D_{2-4}).

(1) D_{2-1} subarea: This subarea lies in the Tertiary–Quaternary fold zone in the northwestern part of the basin. A series of playas such as the Dalangtan Playa are distributed in some of its NNW–trending synclinal depressions. Starting from the initial Early Pleistocene, the Quaternary salt–forming epoch of this subarea continued intermittently into the terminal Pleistocene, at about 10 000 a B.P. In the Dalangtan Playa, for instance, the Quaternary salt beds total 242 m or more in thickness (see Table 1.4). Q_1–Q_3 consist of halite beds occasionally intercalated with mirabilite beds and are abundant in K–bearing intercrystalline brine of sodium sulfate subtype. At the late sage of the salt–forming epoch (Q_3^2), the saline lake at Dalangtan turned to be of the magnesium sulfate subtye and large quantities of carnalite, sylvite, schoenite and bischofite were deposited. The saline lakes in the subarea were formed under the control of the Altun left–lateral major fault and

WNW basement structere. Under the compression from opposing directions, the synclinal depression swarm tended to contract and the brine in the saline lakes migrated towards the northeast or southeast. At the late stage of the Quaternary, this subarea was gently uplifted and the overwhelming majority of the saline lakes tended to extinguish, giving rise to playas and under—sand lakes. Therefore, future attention should be given to the search for concealed Quaternary saline lakes in the northeastern and southeastern parts of this subarea.

(2) D_{2-2} subarea: This subarea lies to the northeast of the D_{2-1} subarea. It extends in WNW direction in the western part of the Cha—Huo fault block in the slope zone of the central depression, belonging to the same WNW—trending fault block as the D_{2-4} subarea to be described does. In this subarea there occurs the Chahansilatu Playa, whose salt—forming epoch is approximately the Pliocene—Early Quaternary. As a result of the tilting of the fault block from the west, the salt basin inclined to the southeast, possibly leading to the southeastward migration of the K—Mg—bearing brine. Therefore, the Chahansilatu salt basin is primarily dominated by halite and mirabilite and is less prospective for K—Mg—bearing brine.

(3) D_{2-3} subarea: Lying to the east of the D_{2-1} subarea, the D_{2-3} subarea extends in a WNW direction in the middle of the central depression zone and is limited on the east by the N—S concealed fault to the south of the Eboliang anticlinal zone. There occur enormously thick Quaternary sediments in the Yiliping area, which used to be the centre of the Qaidam Early Quaternary palaeo—lake basin. Known in the subarea are the Yiliping Playa, Dong Taijnar and Xi Taijnar semi—playas, etc., all being of magnesium sulfate subtype. Salts in these lakes began to form at 28 000 a B.P. This saline lake subarea developed inheritantly from the termial Pleistocene ancient saline lakes in the D_{2-1}. Because of the subsequent inclination of the surface of the Chahansilatu salt basin of the D_{2-2} subarea toward this subarea and the replenishment of B, Li and K—bearing meteoric water from the Hoh Xil Cenozoic volcanic terrane on the south (through the Narin Gol River)[1], the present subarea shows much increased K, Mg, B and Li contents as compared with the D_{2-1} subarea, and the Dong Taijnar and Xi Taijnar Lakes, which are abundant in sylvite deposits, have become saline lakes with the highest contents of B and Li in brine in the Qaidam Basin.

(4) D_{2-4} subarea: Extending also in WNW direction, this subarea lies to the east of the D_{2-2} subarea and to the west of the Xiao Qaidam—Hulsan nearly N—S concealed fault and belongs to the lowest and largest sag area of the central depression zone. It is a Late Quaternary saline lake area developed slightly later than the D_{2-3} subarea as well as the area of termination of the prolonged evolution of the Cenozoic ancient saline lake basin in the Qaidam Basin, and thus has become the

---

[1] Based on the investigations carried out by the Qinghai No. 1 Geological and Hydrogeological Party and Wu Bihao of the Institute of Mineral Deposits, CAGS.

concentration centre of salt—forming elements such as K and Mg that have the strongest ability of geochemical migration① and the largest area of saline lakes on the Qinghai—Tibet Plateau. There are scattered, from west to east, a number of saline lakes such as the Suli Lake, Da Biele Lake, Xiao Biele Lake, Dabsan Lake, Qarhan Playa, Tuanjie Lake, Xiezuo Lake, Nan Hulsan Lake, Bei Hulsan Lake and Dongling Lake. In fact, these saline lakes (excluding the Dongling Lake) used to be one unified lake. The whole area was dried up to become a unified playa, the Qarhan Playa which covers an area of about 5900 km^2, approximately at the early stage of the Holocene. Later on, a series of marginal saline lakes mentioned above emerged as a result of abundant recharge of fresh water mainly from the rivers in the south. At present, the saline lakes in the northern and eastern parts of the Qarhan Playa are of the chloride type, while those in the southwestern part are of the magnesium sulfate subtype. This fact hydrochemically justifies the afore—mentioned evolution process of the Qarhan Playa. This subarea is the principal zone of K—Mg salts and halite. It is the main mineralization area of K—Mg—bearing saline lakes on the Qinghai—Tibet Plateau. In this subarea there also occur mineral resources of Li, B and Sr which are of significance for comprehensive utilization as by—products. Because the evolution of the saline lakes in this subarea is characterized by their contraction from WNW to ESE as well as from SSW to NNE, there occur, too, abundant resources of low—grade K and Mg—bearing brines under the S_4 salt—bearing formation in the peripheral parts, particularly the southern and western parts, of the Qarhan Playa. Should attention be paid to the studies on mining geology and comprehensive utilization (focussing on the lowering of the workable grade of K), apart from the exploitation of the K and Mg bearing brine resources in the S_4 bed in salt flat areas, the prospects of salt resources in the Qarhan Playa will most probably be expanded in the future.

North Qaidam K—Mg *and* B—Li—K *Minerogenic Subzone* (D_3) This subzone is located in the northern block—faulted zone of the Qaidam Basin (Yang Shaoqing, 1985), with a rigid—flexible mixed basement essentially of Paleozoic rocks. During the Mesozoic, relatively active block—faulting occurred, leading to the deposition of Middle—Lower Jurassic sediments. In the Cenozoic period the tectonic activities remained to be characterized by block—faulting, which played an important role in controlling the salt—forming process in the Quaternary lake basin. Based on the mineralization characteristics of the saline lakes, this subzone is further divided into three subareas: Kunteyi—Mahai prospective subarea of K, Mg, (Li, B) and halite (D_{3-1}), Da Qaidam—Xiao Qaidam subarea of B—Li—K bearing brines and N, Ca and Mg borates (D_{3-2}) and Delingha—Hoh halite subarea (D_{3-3}).

---

① Mg is also an element with the strongest ability of migration to lower position when the water is poor in the carbonate (bicarbonate) root.

(1) D_{3-1} subarea: Lying to the northeast of the D_{2-1} subarea, it is roughly coincident with the Sai-Kun downfaulted zone and the middle and western portions of the Mahai-Dahonggou uplift classified by Yang Shaoqing. Though similar to the D_2 subzone as a Cenozoic ancient saline lake basin, this subarea has a basement of different character, which was downfaulted to a greater depth, and it is seperated from the D_{2-2} subarea by the Eboliang No. 2 basement fault. In this subarea, there are the NW and WNW-trending Kunteyi Playa and Mahai Lake area (Mahai Playa and Dezong Mahai Lake), and the extent of the saline lake basins is strictly delimited by NW and NNW-trending faults. The Quaternary salt-forming epoch of this subarea continued intermittently from the initial Early Pleistocen into the Late Pleistocene but the Quaternary salt beds here are less thick than those in the Dalangtan Playa. In the Kunteyi Playa, for instance, the Quaternary salt beds have an aggregate thickness of about 70-80 m, but in the Upper Pleistocene sediments above the borehole depth of 50 m in its upper part the individual halite beds are relatively thick and there occur K-Mg and Li-bearing brines of commercial grade. The Kunteyi Playa and Balun Mahai Lake are both of chloride type, suggesting higher degrees of differentiation and metamorphism of their brines, and therefore are prospective for the search of K and Mg salts. In comparison with the Kunteyi Playa,phowever, the Mahai Lake area has a fairly stable basement and is located in an uplifted position, and so is possibly less prospective for K-Mg salts. Besides, in the mud-flat in the eastern part of the Mahai Lake area there also occur salt-pit type borax and ulexite mineralizations. Although they are known to be small in distribution extent and are of insignificant economic value, these occurrences should be traced with a view to further prospecting for boron deposits.

(2) D_{3-2} subarea: Lying to the north of the D_{2-3} subarea, this subarea extends in NNW direction approximately in conformity with the eastern Mahai-Dahonggou uplift zone and the Da Qaidam-Hongshan depression zone. This minerogenic subarea, where the Da Qaidam aand Xiao Qaidam saline lakes are located, is made up of intermountain basin type sediments and is confined by pre-Sinian "old mountains" on the south and north. The Tertiary and Quaternary sediments in this subarea are relatively thin and remarkably differ from those in the afore-mentioned salt basins. Hydrochemically the lakes are mainly of the magnesium sulfate type (in modern times) or sodium sulfate subtype (at the early stage of the salt-forming epoch). The principal salt-forming epoch is the terminal Late Pleistocene-Holocene (21 000 a to the present). Owing to the recharge of the B and Li bearing geothermal water from the South Qilian fault belt on the northern edge and the river water of the Tatalin Gol over a long period of time (about 24 420 a), there have formed sizable B and Li-bearing brine and ulexite-magnesium borate deposits. The saline lakes in the subarea are also characterized by their migration from ENE to ESE and from SSW to NNE. For instance, the depocentre of the Da Qaidam saline lake at the Holocene lay to the east of the early-stage (about 19

500–16 800 a B. P. and 12 000–9 600 a B. P.) salt depositional centre; at the beginning of the Holocene, the Tatalin Gol (river), an important recharge water source of B and Li, changed its course to flow into the Xiao Qaidam Lake on the southeast instead of the Da Qaidam Lake. This geological characteristic feature should be kept in mind in the process of future mineral prospecting and development in this subarea. Moreover, with a view to search for more boric mineral resources, the area north of Iqe to Yashatu should not be neglected.

(3) D_{3-3} subarea: Lying to the east of the D_{3-2} subarea, this subarea is located in an uplift zone south of the Delingha depression and extends also in the NNW direction. In this subarea the basement is made up of Paleozoic strata. Here, Mesozoic strata are relatively poorly developed, mostly occurring in medium to small–sized lake basins of the intermountain basin type. With the abundant recharge of fresh water from the northern semi–humid area, the subarea shows an interwoven pattern of saline and fresh–water lakes. The main saline lake is the Hoh Saline Lake of the sodium sulfate subtype where mining is under way. The principal salt–forming epoch of this lake is the Holocene. Halite beds are the predominant solid mineral resources, whereas mirabilite beds are of secondary importance; the brine here contains Li and K, which can be extracted as by–products. Common salts are the principal mineral resources in the subarea.

Kumkol Inferred Minerogenic Subzone of Li, B and Halite (D_4)   The Kumkol Basin, which lies between the C and D_1 zones, is a large intermontane basin which is similar to the Qaidam Basin in geological–geomorphological and climatic conditions. In this basin, which is confined by deep faults on all sides, there was a long period of dry climate and salt generation during the Tertiary, and salt springs have been found to occur around lakes. Particularly, the southern part of the Kumkol Lake is concurrently recharged by the B, Li and K bearing meteoric water from the volcanic–geothermal area in the Hoh Xil Mountains. Preliminary analyses indicate that large lakes in this subzone such as the Ayakkum Lake, Aqqikkol Lake and Jingyu Lake are all salt water lakes with salinities in the range of 62–145 g / l but commerical or marginal grades of Li and B and relative contents of K no less than those of the lakes in the Qaidam Basin. Moreover, the rare earths and other trace elements contained in this subzone are similar with those in the A_1 subzone (see Chapter 1). Hence, it is considered that the saline lakes in the subzone are, to some extent, prospective for Li and K, and possible occurrence of burried mineralizing brine in the Cenozoic formations is worth attention. Judging from the hydrochemical indicies of brines, it seems that attention should also be paid to the search for Zn–Pb deposits around the lakes.

Qilian Aarea Poor in Saline Lakes (D_5)   The Qilian mountains area to the east of the Qaidam Basin is an area of high–mountain lakes recharged by glacial water, of which the Har Lake is the largest in size. Because of the low salinity of water in all these lakes, the area is not considered as an area of saline lakes. Lateritic sinter type fresh–water facies ulexite deposits have been found along the South Qilian B–Li–bearing deep fault belt in this area. Attention should also be paid to the search for fresh–water facies boron deposits in local depressions along that belt.

11.5. Third–Class Minerogenic Area of Halite and Mirabilite–Bearing Saline Lakes of Sodium Sulfate Subtype (E)

This minerogenic area is located to the east of the Yang–Du major fault and its most part belongs to the Huanghe (Yellow River) and Sanjiang (Nujiang, Jinsha and Lanchang Rivers) exorheic regions. In this area many of the ancient lakes have been drained up by rivers and new small lakes have emerged in local depressions. Since no accumulation of salt components can take place in the relict exorheic lakes such as the Gyaring Lake and Ngoring Lake, fresh–water lakes are predominant in the vast area. However, some interbarrier lakes formed by lake contraction on the edges of large lakes, residual depressions of ancient exorheic lake basins, and lakes remaining on the plateau surface have turned into brackish lakes and medium to small–sized saline lakes. Based on the differences in closing conditions of lakes, the present area is divided into 3 subareas: South Tibet subarea of brackish lakes (E_1), Sanjiang–Huanghe subarea of exorheic lakes (E_2) and Qinghai Lake–Gonghe subordinate prospective subarea of halite and mirabilite (E_3).

In the E_2 subarea thin–bedded halite deposits have been found in two small interbarrier saline lakes (e.g. the Hajiang Yanchi Lake) on the periphery of the Ngoring Lake. The E_3 subarea originally belonged to the Qinghai Lake–Gonghe basin of the Yellow River drainage system. This largest brackish lake on the plateau and the Caka Lake were formed there as a result of the lake basin contraction into local closed–flow lake at the late stage of the Late Pleistocene. According to the satellite image interpretations, the Gengga Lake to the southwest of Gonghe County is probably a small saline lake, too. The Caka Lake has a great prospect of halite and is now under exploitation, and its brine whose K content approaches the commercial grade is worth future attention. Hydrochemically, lakes of various kinds in the E area are predominantly of the sodium sulphate subtype, except for individual lakes that are of magnesium sulphate subtype (Caka Lake) or carbonate type (Xiaojian Lake of the Jianhaizi Lake at Qamdo, which is of geothermal origin).

CHAPTER 12 PROSPECTS OF DEVELOPMENT OF RESOURCES OF SALINE LAKES AND THEIR ENVIRONMENTAL PROTECTION

As stated above, saline lakes are a kind of valuable synthetic natural resources, and contemporaneous saline lake resources not only break through the conventional concept of mineral resources but also include resources of various peculiar organisms (halophilic algae and *Artemia*) and tourism resources. Besides, some physical properties and environmental conditions of saline lakes have been utilized. For example, the heat—storing property of brines of saline lakes has been used to generate electricity by means of "solar energy salt water ponds". People also use dry salt to construct roads, railways, automobile race courses or airfields. Information of past and modern nature evolution is preserved in saline lakes, so they are a "nature laboratory" for the study of paleoclimatic and environmental evolution and prediction of the future.

The large—scale exploitation of saline lakes on the Qinghai—Tibet Plateau did not begin till the end of the 1950s. Now mining in several saline lakes of the plateau has become an important industry and made more and more outstanding contributions to the country. For example, potassium chloride was produced using the floatation method from the Qarhan Salt Lake in the Haixi Prefecture, Qinghai Province, in 1958—1983. For 25 years a total of 496000 tons of potash fertilizer (KCL) had been produced. At present a potash fertilizer factory with an annual production of 200000 tons has been built, and an agreement of joint development of the salt lake has been reached with Israel and a potash fertilizer (KCL) factory with an annual production capacity of 800000 tons is to be built by the year 2000. The annual production of halite in the Caka Lake and Hoh Yanhu Lake attains about one million tons. During the period from 1958 to 1963 about 150000 tons of raw borax were produced in Bangkog Co and Koggyaling Co, Tibet, which had once become the main source of revenue of the Tibet Autonomous Region. During 1986—1989 borax and magnesian borax mined from the Chalaka Lake and Chagcam Caka amounted to nearly 100000 tons, which not only alleviated the shortage of borax but also increase income of the local government and people. Meanwhile marked advances have also been made in studies of exploration and processing of resources of saline lakes. A group of saline mineral deposits have been explored and several research results have been up to the advanced level of the world. Nevertheless there still exist the following problems concerning the exploi—

tation and utilization of saline lakes on the plateau that await urgent solution or merit great attention.

12.1. Problems in Exploitation of Saline Lakes

1. Unplanned mining with indigenous methods, mining high—grade ore and giving up low—grade ore and low recovery cause serious waste of resources of saline lakes and disturb the production and sale plan and tax revenue of salt products of the state. This is of common occurrence in some borax—, mirabilite— and halite—producing areas. One example is given by Bangkog Co where the mining ratio of high—grade borax is less than 1 / 1 and hundreds of thousands of pure mirabilite have been polluted and depleted. Another example is the Da Qaidam lakeshore borax deposit where the ore that has been mined only accounts for 1 / 4 to 1 / 5 of the explored reserves.

2. Large amounts of useful components are abandoned, thus not only causing waste of resources and increasing production cost and consumption of raw materials but also making mineral deposits susceptible to metamorphism. For example, in Qarhan the production of one ton of KCl will yield 8 tons of $MgCl_2$ as a by—product. But as its utilization has long been neglected, about 1.6 million tons of $MgCl_2$ are given up each year if the calculation is based on the production of 0.2 million tons of KCl per year. From this amount of MgCl, 408500 tons of magnesium metal may be extracted. If the recovery rate is 1 / 3, this amount of metal magnesium is worth 458 million U.S. dollars. Another example is given by the technological test of a saltfield with a designed annual production capacity of 200 000 tons of KCl performed by the Chemical Mine Design Institute of the Ministry of Chemical Industry. 300 000 tons of halite with over 85% NaCl can be removed each year from the NaCl crystalization pools and they are valued at about 500 000 U.S. dollors. Besides, the brines yielding KCl also contain B and Li which can be recovered as by—products. If 0.2 million tons of KCl are produced annually, 2000 tons of LiCl and 1 000 tons of borax may be produced. On the basis of the 1 / 10 recovery rate, they are valued at about 4.3 million and 0.24 million U.S. dollars respectively. The above by—products amount to about 23 million U.S. dollars, which is about 3 times more than the value of KCl. It is a pity that the available resources are not utlized and it is also unreasonable from an economic analysis. The investment of the factory has amounted to 45 million U.S. dollars with a unit investment of 220 U.S. dollars / ton KCl, which is 2.6—6.3 times that of the Dead Sea and Chile (on the basis of the price of 1985), and large amounts of $MgCl_2$ and NaCl will be drained into the lake; as a result, the composition of brines would change and finally carnallite would fail to crystallize out.

3. In the course of mining and producing salt chemicals, more soluble salts and brines are liable to be lost, and exposed mirabilite and alkalis are readily pulverized

and fly with wind, thus causing pollution. Meanwhile, saline lakes are the ultimate drainage area of a catchment basin. When a factory, mine or town is built in a saline lake area, some saline lakes will be polluted more or less because of improper disposal of waste water and garbage or mining of salts with indigenous methods. For example, in the above-mentioned Qarhan potash fertilizer factory, the by-products MgCl and NaCl have been hardly used but are piled in large amounts or allowed to be lost. As the promoter of aliphatic ammonium used in the floatation method is slightly poisonous, its toxicity has increased daily with increasing discharge; so if the by-products are not used or treated, the entire lake area will be seriously polluted and so will water for industrial uses, and the health of local people will be jeopardized. Saline lakes will also be polluted when livestock transport halite or borax in lakes. This is of common occurrence in Chagcam Caka, Zabuye Caka and Xia Caka, where some salt beds are often polluted by droppings and mud and sand of flock and herd.

The above-mentioned environmental pollution is all caused by artificial factors. The following two natural factors that endanger resources of saline lakes merit attention in exploitation and utilization of resouces of saline lakes.

(1) Sandization of saline lakes   In arid areas, lakes change from saline lakes through playas to undersand lakes, which is a natural law. As far as human activity is concerned, it is usualy a relatively slow process and periodic change. However, digging sand plants and reclaiming wasteland blindly may intensify sandization of saline lakes. This situation has happened in the Qaidam Basin. Now sand plants in an areal extent 240 km long from east to west and 25 km wide from north to south along the Golmud highway have been mostly dug up. In the basin the sand vegetation covers an area of about 2 million hactares, 1/3 of which have been destructed; in addition, the land that was reclaimed blindly and the arable land that was abandoned are up to 4.5 million hactares. As a result, soils are salinized, the salts that recharge the saline lakes are lost and sandization of the saline lake area is intensified.

Acording to the statistics, the areas of some saline lakes of Qaidam have been reduced by about 10% because they are covered with silts.

(2) The influence of change of water recharge on saline lakes   Long-term excessive recharge of water may lead to dilution of saline lakes and dissolution of salt beds. Sudden seasonal flood usually bring large amounts of silts, which not only causes salt beds to be dissolved but also results in increase of impurities. But water is the life line of saline lakes, and, under the specific conditions of plateau arid climate, more attention should be paid to the serious consequences arising from abrupt reduction of waters feeding saline lakes, e.g. sharp reduction of recharge waters leads to deep burial of surface brines; thus the production of halite and potassium salt will meet great difficulty. This situation has occurred in Lop Nur neighboring the plateau. In Lop Nur, owing to construction of dams and reclamation of waste land in the middle and upper reaches of the lake district and excessive

use of fresh water, surface brines as vast as 1500 km² in area had submerged at depths of 2–4.5 m in a short period of time in 1959, bringing about sandization over large tracts in the lower reaches of the Konqi River and affecting the Lop Nur Salt Lake. The proper authorities of the plateau, particularly the Qaidam Basin which lies in a very arid area should remember this lecture.

12.2. Suggestions about the Utilization of Resources of Saline Lakes and Environmental Protection

Saline Lakes are dominant resources of the Qinghai–Tibet Plateau. K, B, halite, alkalis, Mg, Br and I are important mineral materials needed urgently by departments of chemical industry, light industry and metallurgical industry of China now; while Li, Cs and Rb have relatively high economic value. Some saline lakes may also serve as bases for producing food pigments and protein feed and bait as well as tourism bases. So much can be accomplished in saline lakes of the plateau. In order to bring benefit to mankind unceasingly by use of these resources, the following suggestions are put forward.

1. Build a salt–chemical complex   According to the geological data available, saline lakes of the plateau possess rich resources for the construction of a salt–chemical complex. There is no problem for the supply of water for domestic and industrial uses and auxiliary materials in some major saline lake districts, while conditions of transportation have also been improved somewhat. Based of salt industry have been established in some saline lakes and studies of exploration and exploitation have been carried out for many years; so a foundation is available for further development. Such a complex should have unified leadership and management, break down the existing situation of separate management of salt products by the Ministries of Chemical Industry, Light Industry and Metallurgical Industry and overcome various malpractices such as use of only one useful component, disharmony, overlapping of administrative organizations and low economic effects. The establishment of such a complex not only can raise economic effects double to even over ten times but also will be beneficial to the rational utilization of resources and protection of environment.

Analysis of conditions of salts, water resources and transportation as well as economic–tectonical conditions shows that Golmud is provided with the basic conditions for the establishment of a salt–chemical complex. A pressing matter of the moment is to carry out more intensive experimental studies of the feasibility of comprehensive development such as processing, and then the processing will develop from rough finish to finish and attention will be paid to the study of expansion of the application of salt chemicals. For example, $MgCl_2 \cdot 6H_2O$ is now applied in magnesium cement, but as it is apt to deliquescence and difficult to transport, it is advisable to study or introduce techniques as soon as possible and treat it into

$MgCl_2$ with 3–4 H_2O. This product is applied in making magnesium cement which can replace wood and steel and also in manufacturing man–made marble, dust–proof and sugar industry. The magnesia made by $MgCl_2$ is higher in quality than that made by magnesite and its cost is lower. The country requires at least 400 000 tons of pure magnesia each year.

2. Develop the synthetic industry of halite and fish farming   Halophilic algae and *Artemia* in saline lakes have great economic value and have become an important industry in the world. From halophilic algae we can extract β–carotene ($ 300–1500 / kg), glycerine ($ 1.6 / kg), algal protein, fatty acid ($ 900–2000 / t), chlorophyll ($ 100–200 / kg) and others. *Artemia* has very high nutritive value. It contains 57.6% protein and 18.11% fat as well as minerals required for the growth of fish, shrimps and crabs and hormone that can promote the development of sex gland of prawns. So it can serve as high–quality feeder and bait. The key to developing aquatics breeding is to solve the problem of the source of living baits— *Artemia*—for the larvas of prawns. At present the demand of the ova of *Artemia* exceeds their supply in the world. High–quality *Artemia* ova are valued at over 50 000 U.S. dollars. Besides, *Artemia* can take in organisms in saline water realms, thus raising the quality of halite. *Artemia* occurs in saline water realms (salt lakes, salt fields) of different types in the world except the Antarctic. In particular, some saline lakes with appropriate salinity are most favorable for natural and artificial intensive production of *Artemia*. Intensive production of *Artemia* in saline lakes needs less investment and yields quick results. Therefore, saline lakes are not only "treasure bowls" of inorganic salts but also "treasure houses" of natural pigments, fodders and baits. On the Qinghai–Tibet Plateau there are many saline lakes and playas that are not suitable for farming. So while exploiting saline lakes, we can develop such "saline lake agriculture", which is an important supplement to China's pasture land and arable land and can also raise the quality of halite, purify salt water realms and improve environment. For this reason, the following suggestions are put forward. First tests will be performed in saline lakes which are easily accessible (e.g. breeding test of *Artemia* in Caka and Gahai and breeding test of halophilic algae in Caka); then a synthetic enterprise of halite and aquatic breeding will be set up. Tests may also made in some areas where conditions are available.

3. Include the exploitation and utilization of key saline lakes in Tibet in the plan   There are more abundant "high–value" minerals such as Li, B, K and Cs and trona, halite and niter in saline lakes of Tibet than in other saline–lake districts of China. For example, it is estimated that in the Zabuye Lake there are 68 million tons of KCl, 32 million tons of B_2O_3 as well as large amounts of Li, Cs and halite, alkalis and niter. If 1 / 3 of them are minable, their potential value amounts to more than 100 billion U.S. dollars. Besides the lake also yields a large amount of seasonal halophilic algae rich in β–carotene. Detailed geological and mineral resources investigations as well as preliminary geoecological studies have been conducted in the lake and experiments of extracting Li, B and K yielded good results.

The special character of this saline lake has aroused attention both at home and abroad. It possesses the resoruces conditions for establishing a large—scale salt—chemical enterprise. But there are a number of unfavorable factors such as inconvenient communication, long distance of transport, frigid climate; insufficiency in oxygen and lacking in fuel. It is advisable to mine Li, B and K first as is the case with the Afacama mine in Chile and then to establish a complex producing all kinds salt products. Tibet adjoins Nepal and Pakistan of South Asia. If the problem of transport to these countries should be solved earlier, the cost of transport could be reduced greatly and the trade with South Asia would be promoted.

4. Draw up an 11—year and long—term overall programme for the scientific research and development of saline lakes of the Qinghai—Tibet Plateau   In 1963 the Science and Technology Commission set up a group of saline lake surveys and comprehensive utilization including such disciplines as geology, exploration, chemistry and mining and dressing and worked out a ten—year program, which promoted the development of saline lakes of the plateau. The contents of research in the new program should be widened and also include organism resources of saline lakes, environmental protection of lake districts and brine solar cells, and the application scope of resources of saline lakes of the plateau should be broadened. Besides, prominence should be given to researches on saline lakes and their related resources and harmonization between human activities and environment, and multidisciplinary coordination should be strengthened. The disciplines and specialities taking part in the researches should also include ecology, environmental protection, breeding of aquatics, social economics and systems engineering in addition to hydrology, engineering geology, sedimentary gelogy, physicochemistry, chemical technology, salt field engineering, mining and ore dressing.

5. Implement the Mineral Resources Law and relevant regulations of protection of resoruces and environment seriously   Check unplanned mining of borax and mirabilite deposits and reduce the loss of resources and pollution of environment. Authorities concerned should support departments in charge of mining and environmental protection in the implementation of relevant laws and regulations, adopt powerful measures and organize people to mine and maintain ecological environment according to the laws.

As the plateau covers a vast area, only a small part of saline lakes have been studied and explored in detail, while most lakes are less studied and there are also some blank areas. Substantial regional surveys and intensive studies of key saline lakes still remain to be made in the future. Through the efforts of the Chinese people, we believe that the Qinghai—Tibet Plateau will become a China's major base of such inorganic salts as K, B and Li salts and pigments and fodders and baits. Moreover, the plateau is also an ideal and attractive region for researches on the gelogy of saline deposits, chemistry and biology.

CONCLUDING REMARKS

This monograph reflects the results of the investigations and studies of saline lakes on the Qinghai–Tibet Plateau since 1956. Representing the continuation and deepening of previous work, the book highlights the data of comprehensive surveys of saline lakes in the Qaidam Basin and Tibet performed by the authors along with other scientists from the Ministry of Geology and Mineral Resources, the Ministry of Chemical Industry and the Chinese Academy of Sciences since 1956. In this book, proceeding from studies of the material compositions of mineral deposits and their structural–geological settings, the authors have made a synthetic analysis of a wealth of data on the regional geology, geophysics, Quaternary geology, sedimentology, hydrogeology, solid–liquid phase geochemistry, bioecology and physicochemistry of the saline lakes on the Qinghai–Tibet Plateau and, by applying the theories and methodologies of some related sciences and taking these saline lakes as an interlinked whole, studied from an intergrated point of view the formation and temporal–spatial evolution of these lakes, their prospects for B, Li and K as well as organisms of saline lakes and the geoecology of saline environment of these lakes. Based on these studies, they have arrived at the following points of knowledge.

1. The saline lakes on the Qinghai–Tibet Plateau—derivatives of the new-born and independent "Qinghai–Tibet tectosome"   The Qinghai–Tibet tectosome is a new tectonic unit that began to emerge as a result of the latest event of the Indus–Yarlung Zangbo suturing in the Eocene on the basis of the block–faulting of the fault blocks (fault–fold systems) of different ages in the Eocene. In such a unique tectonic setting, there formed four types of orderly arrayed tectonic basin systems, which not only provided the spaces and driving forces of water–salt accumulation, migration–evolution and differentiation–mineralization for the formation of saline lakes and controlled, to a considerable extent, their depositional processes, but also furnished the saline lakes with basic material sources.

The Cenozoic saline lakes in the plateau region are primarily tectonic lakes initiated also in the Eocene. They have experienced a process of accordion–type reciprocating spreading–contractions and migrations, generally with the western Qaidam as their base. Along with the drastic uplifting of the Qinghai–Tibet Plateau and regional climatic changes in the Quaternary, the atmospheric circulation on the plateau underwent repeated vertical and horizontal differentiations. This not only led to the evolutionary changes in salinity of water bodies and character of sediments in lake basins but also caused the formation of glaciers; marked N–S zonation and E—W regionalization of the lake basins on the plateau; frequent vertical and lateral variations of sediments in lake basins and distinct migrations of depocenters and salt—forming processes; warm and cold alternations, exorheic to endorheic transformations and fresh to brackish variations of water bodies in the lake basins as well as evolution from the repeated spreading—contraction changes of vast water bodies to the mutation of complete drying—up of some lake basins, etc. On the Qinghai—Tibet Plateau there occurred time and again the alternation of high lake levels and drying—ups. The Quaternary last high lake level, i.e. the "panriver—lake period", corresponds to the "Dali interglacial stage" of 36—25 ka B.P., while the recent saline lakes in Qinghai and Tibet have developed from north to south, following the arrival of the "late Dali interglacial stage" generally since 25 (31 locally)—10 ka B.P.

2. The tectono—geochemical point of view on the origin of B, Li, Cs and K on the Qinghai—Tibet Plateau   The hydrochemical characteristics of the natural waters and the trend surface analysis of B, Li, K and other components of lake waters indicate that the saline lake region of the Qinghai—Tibet Plateau is a region with anomalous abundances of elements such as B, Li, Cs and K in modern lakes in China and the Gangdisê Mountains— Yarlung Zangbo River area is the centre of their high abundance values, which just coincides with the latest plate convergence zone and deep—seated thermal anomaly zone (high—temperature geothermal zone) of the plateau; the chemical components of the lake water and geothermal water mostly belong to the group of lithophilic elements primarily derived from the crust; the B, Li and Cs contents in rocks of regional extent tend to increase toward the Cenozoic rocks; the crustal structure of the Qinghai—Tibet Plateau demonstrates that at the depth of 20—50 km in the upper and middle crust there exist a low—velocity layer and a layer of low Q values, which are interpreted as the expressions of the presence of a high—temperature melt in the crust of the plateau. This may be also proved by the fact that Cenozoic ring structures, geothermal zones and Cenozoic volcanoes are better developed in the region.

All the above indicates that at the convergence stage of the Qinghai—Tibet plate, the plate subduction, N—S compression and shearing caused intensified thermal expansion, which in turn led to the fusion of crustal layers and the formation of partially remelted magma pockets as well as the redistribution of lithophilic elements in the crust and facilitated the concentration of light elements and volatile components such as B, Li, K, Cs, S, As and F upward into the surficial layers, their diffusion in various wyas, through a series of fault systems, toward the surface of the crust to influence, in varying degrees, the material composition of the rocks in the various fault blocks (faulted—fold systems) of the region, and then their emanation onto the surface in the form of geothermal water or volcanic lava flow to become principal supply sources of special components for the lakes in the region. Therefore, despite the fact that the salt materials on the Qinghai—Tibet Plateau have a multiple of sources, those that constitute the numerous special saline lakes abundant in B, Li, Cs and K were mainly derived from the remelted magma at the depth of the crust and the surrounding rock series with relatively high B and Li contents and were carried up onto the surface by geothermal water and leaching water. However, the potash and magnesian salts in the Qarhan Playa of the Qaidam Basin, apart from their deep source, are essentially related to the long history of the fault–block subsidence and corresponding large Cenozoic basin evolution.

3. Salt–forming model of multi–step saline–lake subbasins and their minerogenic sequence   According to the geological and geomorphological conditions of their formation, the numerous lake basins of the Qinghai–Tibet Plateau can be generally divided into two groups:"medium to shallow down–faulted basins with moderate to low mountains" and "intermontane deep block–faulted basins", which are distributed on the first and second steps (4750–5100 m) and third step (2670–3270 m) of the plateau respectively. The saline–lake basins of both groups are characterized by multi–step subbasin salt–forming deposits. The saline lakes on the plateau have been universally evolved from pan–river lakes or pan–lakes. In the process of contraction and disintegration of fresh–water and exorheic lakes in various major lake districts (domains), a lake series–lake swarm–lake chain system that is disconnected in appearance but interconnected in essense was formed under the control of tectonic–geomorphological conditions. Under the influence of the gravity field, magmatic field and selective chemical differentiation and due to the difference in time of salt formation in various subbasins of the system, the salt components are bound to form different solid–liquid phase minerogenic series from the high–terrace lakes to low–terrace ones. Although salt assemblages formed in lakes of different hydrochemical types are somewhat different, yet less soluble salts would first assemble in high–terrace lakes and the salinity of lake water would be the lowest, whereas more soluble salts would gather in low–terrace lakes (subbasins of the lowest or lower steps) and the lakewater salinity would be the highest, thus giving rise to various corresponding mineral sequence of saltern carbonates, diatomaceous earth or clay→magnesian borate or mirabilite–borax or halite→K–Mg salts or K, Li, B and Cs salts.

The multi–step saline–lake subbasin model is a quite common salt–forming process in both ancient and modern inland lakes as is manifested in the Searles Lake chain in the United States and the numerous Eogene salt–forming basin systems in eastern China.

4. "Fault–block upwarping" model controlling the evolution of saline lakes and the migration direction of depocenters   The block differential movement causing the uplifting of the Qinghai–Tibet Plateau is the inevitable result of collision in the third plase of intracontinental convergence. Because of the intensified northward subduction of the Indian plate, the downward wedging of the Tarim,

Sino–Korean and Yangtze blocks into the lower crust and their confinement, the piling–up and thickening of the thrust slices of the Qinghai–Tibet crust and the thermal expansion of the lower crust, the horizontal compression stress was converted into a huge vertical stress and a nearly E–W (especially ESE) horizontal tensile stress. When the two stresses acted on the different fault blocks of the plateau, there occurred a differential movement of the fault blocks on the background of overall uplifting, thus resulting in the formation of first–order block areas (fault–fold systems) of various sizes, in which isolated lake districts (domains) were formed. Under the action of the N–S tectonic compressive stress of the Qinghai–Tibet Plateau, the various fault blocks and even one and the same block showed some differences in uplifting speed, dipping direction and horizontal torsion, thus causing a regular migrational movement of the depocenters of saline lakes in various lake districts of the plateau. Therefore, based on the movement pattern of each fault block in the general stress field, the migration directions of the saline lakes on the plateau can be effectively inferred and expounded; reversely, these lake districts can serve as a sensitive "levelling instrument" to indicate the vertical fluctuation and upwarping direction of the hosting block.

5. Supply source—the leading factor controlling the type of water in a lake basin   A comparative analysis of the hydrochemical characteristics of natural waters, climatic and structural features, and dominant lithological–geochemical associations in lake districts indicates that the rock types and related surface and subsurface waters in the eroded source areas are the basic controlling factors for the hydrochemical types of lake basins in arid zones whereas the climatic conditions are the secondary factor for the lake water types.

6. Geoecology and biomineralization of boron in supersaline environments
More and more facts have proved that activities of algae and bacteria and biochemical processes occur almost throughout the process of deposition in the saline–lake basin. In the course of deposition and mineralization of the saline lake, the biochemical reaction plays a negligible role, and biotic species and the composition of organic matter are of great significance as indicators. For instance, algae and bacteria had a decisive part in the formation of the dolomite on the beach of the Xiao Qaidam Lake, whereas the formation of some styles of borates was closely related to the action of halophilic algae and bacteria. Therefore, the authors suggest that an integrated viewpoint be applied to the study of the geoecology of salt–forming environments, which is a science dealing with the study of the interaction between the activities of organisms and the geological environments and its geological and minerogenic implications.

7. Isomorphous replacement of Li and Mg in carbonates of exogenic deposits
The replacement of Mg by Li under exogenetic conditions is referred to as an ex-

ample of the Goldsmith's diagonal law (replacement of 2 Mg^{2+} by Li^+ and Fe^{3+}). The discovery of Li—bearing magnesite and Li—bearing dolomite suggests that the replacement of Li and Mg in carbonates deposited at normal temperatures and pressures may also take place in closed evaporite basins, provided the Li concentration is high and the Mg and Ca concentrations are appropriate in the solution, thus demonstrating a new realm for the deep—going study of the microscopic world of Li—Mg—Ca carbonates.

In the future, a great deal of regional investigations and studies are still to be vigourously carried out on this vast plateau, since only a minor amount of saline lakes there have been surveyed in detail and the majority of saline lakes studied in relatively low degrees, while still others remain uninvestigated at all. In the future studies, temporal and spatial studies must be unified in the light of the development of geological history; "the saline lakes in the Qinghai, Tibet and Xinjiang Plateau region must be studied as an interrelated whole"[1]. The synthetic analysis should be continued by applying multi—disciplinary theories and methodologies, importance should be attached to the studies of basic theories of saline lakes, and the studies on structural geology, tectonochemistry, comparative sedimentology and biochemistry of saline lakes should be strengthened simultaneously with the studies on geology of mineral deposits, mineralogy, geochemistry and physicochemistry of saline lakes. Great attention should be paid to the accumulation of primary data, and data banks of basic resources of saline lakes, and deposits of material samples from saline lakes should be established step by step, so as to master more effectively the knowledge of salt—forming processes and mineralization regularities with a view to the establishment of the theoretical system of the unique halolimnology of the Qinghai—Tibet Plateau.

The studies on the geology and chemistry of key saline lakes and full utilization of all their useful components must be further carried out and well accomplished. In foreign countries, for a few large—sized saline lakes of high economic value, sustained and unremitting basic and applied researches such as studies on basic geology, hydrogeology, chemistry, and exploitation and processing of chemical raw materials are usually persevered in order to ensure the rational development and comprehensive utilization of useful mineral resources in the saline lakes, to protect the natural ecological environments of the lake areas and to gain greater economic and social benefits. For a few key saline lakes on the Qinghai—Tibet Plateau, it is necessary to carry out researches on two groups of subjects. One group of research subjects that have universal and long—term significance include the studies of continental facies K—mineralization model and theory, deposition mechanism and mineralization process of borates, Li—salts and carbonates in recent times,

---

[1] Zheng Mianping et al., 1974. Research Report on Boron Deposits in Saline Lakes of Tibet. Chinese Institute of Science and Technology Information. p.140.

paleoclimatic and geological events of saline lakes, inorganic chemistry of salt—forming elements, and physicochemistry of salt solutions; the other group of research subjects are subjects of applied and basic researches related to rational development of resources, involving mining hydrogeology, phase equilibrium of brine system, theory and technology of separation and extraction of salt minerals, biotechnology, etc. Specific research subjects of this group may include, *inter alia*, the experimental study on the utilization of the low—grade KCl—bearing brine and solid potash salt in the Qarhan Playa; the study and application of mining hydrogeology, mining techniques and mining equipment for the sublacustrine boron deposits in the Da Qaidam and Xiao Qaidam Lakes; the comprehensive utilization of B, Li and K in brine resources of the Zabuye Caka, Bangkog Co, Dujiali Lake and Chagcam Caka; the processing and utilization of the solid—phase Li—salt in the Zabuye Caka and Bangkog Co; the experimental development of the algal resources from the Zabuye Caka. Among the subjects of the above two groups there are some basic researches of common nature, e.g. the establishment of long—term hydrogeological observation stations, which is necessary for studying the mineralization regularities of plateau saline lakes and the development direction of their solid—liquid phase equilibrium as well as for providing plant construction and production with basic scientific data and which, at the same time, is of great importance to the rational protection of saline lake resources, the prediction of climatic changes and the preservation of ecological equilibrium of lake areas. Therefore, for the above—mentioned key saline lakes, the hydrogeological long—term observation stations already established should be strengthened (in Qaidam), and such stations need to be set up as soon as possible where they are not yet available (in Tibet).

The Qinghai—Tibet Plateau saline lake region is an enormously prospective minerogenetic province of saline lakes characterized by the occurrence of B, Li, K, Mg, Cs and Na salts, which ranks first in China in terms of the potentials of Li, K and B resources. Moreover, it is also possessed of favorable conditions of exploitation, processing and utilization of natural energy resources. Besides, abundant halophylic organisms have been discovered successively on the plateau since the late 70's. The authors are confident that, with the efforts of the Chinese people and vigorous international collaboration, the Qinghai—Tibet Plateau will become China's important production base of mineral raw materials of K, B and Li, chemical raw materials of inorganic salts, and halophilic organisms. At the same time, the plateau region also represents an exceptionally advantageous and attractive region for researches in the field of geology of saline mineral deposits and biological and chemical sciences, and its continued and deep—going studies will enable China to make great contributions to salt sciences.

REFERENCES

Editorial Committee of Chinese Chemistry for 50 Years, CCS (1985). *Chinese Chemistry for 50 Years.* Chapter I, part 7 'Chemistry of saline lake', 37–43, Science Press, Beijing (in Chinese).

Investigation Group of Radiation (1982). Climatic of radiation on Qinghai–Xizang Plateau. *Proceedings of the Symposium on Glaciology and Cryopedology*, held by Geographical Society of China. 46–51. Science Press, Beijing.

Ami Ben–Amotz (1982). Accumulation of β–Carotene in halotolerant algae: purification and Characterization of β–Carotene–rich globules from Dunaliella bardawil (chloro–phyceae). *J. of Phycology*, 18, pp.529–537.

An Zhimin, Yin Zhesheng and Li Bingyun (1979). Paleolith and microlith in Shuanghu and Xainza, *Northern Tibet Archaeology*, No. 6, 481–491.

An Zhisheng, Porter, S.C., Wu Xihao et al. (1993). The Climatic optimum in Middle–East China and the East Asian monsoons in summer. *Review of Science (Kexue Tongbao)*, Vol. 23, 1810–1813 (in Chinese).

Barsukov, V.T. and Dmitriev, L.V. (1972). Inference of source of ore from surface sail. *Geochemistry*, No. 12, 1558 (in Russian).

Bayley, S.T. et al. (1964). The Ribosomes of the Extremely Halophilic Bacterium Halobacterium Cutirubrum. *J. Mol. Bio.*, 9, pp.654–669.

BeCheyeva, K.E. (1981). *Hydrogeochemistry.* Geological Publishing House, Beijing (Translator: Pan Liren) (in Chinese).

Borowitzka, L.J. et al. (1977). The salt relations of Dunaliella, *Arch. Microbiol.*, 113, pp.131–138.

Borowitzka, L.J. et al. (1981). The microflora–adaptations to life in extremely saline lakes. *Hydrobiologia*, 81, pp.33–46.

Borowitzka, L.J. et al. (1983). The mass culture of Dunaliella salina for fine chemicals: from laboratory to pilot plant, xith International Seaweed Symposium, Abstracts, pp.21.

Brock, T.D. (1978). Ecology of saline lakes. *Strategies of Microbial Life in Extreme Environments* (Ed, Moshe Shilo), pp.31–47.

Brown, A.D. et al. (1979). Halotolerance of Dunaliella. *Biochemistry and physiology of protozoa*, Second edition, Vol. 1, pp.139–190.

Ce Xiande, Qian Zhiqiang and Liu Laibou (1996). Carborite—new boron carbonate. *Scientia Geologica Sinica*, No. 1. (in Chinese).

Ce Xiande and Zheng Mianping (1963). The initial study of indelite in China. *Scientia Sinica*, No. 12, 1246–1248 (in Chinese).

Ce Xiande and Zheng Mianping (1963). The study of kurnaporite crystal. *Acta Geologica Sinica*, Vol. 43, No. 2, 184–192 (in Chinese).

Ce Xiande, Zheng Mianping and Liu Laibou (1965). *Minerals of Borate.* Science Press, Beijing (in Chinese).

Chen Deqian (1982). *Geological Introduction of Rare Element.* Geological Publishing House, Beijing,

REFERENCES

32–35 (in Chinese).

Chen Jingqing, Liu Ziqiu, Fang Chunhui and Xu Xiaodong (1986). 25°C Isothermal evaporation of the Xiao Qaidam salt lake brine. *Geological Review*, Vol. 32, No. 5, pp. 470–480 (in Chinese).

Chen Kezao and Bowler, J.M. (1985). Preliminary study of sedimentary characteristics and paleoclimatic evolution of the Qarhan Salt Lake in the Qaidam basin. *Scientia Sinica*, Series B, No. 5, 463–473 (in Chinese).

Chen Kezao and Bowler, J.M. (1987). Evolution of salt lakes in late Pleistocene in Qaidam Basin, Qinghai Province, China. *Proceedings of the Symposium on Quaternary between China–Australia*. Science Press, Beijing, 83–91 (in Chinese).

Chen Kezao, Bowler, J.M., and Kelts, K. (1990). Palaeoclimatic evolution within the Qinghai–Xizang (Tibet) Plateau in the last 40,000 years. *Quaternary Sciences*, No. 1, 21–31 (in Chinese).

Chen Zhiming (1981). The origin of lakes on Xizang Plateau. *Oceanologia Et Limnologia Sinica*, Vol. 12, No. 2, 178–187 (in Chinese).

Cheng Kezhao, Yang Zhaoxiu and Zheng Xini (1981). The salt lakes on the Qinghai–Xizang Plateau. *Acta Geographica Sinica*(in Chinese).

Chiang Sieh–chih, Shen Yunfan and Gong Xunju (1983). *A Quatic Invertebrates of the Tibetan Plateau*. Science Press, Beijing, pp.443–451 (in Chinese).

Chu YiHua, Han Weitien, Chien Tsechang, Liu Laipao and Ming Lingsen (1965). Trigonomagneborite —a new borite mineral. *Acta Geologica Sinica*, Vol. 45, No. 3, 298–305. (in Chinese).

Coulson, A.L. (1933). Geological notes on traveses in Tibet. *Records, Geologica Surv., India*, Vol. 67, No. 3, 338.

Delost, W. et al. (1974). Isotope method in groundwater hydrology. Isotope Techniques in Groundwater Hydrology, *Proceedings of Symposium*, Vienna, 1974 & 1978.

Dobrovonisky, V.V. (1983). Geography of Microelement. *'Thought'*, Moscow.

Dong Lumau, Dai Aiyun and Jiang Xiezhi (1982). *The atlas of Chinese animal*: Crustacean. book I 6, 22, Science Press, Beijing (in Chinese).

Du Naiqiu and Kong Zhaochen (1983). Palynoflow of the Qarhan saline lake and its significance in geography and botany— the sporo–pollen assemblages from CK2022 drilling core at the bieleton. *Acta Botanica Sinica*, Vol. 25, No. 3, 275–282 (in Chinese).

Du Naiqiu, Kong Zhaochen and Shan Fashou (1989). Sporopollen analysis of samples from hole QM85–14C of the Qinghai Lake and preliminary study of the paleoclimate–paleoenvironment of the lake. *Acta Botania*, Vol. 31, No. 10, 879–890 (in Chinese).

Durov, S.A. (1948). On origin of native water. *Laboratorial Record for Hydrogeologic Problem*, III, 104–113.

Ellis, A.J. and Mohon, W.A.J. (1977). *Chemistry and Geothermal Systems*. New York, Academic Press, Inc.

Eugster, H.P. (1979). Lake Magadi, Kenya, and its precursors. *Hypersaline Brines Evaporite Environments—Development in Sedimentology*, Vol. 28, Chapter 15, 195–230.

Eugster, H.P., and Mardie, L.A. (1978). Chapter 8, Saline Lakes.—Lakes: Chemistry, Geology and physics, Edited by Abrabam Lerman, Springer–Verlag, New York Inc. 237–293.

Fonte, J.Ch., Mélières, F., Gibert, E., Liu Qing and Gasse, F. (1993). Stable isotope and radiocarbon balances of two Tibetan Lakes (Sumxi Co, Longmu Co) from 13,000 B.P., *Quaternary Science Re-*

views, Vol. 12, pp. 875—887.

Fott, Bohuslav (1971). *Algenkunde*. VEB Gustav Fischer Verlag.

Gao Shiyang and Li Bingxiao (1982). Borate minerals in saline lakes on the Qinghai—Xizang Plateau. *Acta Mineralogica Sinica*, No. 2, pp.107—112 (in Chinese).

Gau, Zhanghong. Preliminary study for cabonate of clastic sediment in some salt lakes, Qaidam Basin. *Proceedings of the Symposium on Quaternary between China—Australia*, Science Press, 142—152 (in Chinese).

Godbole, N.N. (1972). Theories on the origin of salt lake in Rajasthan. *India International Geological Congress Twenty—Fourth Session*, Section 10, Geochemistry Mountreal. 354—357.

Gorbov, A.F. (1976). *Boron Geochemistry 'Leningrad Mineral Resources'* (in Russian).

Gu Zhaoyan, Liu Jialing, Yuan Baoyin, Liu Dongsheng et al. (1993). Changes of monsoons on the Qinghai—Tibet plateau since 12,000 a B.P.—Geochemical evidence from the deposits of the Siling Co. *Kexue Tongbao*, Vol. 38, No. 1, 61—64 (in Chinese).

Guan Zhihua, Chen Chuanyu, Fan Yunchi et al. (1984). *River and Lake in Tibet*. Science Press, Beijing (in Chinese).

Han Tonglin (1984). Microgeomorphic features of ancient lake erosion in Xizang (Tibet) and their significance. *Himalayan Geology* II — part of achievements in Geoscientific Investigation of Sino—French Cooperation in the Himalayas in 1981. 45—58 (in Chinese).

Han Tonglin (1984). Microgeomorphic features of ancient lake erosion in Xizang (Tibet) and their significance. *Himalayan Geology* II (part of achievements in Geoscientific Investigation of Sino—French Cooperation in the Himalayas in 1981). Geological Publishing House, Beijing, China. 272 (in Chinese).

Han Weitian (1981). The experimental study of the formation condition of borate minerals on the Qinghai—Xizang Plateau. *Kexue Tongbao* (Review of Science), No. 21, pp.1315—1318 (in Chinese).

Hayden, H.M. (1921). *Record of the geological survey of India*. Vol. 52, pp.262—265.

Hsti Ginntze (1937). The northern heaven lake of Tibet. *Acta Geographica Sinica*, Vol. 4, 891—904 (in Chinese).

Hsti Ginntze (Xu Jinzhi) (1960). *The Physicogeographical Material on the Qinghai—Tibet Plateau*. Geographic Institute C.A.S. (in Chinese).

Huang Jiqing (T.K. Huang) (1984). New researches on the tectonic characteristics of China. *Bulletin of the Chinese Academy of Geological Sciences*. No. 9, 5—18 (in Chinese).

Huang Qi, Cai Biqin, Yu Junqing, Chang Hong and Pu Hong (1981). The determination of the age of the salt lakes in China. *Oceanologia Et Limnologia Sinica*, Vol. 12, No. 1, 41—47 (in Chinese).

Huang Sanyou (1986). *Volcano, Hot Spring and Geothermal Energy*. Geological Publishing House, Beijing, 102—103 (in Chinese).

Ji Hungxiang, Hsu Chinqi and Huang Wanp (1980). The Hipparion Fauna from Guizhong Basin, Xizang. *Palaeontology of Xizang*, Book 1, 18—21, Science Press, China (in Chinese).

Jiao Shupei (1984). On characteristics of the evolution of geological history Tarim Qaidam Piwa region and the problem on its belongingness. *Contribution to the Geology of the Qinghai—Xizang (Tibet) Plateau*, part 15, CGQXP Editorial Committee Ministry of Geology and Mineral Resources PRC, Geological Publishing House, Beijing, 121—130 (in Chinese).

Jones—Lidofsky, A.E. (1957). *Method of synthetic survey and exploration for saline lake deposit*. National

REFERENCES

Chemical Publishing House, Leningrad (in Russian).

Joseph Needham (1954). *Science & Civilisation in China*. Vol. 5, No. 2.

Kornevsky, C.M. (1973). Assemblage of Salt Mineral Deposits. *'Resources'*, 145—148 (in Russian).

Khorana, H.G. (1979). Amino acid sequence of bacterior—hodopsin. *Proc Natl. Acd. Sci., USA*, V. 76, 10. pp.5046—5050.

Kudhner, D.J. (1978). Microbial Life in Extreme Environments. Academic Press INC (London) Ltd., pp.319—325.

Li Bingyuan, Wang Fubao, Yang Yichou et al. (1983). *Quaternary in Tibet*. Science Press, Beijing (in Chinese).

Li Bingyun, Zhang Qingsong and Wang Fubao (1991). Evolution of the lakes in the Karakorum—West Kunlun mountains. *Quaternary Sciences*. No. 1, 64—71.

Li Hongji (1993). Preliminary estimation to the crustal thermal state of the Himalayan region. *Acta Geophysica Sinica*, Vol. 26, No. 3, 249—255 (in Chinese).

Li Jijin et al. (1986). *Tibetan Glaciers*, Science Press, Beijing (in Chinese).

Li Shijie and Li Shude (1991). Significance and research on the two boreholes of Tianshuihai in the West Kunlun mountains, *Journal of Glaciology and Geocryology*, Vol. 13, No. 2, 187—188 (in Chinese).

Li Shuanke (1992). Fluctuation of closed lake—level and its climatic significance on the Central Kunlun Mountains. *Journal of Lake Sciences*, Vol. 4, No. 1, 19—30 (in Chinese).

Lin Zhengyao and Wu Xianding (1981). The climatic divisions of the Qinghai—Tibet Plateau. *Acta Geogr. Sinica*, No. 36, 22—33 (in Chinese).

Liu Yingjun, Zhou Liming and Wang Hongnian et al. (1984). *Element Geochemistry*. Science Press (in Chinese).

Liu Zengqian, Jiang Chunfa and Liu Baotian (1984). New thoughts on the geology of Tethys—Himalaya tectonic domain. *Contribution to the Geology of the Qinghai—Xizang (Tibet) Plateau*. part 15, Geological Publishing House, Beijing. 131—146 (in Chinese).

Mao Yanshi, Wei Guanyi and Zhang Bonan et al. (1984). Geologic— Tectonic Characteristics of Precambrian metamorphic rock series in China's Himalayan Region. *Contribution to the Geology of the Qinghai—Xizang (Tibet) Plateau*, part 15, CGQXP Editorial Committee Ministry of Geology and Mineral Resources PRC, Geological Publishing House, Beijing, 15 (in Chinese).

Mastui, M. et al. (1979). *Halophilic microorganisms......ecology, physiology, food hygine and exploition in fod industry* (Japanese). Medicinal Publishing House (Japan).

Mutchinson, G.E. (1937). Limnological studies in India Tibet. *Int. Rev. Hydrobiol*, No. 35, pp.134—177.

Nissenbaum, A. (Editor) (1980). Hypersaline brines evaporite environments, *Developments in Sedimentology*, 28, pp.23—29.

Noriu, E. (1946). Geological explorations in Western Tibet. III, *Geology* 7, pp.26—29.

Pan Baotian, Xu Shuying et al. (1991). Discussion on the sequence and extent of changes of environment in the eastern Qinghai—plateau during the past 30 000 years. *Arid Land Geography*, Vol. 12, No. 2.

Pan Guitang, Jiao Shupei, Wan Peisheng et al. (1984). The Cenozoic tectonics and its orogeny in Altum. *Contribution to the Geology of the Qinghai—Xizang (Tibet) Plateau*, part 15, CGQXP Editorial Committee Ministry of Geology and Mineral Resources PRC, Geological Publishing House, Beijing, 113—120 (in Chinese).

Prell, W.L. and Kutzback, J.E. (1987). Monsoon variability over the past 150,000 years. *Journal of*

REFERENCES

Geophysical Research, Np. 92, 8411—8425.

Qian Dinyu, (1982). An attempt at exploration uplifting of the Qinghai—Xizang Plateau, on the basis of geological data on geosuture along Indua—Yarlung Zangbo Rivers. *Contribution to the Geology of the Qinghai—Xizang (Tibet) Plateau*, part 1, CGQXP Editorial Committee Ministry of Geology and Mineral Resources PRC, Geological Publishing House, Beijing, China, 104—123 (in Chinese).

Qian Fang and Ma Xinghua (1979). The Preliminary discussion on problems the Quaternary magnetic stratigraphy of China. *Science Bulletin*, Vol. 24, No. 23, 1087—1090 (in Chinese).

Qian Fan, Ma Xinghua, Wu Xihao and Pu Qingyu (1982). Study of magnetic strata from the Qangtang and Qugua Formations. *Contributions to the Geology of Qinghai—Xizang Plateau*, part 4, CGQXP (in Chinese).

Qian Zhiqiang (1985). Borate minerals in salt lake deposit at Qaidam basin. *Sixth Symposium on Salt*, Vol. 1, 185 (in Chinese).

Qi Wen and Zheng Mianping (1995). Sedimentary characteristics of ZK91—2 drilling hole and climate—environment evolution of Zabuye Lake, Tibet. *Journal of Lake Sciences* No. 2 (in Chinese).

Qu Yihua, Ce Xiande, Qian Zhiqiang and Liu Laibou (1964). Hungtsaoite— A new water—bearing Mg—borate. *Acta Geologica Sinica*, Vol. 44, No. 3 (in Chinese).

Rubanov, E.V. (1985). Geochemistry of recent continental eveporite. Fundamental Problem of Eveporite, *'Science'*, 74—82 (in Russian).

Searle, M.P. (1983). On the tectonics of the western Himalaya. *Episodes*, No. 4, 21—26.

Shan Fashou, Du Naiqiu and Kong Zhaochen (1993). Vegetational and environmental changes in the last 350 Ka in Erlangjian, Qinghai Lake. *Journal of Lake Sciences*, Vol. 5, No. 1, 9—17 (in Chinese).

Shen Xianjie (1985). Crust and upper mantle thermal structure of Xizang (Tibet) inferred from the mechanism of high heat flow observed in south—Xizang. *Acta Geophysica Sinica*, Vol. 28, supplement I, 93—107 (in Chinese).

Shen Zhanli (1986). *The Foundation of Hydrogeochemistry*. Geological Publishing House, Beijing (in Chinese).

Shen Zhenshu, Cheng Guo, Le Changshuo, Liu Shuqin et al. (1993). *The Division and Sedimentary Environment of Quaternary Salt—bearing Strata in Qaidam Basin*, Geological Publishing House, Beijing (in Chinese).

Smith, G.I. (1979). Subsurface stratigraphy and geochemistry of lake Quaternary evaporite, Searlies Lake, California. *U.S. Geological Survey*, Prof. 1043, pp.98—100.

Stanov, V.P. (1962). *Petrogenic Fundamental Principle VII*, AN U.S.S.R. (in Russian).

Stoeckenius, W. et al. (1979). Bacteriorhodapsin and purple membrane of halobacteria. *Biochimicact Biophysica Acta*, 505, pp.215—278.

Sun Kezhong and Teng Jiwen (1985). The velocity distribution in the crust and upper mantle beneath the Xizang (Tibetan) Plateau from long period surface wakes. *Acta Geophysica Sinica*, Vol. 28, supplement I, 193 (in Chinese).

Sun Tapeng (1974). The origin of recent potash deposits in a certain salt lake, China. *Geochemica*, No. 4, 230—248 (in Chinese).

Sun Xiangjun and Wu Yushu (1987). Holocene vegetation history and environmental changes of the Dianchi lake area, Yunnan Province. *Proceedings of the Quaternary Symposium between China and Australia*, Sciencces Press, Beijing, 28—42 (in Chinese).

REFERENCES

Sven Hedin (1901). *Central Asia, 1899–1902*, Maps (III). STOCKHOLM.

Sven Hedin (1909). Journal in Tibet, 1906–1908. *The Geographical Journal*, Vol. 33, No. 4, Sketch map of the part of Tibet.

Tapponnier, P. (1982). Propagating extrusion tectonics in Asia: New insights from simple experiments with plasticine. *Geology*, Vol. 10, No. 12, 611–616.

Tong Wei, Zhang Mentou and Zhang Zhifei et al. (1981). *Geothermal Field in Tibet*. Science Press, Beijing (in Chinese).

Truesdell, A.H. (1975). Summary of Section III, Geochemical technique in exploration: UN symposium on the Development and Utilization of Geothermal Resources, San Francisco, Proceedings, Vol.,liii–lxxix.

Valyashk, M.G. (1955). Basic chemical types of natural waters and the conditions producing them. *Record of Academy*, U.S.S.R., No. 102, 315–318 (in Russian).

Vine, J.D. (1979). Lithium— Nature's lightest metal. U. S. Department of the Interior Geological Survey. *Journal Research* U. S. Geological Survey.

Walwer, D. (1987). The regional significance of the palynological data from Yunnan Province. *Proceedings of the Quaternary Symposium between China and Australia*. Science Press, Beijing, 21–27.

Wasson, R.J., Smith, G.I., and Agarawal, D.P. (1984). Late Quaternary sediments, minerals and inferred geochemical history of Didwana lake, Than Desert, India. *Palaeogeography, Palaeoclimatology and Palaeolimnology*, Vol. 46, 345–372.

Wen Shixuan, Zhang Binggao and Wang Yigang et al. (1984). *Stratigraphy of Xizang (Tibet) Plateau*— The series of the comprehensive scientific expedition to the Qinghai–Xizang Plateau. Science Press, China (in Chinese).

Wu Bihao, Duan Zhenghao, Guan Yuhua and Lian Wei (1986). Deposition of potash–magnesium salts in the Qarhan Playa, Qaidam basin. *Acta Geologica Sinica*, Vol. 60, No. 3, 289–296.

Wu Yimin (1983). The Tertiary of Xizang (Tibet). *Contribution to the Geology of the Qinghai–Xizang (Tibet) Plateau*, part 3, CGQXP Editorial Comittee Ministry of Geology and Mineral Resources PRC, Geological Publishing House, Beijing. 224–232 (in Chinese).

Xia Wenjie and Li Xiuhua (1984). The discovery of primary dolomite from beach rock in the Xiaochadan salt lake of Qinghai and its significance. *Acta Sedimentologica Sinica*, Vol. 4, No. 2, 19–25.

Xia Wenjie and Li Xiuhua (1986). The discovery of primary dolomite from beach rock in the Xiaochadan salt lake of Qinghai and its significance. *Acta Sedimentologica Sinica*, Vol. 4, No. 3, 20.

Xu Chang (1985). Primary study of clay minerals and its significance in salt lake sediments of the Qinghai–Tibet. *Scientia Geologica Sinica*, No. 1, 87–96 (in Chinese).

Xu Guowen (1992). Evolution and developing trend of Qinghai Lake since 8000 a B.P. *Qinghai Geology*, No. 1, 70–72 (in Chinese).

Xu Jiasheng, Gao Jianxi and Xie Fuyuan (1981). Yellow Sea during the last glacial epoch. *Scientia Sinica*, No. 5, 605–613.

Xu Shuying (1985). On the palaeogeographical environmental evolution of Pliocene–Pleistocene epoch of the Tanglha Mountains. *Quaternary Sciences*, 73–81 (in Chinese).

Yang Hua (1985). Geophysical features demonstrated by aeromagnetic maps of Qinghai–Xizang Plateau and their geotectonic significance. *Acta Geophysica Sinica*, Vol. 28, Supplement I. 193 (in Chinese).

REFERENCES

Yang Lihua and Liu Tungsheng (1974). On the neotectonic movements in the Mt. Jolmo Lungmu region. *Scientia Geologica Sinica*, No. 3, 209–219 (in Chinese).

Yang Liufa, Fan Yunqi and Huang Baoren (1982). Relation between ostracode distribution in surface deposits and water salt of recent lakes in Xizang Plateau. *Transactions of Oceanology and Limnology*, No. 1, 20–28.

Yang Qian (1982). The sedimentation mechanism of potash deposits on the Qarhan inland salt lake. *Acta Geologica Sinica*, Vol. 56, No. 3, pp.281–292 (in Chinese).

Yang Yichou, Li Bingyuan, Yin Zhesheng et al. (1983). *Geomorphy in Tibet*. Science Press, Beijing (in Chinese).

Yaoduoyandan Gongbu, AD 720, *The Four Ancient Books of Medicine* (in Tibetan).

Yu Shengsong (1984). The hydrochemical feature of salt lakes in the Qaidam basin. *Oceanologia Et Limnologia Sinica*, Vol. 15, No. 4, 341–359 (in Chinese).

Yu Shengsong and Tang Yuan (1981). The hydrochemical characteristics of the saline lakes on the Qinghai–Xizang Plateau. *Oceanologia Et Limnologia Sinica*, Vol. 12, No. 6, 498–511 (in Chinese).

Yu Zhihong, Liu Zhongping, Wan Defang and Fu Zijie (1981). *Tectonic Map of the Linear Structures on the Territory of China* (By using of the satellite images), Cartographic Publishing House, Beijing (in Chinese).

Yu Zhihong, Liu Zhongping and Wan Defang et al. (1981). *Some tectonic features of the linear structures on the territory of China*. Geological Publishing House, Beijing (in Chinese).

Yuan Jianqi (1946). Caka Lake, Qinghai province. *Selected Papers on Geology of Salt Deposits* (1989), Palaestra Press, Beijing, 33–38 (in Chinese).

Yuan Jianqi, Huo Chengyu and Cai Keqin (1983). The high mountain–deep basin saline environment—a new genetic model of salt deposits. *Geological Review*, Vol. 29, No. 2, 159–164 (in Chinese).

Yuan Jianqi, Huo Chengyu and Cai Keqin (1985). Characteristics of salt deposits deposits in the dry salt lake and the formation of patash beds. *Earth Science*. 1–9 (in Chinese).

Yuan Jianqi, Huo Chengyu and Cai Keqin (1985). Characteristics of salt deposits deposits in the dry salt lake and the formation of patash beds. *Earth Science—Journal of Wuhan College of Geology*, Vol. 10, No. 4, 1–9 (in Chinese).

Zhang Chao and Ma Pinqu (1989). *Geographic climatology*. Meteorologic Publishing House, Beijing (in Chinese).

Zhang Hongzhao (1954). *The Record of the Ancient Ore Deposit*, Geological Publishing House, Beijing, (in Chinese).

Zhang Pangxi, Zhang Baozhen and Yang Wenbo (1988). The evolution of the water body environment in Qinghai Lake since the postglacial age. *Acta Sedimentologica Sinica*. Vol. 6, No. 2, 1–14 (in Chinese).

Zhang Pengxi, Zhang Baozhen, Qian Guimin, Li Maijun and Xu Liming (1994). The study of paleoclimatic parameter of Qinghai Lake since Holocene. *Quaternary Sciences*, No. 3, pp. 225–238 (in Chinese).

Zheng Mianping (1989). Prospect for geological research of the global saline lake. *Abroad Geology of Mineral Deposits: Special Issue of the Geology for Saline Lake*, No. 3–4, 1–34. Publishing by Institute of Mineral Deposits, MGMR (in Chinese).

Zheng Mianping (1991). Saline–lake resources on the Qinghai–Xizang Plateau and their developed

REFERENCES

prospect. *Resources, Environment and Development of the Qinghai–Xizang Plateau: Proceedings of the First Symposium on the Qinghai–Xizang Plateau*, 91–97 (in Chinese).

Zheng Mianping and Jin Wenshan (1964). The initial study of a new type—magnesian borate deposit in China. *Chinese Geology*, No. 2, 24–30 (in Chinese).

Zheng Mianping and Liu Wengao (1982). The discovery of a lithium–rich magnesian borate deposit in Xizang (Tibet). *Geological Review*, Vol. 28, No. 3, 263–266 (in Chinese).

Zheng Mianping, Liu Wengao and Jin Wenshan (1981). A study on the Xizang (Tibet) salt minerals. *Bulletin of the Chinese Academy of Geological Sciences*, Series III, Vol. 2, No. 1, 75–101 (in Chinese).

Zheng Mianping, Liu Wengao and Xiang Jun (1985). Study of Zabuye saline lake, Tibet. *Scientific Papers on Geology for International Exchange*—Prepared for the 27th International Geological Congress. 4. Geological Publishing House, Beijing. 173–184 (in Chinese with English abstract).

Zheng Mianping, Liu Wengao and Xiang Jun (1985). The discovery of halophilic algae and Halobacteria at Zabuye salt lake, Tibet and preliminary study on the geoecology. *Acta Geologica Sinica*, Vol. 59, No. 2, pp.162–171 (in Chinese with English abstract).

Zheng Mianping, Liu Wengao, Xiang Jun and Jiang Zhongti (1983). On saline lakes in Tibet, China. *Acta Geologica Sinica*, Vol. 57, No. 2, 184–194 (in Chinese with English abstract).

Zheng Mianping, Qi Wen, Wu Yushu and Liu Junying (1992). Preliminary study on sedimentary environment of the Nop Nor salt lake and its prospect for potassium. *Chinese Science Bulletin*, Vol. 37, No. 11, pp. 935–939.

Zheng Mianping, Xiang Jun et al. (1989). *Saline Lakes on the Qinghai–Xizang (Tibet) Plateau*. Beijing Scientific and Technological Publishing House, Beijing (in Chinese).

Zheng Mianping, Zheng Yuan and Liu Shuqiu (1989). Shoreline of Zabuye salt lake in Tibet and its climatic and environmental significance. *Progress in Geosciences of China*(1985–1988)— papers to 28th IGC, Vol. III, Geological Publishing House, (Beijing, China), 181–184.

Zheng Mianping and Xiang Jun (1990). Lake basin evolution and plateau uplifting. *Geology of the Himalayas*, Vol. 2, Papers on Geology, Edited by Li Guangcen et al., Geological Publishing House, Beijing, 219–233.

Zheng Xiyu (1981). Formation of the salt lakes resources and its utilization on the Xizang Plateau. *Scientia Geographic Sinica*, Vol. 1, No. 1, 66–76 (in Chinese).

Zheng Shaohua (1980). The Hipperion Fauma of Bulung Basin, Birn, Tibet. *Tibetan Paleontology*, Book 1. Science Press, 32–34 (in Chinese).

Zheng Xiyu and Yu Shengsong (1982). The distribution characteristics of B and Li in the brine of Zhacang Caka (Zhangzang Caka) saline lake, Xizang autonomous region, China. *Oceanologica Et Limnologia Sinica*, Vol. 13, No. 1, 26–34 (in Chinese).

Zhu Meixiang, Tong Wei and You Maozheng (1982). Effloresence in geothermal areas of Xizang (Tibet) and its geological significance. *Acta Scientiarum Naturalium Universitatis Pekinesis*, No. 1, 109–110 (in Chinese).

Zhou Kunshu, Li Wenyi and Kong Zhaochen (1984). The main results of the Quaternary sporo–pollen analyses in China. *Sporo–pollen Analyses of the Quaternary and Paleoenvironments*, 1–14 (in Chinese).

Zhu Xia (1983). *Tectonics and Evolution in Mesozoic and Cenozoic Era of China*. Science Press, Beijing (in Chinese).

Zhu Zhonghau (1985). Tertiary climate in Chaidam Basin. *Oil Geology in Qinghai*, No. 2 (in Chinese).

APPENDIX 1
RULES FOR TRANSLATING GEOGRAPHIC NAMES RELATED TO SALINE LAKES ON THE QINGHAI–TIBET PLATEAU

A. In translating Chinese geographic names, their transliterations in the Chinese phonetic alphabet in the GAZETTEER OF CHINA (published by the Cartographic Publishing House in 1983) are unexceptionally followed; for geographical names that are not available in the GAZETTEER OF CHINA, their transliterations in the Chinese phonetic alphabet in the GAZETTEER OF THE TIBET AUTONOMOUS REGION, Series B, edited and published by the Bureau of Geodesy and Cartography of the Revolutionary Committee of the Tibet Autonomous Region and the Institute of Geodesy and Cartography of the State Bureau of Geodesy and Cartography in 1978 are taken as standards; for those geographical names that are not included in the above two gazetteers and the names of individual key saline lakes (e.g. Zabuye), they are transliterated into the Chinese phonetic alphabet.

B. Common component parts in geographical names such as mountain range, mountain pass, river, lake, basin and playa are all put into English, but those in Tibetan such as caka (saline lake) and co (lake) that are commonly accepted abroad are kept as they are in the Chinese phonetic alphabet as caka, co, etc.

C. The proper and common components of a geographical name are generally seperated in writting for their distinction (e.g. Gomo Co), except for those geographical names with their proper and common components joined together in writing as shown in the gazetteers (e.g. Yamzho Yumco). For the name of a river or lake consisting of a monosyllable proper component and a common component, the whole name is transliterated into the Chinese phonetic alphabet and its common component in English is added to it separately for better explicitness (e.g. Nujiang River, Yanghu Lake and Co Nyi Lake instead of Nu Jiang, Yang Hu and Co Nyi, respectively).

D. In case a lake consists of several small lakes, the general name of the lake is used. Should the small lakes need to be specified, the numbers of the small lakes in Roman numerals are inserted between the proper component and common component of the general name of the lake, e.g. Yinhu I Lake and Dubuke II Lake.

E. For names of cities, towns and settlements, their proper components are generally used, with the common components such as city, county, town and prefecture ommitted. In case the proper components of a place name carries the word meaning river or lake which is liable to get confused with the common comnponent, the place name is transliterated in the Chinese phonetic alphabet with all syllables joined together, e.g. Shiquanhe and Lenghu.

APPENDIX 2

APPENDIX 2 List of Lakes on the Qinghai–Tibet Plateau

No.	Lake name	Area (km²)	Hydro-chemical type	Salinity (g/l)	pH	Na	K	Ca	Mg	Cl	SO₄	CO₃	HCO₃	B₂O₃	Note
001	La'nga Co	265	MC	1.02	8.6	142	16	6.0	84	57	84	66	512.6	44.0	(11)
002	Mapam Yum Co	394	MC	0.45	8.2	45	6	27.9	27	11	25	15	274.6	17.0	(11)
003	Kunggyû Co	71	SG	5.40	9.5	1490	62	5.0	82	813	241	1490	18.3	1010.0	(11)
004	Paikû Co	250	NS	2.36	9.5	410	90	2.0	354	47	1068	377	0.0	5.2	(1)
005	Langjang Co	25	SC	2.00	No	515	119	4.2	32	58	41	332	895.7	No	(6)
006	Ngamring Co	23	SC	4.11	9.7	2000	5	4.0	12	211	827	955	0.0	78.5	(1)
007	Lang Co	12	SC	1.97	9.5	511	10	3.3	63	106	18	241	992.1	19.9	(1)
008	Dinggyê Co	12	MC	0.71	8.8	427	20	10.0	18	56	No	35	143.5	No	(2)
009	Como Chamling L.	68	NS	5.97	8.4	1370	198	6.5	101	607	1795	0	215.7	1496.8	(1)
010	Yinhu IL.	0.09	NS	4.86	9.4	1354	95	46.0	105	1470	802	102	286.7	522.9	(2)
010-1	Yin hu II L.	0.15	NS	3.25	8.4	724	30	166.0	76	916	584	No	333.2	372.0	(2)
011	Yinhu III L.	0.74	MC	6.92	9.2	2096	237	12.0	39	2443	413	50	627.0	902.2	(2)
012	Malashan L.	No	NS	116.68	No	10747	9515	581.0	11290	4417	79135	114	497.0	No	(7)
013	Ximen Co	No	C	1.82	No	No	No	3.9	147	31	212	146	949.5	No	(7)
014	Xifeng L.	No	C	0.06	No	No	No	7.9	No	3	9	No	27.4	No	(7)

APPENDIX 2

No.	Lake name	Area (km²)	Hydro-chemical type	Salinity (g/l)	pH	Na	K	Ca	Mg	Cl	SO₄	CO₃	HCO₃	B₂O₃	Note
015	Rabgyai Co **	9.5	C	0.07	7.1	No	No	23.4	16	15	20	No	65.1	No	(7)
016	Amjog Co **	No	C	0.13	7.2	No	No	41.7	19	10	18	No	72.2	No	(7)
017	Cocholung L. **	1	NS	154.10	No	No	No	4.8	82	1	96	1	1.3	No	(7)
018	Daggyai Co	76	SC	3.33	9.4	950	63	11.4	39	220	828	302	756.6	113.8	(1)
019	Dajamang Co	8.2	C	0.12	7.5	17	1	14.3	2	2	19	0	73.2	5.2	(1)
201	Maindong Co	30	SC	12.09	No	3597	291	8.0	328	1077	2083	2392	2309.5	No	(6)
202	Langmari L.	0.01	MS	322.26	7.5	105364	6536	766.0	5805	184934	9382	No	4648.0	3664.0	(2)
203	Rabang Co	32	NS	67.59	9.2	20720	2540	11.6	443	15857	25248	921	1.2	1634.0	(5)
204	Dingzi L.	4	NS	9.16	No	924	88	No	905	2447	4215	551	No	29.5	(2)
205	Mani Co	3	C	1.51	No	310	50	4.0	80	194	487	133	247.6	7.5	(5)
206	Cokeza L.	3	MC	0.97	8.7	188	8	20.0	33	148	263	0	289.7	17.6	(2)
207	A'ong Co	56	MC	58.81	No	19125	1143	39.5	634	18086	14032	3326	0.0	2078.8	(2)
208 *	Xia Caka	3	WC	385.24	1.0	125190	25782	No	409	184190	40284	3099	0.0	4958.0	(2)
208–1	Xia Caka	3	WC	201.47	8.4	68821	9026	96.0	432	108457	6776	740	3538.0	2984.0	(2)
209–1	Dubuke I L.	0.0002	MC	16.60	9.2	5491	734	5.1	100	4320	3357	1672	0.0	774.6	(2)

Unit (mg/l)

APPENDIX 2

Unit (mg/l)

No.	Lake name	Area (km²)	Hydro-chemical type	Salinity (g/l)	pH	Na	K	Ca	Mg	Cl	SO₄	CO₃	HCO₃	B₂O₃	Note
209	Dubuke II L.	1	WC	302.93	9.3	104427	19925	68.9	17	130548	39661	510	0.0	5938.7	(2)
210	Nyêr Co	24	NS	233.99	7.6	46804	16540	610.0	15320	88110	56550	0	4762.0	4482.0	(1)
210–1	Chalaka I L.	0.8	NS	40.02	No	No	No	24.9	184	13183	8810	0	567.9	2807.5	(3)
211	Chalaka II L.	6	MC	107.22	9.5	No	No	16.5	27	19575	28293	4518	419.8	5689.5	(3)
211*	Chalaka II L.	6	MC	332.01	No	No	No	5.3	53	118734	31850	39662	2531.0	7074.6	(7)
211–1	Chalaka III L.	3.7	NS	0.42	No	46	8	30.6	24	24	84	1	183.2	22.9	(2)
211–2	Chalaka IV L.	0.001		1.00	No	No	No	No	No	No	No	No	No	No	(2)
212	Chali Co	48	MC	46.58	9.4	14954	732	0.0	103	8766	12169	5260	0.0	3712.0	(2)
213	Ngangla Ring Co	528	WC	17.48	No	No	No	1.8	35	2838	4553	1384	1234.6	1174.4	(3)
214	Zabuye Caka (north L.)	98	MC	377.21	10.1	125500	24750	0.0	38	154681	58813	23920	0.0	7800.0	(1)
214–1	Zabuye Caka (south L.)*	145	MC	435.66	No	135735	39060	0.0	5	154880	38290	51610	0.0	12240.0	(1)
214–1	Zabuye	145	MC	398.95	No	128699	29490	0.0	31	172090	32350	25000	1424.0	7628.0	(1)
214–2	Swan L.	0.03	SC	4.72	9.6	1194	167	0.0	175	550	387	1014	738.3	424.5	(1)
214–3	Leen L.	0.2	MC	58.30	No	18437	3000	7.2	41	18718	8000	3442	0.0	5685.2	(1)

APPENDIX 2

No.	Lake name	Area (km²)	Hydro-chemical type	Salinity (g/l)	pH	Na	K	Ca	Mg	Cl	SO₄	CO₃	HCO₃	B₂O₃	Note
215	Cam Co	86.7	NS	153.70	9.0	43967	7812	340.0	3627	77090	16130	326	1324.0	1795.0	(1)
216	Qagar Co	23	MC	14.69	9.7	4778	108	0.0	91	3478	2667	1500	1769.5	260.8	(1)
217	Jibucaka Co	8.2	NS	82.72	No	23433	4757	29.5	1369	33830	13210	797	2013.0	2422.0	(1)
218	Lagkor Co	92	WC	59.79	8.8	16674	2618	8.8	794	13150	19430	1213	2914.0	2355.0	(1)
219	Taro Co	485	SC	0.77	8.7	163	14	16.0	19	83	62	32	319.7	55.6	(1)
220	Jiabo Co	6	MC	16.45	9.2	4694	126	3.3	60	4339	1890	1584	0.0	3145.3	(1)
222	Tangra Yum Co	544	C	18.49	9.5	79	3	0.3	16	32	40	18	10.1	590.3	(7)
223	Damxung Co	54	C	1.11	10.1	335	28	11.1	76	89	23	67	478.6	4.5	(5)
224	Gomang Co	55	MC	14.86	No	No	No	17.9	140	1045	5361	1189	2092.5	104.3	(3)
225	Xagbai Co	17	SC	11.00	No	No	No	9.8	39	1359	1112	1807	1935.3	784.8	(3)
226	Kong Co	4	SC	1.22	No	No	No	12.4	34	88	74	80	583.9	26.4	(3)
227	Dagzê Co	240	SC	40.69	9.6	13263	996	2.5	179	2799	10519	8249	3886.0	715.2	(2)
228	Chaxia L.	3	MC	4.61	9.6	1762	198	6.0	40	1188	919	318	No	165.1	(2)
229	Gyarab Pün Co	37	SC	24.09	No	8607	521	4.0	42	2138	2101	6666	2861.0	1017.8	(2)
230	Zhüqu Co	4	C	No	9.5	76	2	1.0	21	21	30	32	16.4	No	(4)

Unit (mg/l)

APPENDIX 2

No.	Lake name	Area (km²)	Hydro-chemical type	Salinity (g/l)	pH	Na	K	Ca	Mg	Cl	SO₄	CO₃	HCO₃	B₂O₃	Note
						\multicolumn{9}{c	}{Unit (mg/l)}	
231	Chagmar Co	5	SC	13.06	9.4	4062	198	5.0	221	1120	2248	1632	3242.0	297.1	(2)
232	Urru Co	336	WC	0.43	No	No	No	18.7	38	25	25	0	265.4	10.6	(3)
233	Baidoi Co	62	C	No	No	91	3	0.1	6	17	37	37	13.5	No	(4)
234	Pongyin Co	48	WC	340.54	9.0	114082	23495	8.0	36	175757	9283	9425	4880.0	3054.7	(2)
235	Pêli Co	17	WC	215.37	9.0	64057	5000	1.6	115	74954	49000	6821	2103.5	1869.5	(1)
236	Norma Co	66	SC	35.42	9.8	12336	622	9.0	115	2488	4288	11119	2655.0	1591.9	(2)
237	Sêrbug Co	60	SC	28.35	9.6	9460	527	5.0	179	2997	4519	6057	3668.0	836.6	(2)
238	Qagoi L.	88	NS	0.38	No	No	No	26.3	25	32	49	0	192.7	8.0	(3)
239	Gyaring Co	480	C	0.26	9.9	33	3	40.6	24	8	10	11	70.8	No	(7)
240*	Koggyaling L.	65	MC	220.50	8.8	74147	11291	0.0	198	64835	23629	33523	5587.2	6052.8	(3)
241	Pongcê Co	15	SC	6.36	9.5	1245	248	6.2	160	1022	1810	775	1023.0	63.5	(2)
241-1	Xiao Pongcê	No	MC	12.32	9.1	3710	419	11.0	281	2179	2408	766	2404.0	121.1	(2)
242	Naiqam L.	44	SC	10.26	9.6	3580	133	8.0	89	1018	915	2217	2175.0	110.1	(2)
243	Siling Co	1658	WC	18.96	9.3	5550	584	2.0	262	3390	6980	412	1441.6	187.0	(1)
244	Yagedong Co	35	WC	67.97	9.8	21521	2306	3.0	242	11639	27230	2855	1427.4	669.8	(2)

# APPENDIX 2

Unit (mg/l)

No.	Lake name	Area (km²)	Hydro-chemical type	Salinity (g/l)	pH	Na	K	Ca	Mg	Cl	SO₄	CO₃	HCO₃	B₂O₃	Note
245	Jiangdongrurui Co	48	WC	19.07	9.3	5780	584	1.0	265	3420	6994	372	1312.9	187.0	(1)
245–1	Shijang L.	3	MC	0.57	No	29	9	6.0	65	30	58	16	347.0	1.9	(1)
246	Gomang Co	94		No	No	No	No	No	No	No	No	No	No	No	( )
247	Goigan Co	24	MC	1.99	No	454	10	15.9	80	43	797	45	528.3	13.0	(2)
248	Mugdê Co	3.5	MC	8.71	No	1892	172	10.2	125	3496	1553	491	941.6	24.3	(3)
248–1	Zhuozha L.	10	WC	3.75	No	560	580	6.0	210	730	1519	0	240.0	71.5	(1)
249	Bobsêr Co	19	SC	12.45	No	4221	106	5.2	84	1206	2010	2301	2412.0	88.4	(2)
250	Ngang L.	5	MC	93.50	No	No	No	4.4	15	22661	26649	5207	3615.5	792.1	(2)
250–1	Ngang L.*	5	MC	98.90	No	No	No	13.1	25	25299	25982	6482	3443.3	748.5	(2)
251	Bangkog III L.	80	MC	258.17	8.7	80241	21279	0.0	111	114387	35256	1448	1327.7	3126.1	(2)
251–1	Bangkog II L.	50	MC	187.07	8.6	60922	11220	0.0	45	54950	37316	16469	3058.0	2555.8	(1)
251–1	Bangkog II L.*	50	MC	402.52	8.6	131708	28942	No	18	150960	38250	40290	3825.0	7650.0	(2)
251–2	Xiaqong L.	10	MC	7.16	8.7	2000	463	4.0	37	2560	970	310	584.0	190.6	(1)
252	Buga Co	34	MC	2.49	8.3	621	62	7.0	160	190	471	120	841.8	12.0	(2)
252–1	Chadongbei L.	0.1	WC	6.31	No	1450	334	10.0	292	1990	879	481	763.0	89.8	(1)

APPENDIX 2

No.	Lake name	Area (km²)	Hydro-chemical type	Salinity (g/l)	pH	Na	K	Ca	Mg	Cl	SO₄	CO₃	HCO₃	B₂O₃	Note
252-2	Punuji L.	0.2	NS	10.11	No	2700	580	11.0	270	3500	2170	231	383.0	223.1	(1)
252-3	Xiangmanan L.	0.05	MC	1.18	No	230	27	8.0	79	200	200	118	296.0	14.8	(1)
253	Ben Co	20	MC	82.80	No	No	No	15.1	130	7951	11128	25276	5995.9	1399.7	(3)
253-1	Hongya L.	0.1	MC	1.37	No	300	30	7.0	65	290	190	62	378.0	40.2	(1)
253-2	Haguo L.	0.24	WC	12.21	No	3650	307	3.0	247	4020	2741	526	517.0	158.4	(1)
254	Angdar L.	34	WC	180.87	9.8	69000	3100	24.0	215	92096	8070	6168	1343.7	706.4	(3)
255	Zainzong Caka	9	NS	412.48	No	94934	17265	5.1	27328	161462	105780	2120	0.0	3069.8	(2)
256	Qêgê Pun Co	3	SC	16.59	No	5683	241	10.2	129	1094	706	5240	3337.7	126.2	(2)
257	Yangnapen Co	12	SC	14.09	No	No	No	No	44	1454	2093	2384	3005.3	50.1	(3)
258	Namka Co	12	WC	14.10	No	4422	310	10.1	231	3920	4221	629	120.6	211.1	(2)
259	Rencogongma III L.	3	C	79.50	No	No	No	26.3	280	10624	22403	10772	8773.6	815.4	(3)
260	Zongmugen Co	2.3	MC	0.61	No	108	7	10.1	46	86	30	71	250.4	No	(4)
261	Kyêbxang Co	142	C	63.35	10.2	23601	807	25.5	163	12850	7099	15344	2635.1	724.9	(2)
262	Xogor Co	23	MC	35.58	No	No	No	14.5	70	11940	6118	1805	2156.7	136.4	(3)
263	Dungqag Co	46	C	36.38	No	No	No	10.6	191	6938	9679	3168	2928.0	442.3	(3)

Unit (mg/l)

APPENDIX 2

No.	Lake name	Area (km²)	Hydro-chemical type	Salinity (g/l)	pH	Na	K	Ca	Mg	Cl	SO₄	CO₃	HCO₃	B₂O₃	Note
										Unit (mg/l)					
264	Sên Co	40	C	23.51	10.2	91	6	0.3	2	15	17	46	22.0	No	(7)
265	Bam Co	190	C	16.96	9.8	6000	493	2.1	136	1894	4470	2016	1499.0	304.8	(7)
266	Nam Co	2376	SC	1.39	9.3	No	No	27.4	24	29	923	No	308.8	No	(6)
267	Daru Co	52	MC	6.94	No	No	No	15.1	278	853	3361	No	504.8	25.7	(3)
268	Jang Co	38	SC	25.67	9.5	8348	704	8.0	78	1753	7127	3996	3244.0	363.2	(2)
269	Zigê Tang Co	188	SC	29.09	9.8	3355	885	110.3	107	2322	9819	7610	4865.6	0.5	(2)
270	Qieli Co	9	NS	7.51	8.8	2096	222	32.0	273	1425	2565	378	499.0	16.5	(2)
271	Pung Co	140	MC	6.01	8.7	1690	222	9.0	27	679	2189	210	807.0	161.9	(2)
272	Bong Co	140		No	No	No	No	No	No	No	No	No	No	No	(2)
273	Dung Co	120	SC	12.38	9.9	2671	827	6.7	151	1301	6972	36	234.0	161.0	(2)
274	Cungu L.	18	NS	1.55	No	189	28	28.9	136	81	806	44	206.0	31.0	(2)
275	Co Nag L.	180	SC	0.99	8.7	259	23	21.0	36	15	160	230	240.0	2.3	(2)
276	Kalong L.	15	SC	0.75	No	159	21	14.3	16	51	104	71	317.0	No	(2)
277	Co Ngoin L.	60	C	0.52	7.0	43	8	12.8	55	14	39	17	327.1	No	(7)
278	Argog Co	64	SC	0.30	8.5	39	4	21.8	10	7	19	15	167.8	12.0	(4)

APPENDIX 2

Unit (mg/l)

No.	Lake name	Area (km²)	Hydro-chemical type	Salinity (g/l)	pH	Na	K	Ca	Mg	Cl	SO₄	CO₃	HCO₃	B₂O₃	Note
279	Ziri L.	No	SC	5.00	No	1440	92	2.0	62	1073	1054	174	1034.0	50.2	(1)
280	Sekazhig Co	16	C	22.74	9.4	9184.0	550.0	25.1	284.9	6863.0	3138.4	2342.9	0.0	1640.8	(14)
281	Donggu Co	9	NS	2.97	8.9	484.00	170.0	27.9	355.0	516.6	476.0	0.00	921.8	160.9	(14)
282	Wuguo Co	24	C	260.84	9.7	39.02	8550.0	39.02	128.5	102941.5	11184.2	50876.1	0.00	5841.9	(14)
283	Cuowa L.	13	C	174.40	9.2	55300.0	7275.0	27.9	204.5	53394.9	13314.2	31986.5	12841.6	3275.9	(14)
284	Chabuluo Co	42	NS	25.17	9.3	4225.0	700.0	67.7	586.5	8552.8	3031.4	2770.5	0.00	215.5	(14)
285	Tungpo Co	29	C	104.88	9.7	8900.0	1460.0	6.3	110.5	600.1	1483.4	7729.3	8629.5	171.1	(14)
286	Puoji Co	13	NS	10.52	9.5	2104.0	1600.0	36.2	316.1	2111.2	2065.9	2014.9	256.1	215.5	(14)
301	Bangong L. (East part)	565.6	MC	0.75	8.7	96	8	36.0	57	87	54	44	345.9	22.0	(2)
301-1	Bangong L. (Middle part)	No	NS	2.72	9.5	613	46	7.9	163	639	533	172	490.6	51.2	(7)
302	Kayi Co	3	MS	167.02	7.8	38000	4048	281.4	10105	77536	34657	384	103.1	1584.4	(5)
303	Changmu Co	6	C	63.14	8.2	16985	768	37.1	3562	25698	14793	554	133.8	590.0	(5)
304	Tai Co	8	C	15.10	8.9	4852	286	1.2	271	4656	1655	1520	1502.0	303.0	(5)
305	Luotuo L.	62	MC	2.26	No	458	18	4.0	31	94	1000	38	255.1	161.0	(4)

APPENDIX 2

No.	Lake name	Area (km^2)	Hydro-chemical type	Salinity (g/l)	pH	Na	K	Ca	Mg	Cl	SO$_4$	CO$_3$	HCO$_3$	B$_2$O$_3$	Note
306	Mêmar Co	134.4	MC	15.97	No	4764	500	2.0	223	1461	5700	1633	1184.0	441.5	(4)
307	Aru Co	36	SC	0.76	No	177	17	3.0	31	41	100	81	267.7	36.1	(4)
308	Chagcam III L.	32	MS	57.25	7.5	29118	2441	166.0	1199	17112	4888	0	183.0	1814.1	(2)
309*	Chagcam Caka II L.	57.5	NS	384.78	8.0	104002	21840	35.0	14700	178970	60510	0	1748.0	2143.0	(1)
309—1	Chagcam Caka II L.	57.5	NS	305.12	No	86507	13830	308.0	10800	143180	45610	0	2286.0	2088.0	(1)
310	Chagcam Caka I L.	24.5	MS	75.35	7.9	23588	2600	83.3	1895	39748	6600	106	66.0	464.6	(1)
311	Duqu Co	2	SC	5.99	9.2	1169	459	10.0	98	800	706	402	2137.0	184.8	(2)
312	Bêro Zê Co	33	NS	29.10	9.2	7170	933	0.0	991	5475	12611	876	103.7	806.4	(1)
313	Oma Co	4	NS	6.97	9.6	1500	217	0.0	466	979	2800	390	538.2	71.5	(1)
314	Chagnag Co	6	MS	349.87	7.6	84572	34183	No	20095	182425	25684	457	0.0	1935.0	(5)
315	Chagbo Co	34	MC	1.79	9.2	480	20	0.0	69	502	200	28	417.4	67.3	(1)
316	Rena Co	16	WC	18.79	9.7	8100	514	8.3	574	4559	1962	790	1749.0	465.6	(4)
317	Gangmar Co	14	WC	369.07	9.5	131480	11720	0.0	95	178970	30710	7354	4051.0	3877.0	(1)
318	Buerga Co	12	NS	135.48	7.9	41723	4713	69.5	3717	65286	18357	365	554.5	571.0	(5)
319	Domar Co	18	MS	117.12	8.5	32550	5000	570.2	2519	72025	2038	116	266.2	1706.8	(4)

Unit (mg/l)

APPENDIX 2

Unit (mg/l)

No.	Lake name	Area (km²)	Hydro-chemical type	Salinity (g/l)	pH	Na	K	Ca	Mg	Cl	SO$_4$	CO$_3$	HCO$_3$	B$_2$O$_3$	Note
320	Riwan Caka (II)	1.5	MS	321.65	No	116592	4350	505.9	3278	188803	7700	16	110.2	175.8	(1)
320-1	Riwan Caka (I)	1.5	MS	342.83	No	109721	7593	324.0	10700	186540	24780	0	630.4	612.5	(1)
321	Zhajieyi Co	9	MS	18.59	8.2	5250	486	293.4	1902	8168	2015	0	364.0	101.3	(4)
322	Cêmar Co	36	MC	214.55	8.8	64942	21500	0.0	0	89054	23900	9357	2777.4	2450.1	(1)
323	Dong Co	88	NS	114.28	8.7	22260	9026	56.0	4164	31147	38455	792	5968.0	1958.0	(2)
324	Zhaxi Co	44		28.73	No	No	No	No	No	No	No	No	No	No	(5)
325	Youbu Co	62	NS	35.84	No	No	No	3.2	881	4100	17234	789	950.7	846.3	(3)
326	Dawa Co	80	NS	35.68	9.3	8412	703	24.3	997	3681	19675	852	323.5	878.0	(5)
327	Qingmuke L.	7	NS	11.52	9.6	2417	300	No	624	1171	5544	708	451.5	264.2	(1)
328	Garing Co	64	NS	277.66	7.4	102512	903	711.0	2579	158897	11887	No	56.0	102.1	(5)
329	Coqên Co	1	NS	1.46	No	277	55	24.0	73	100	500	38	301.0	85.7	(1)
330	Caij Co	24	MC	15.84	9.8	4306	429	0.0	308	1819	6207	1626	475.9	586.4	(1)
331	Qigai Co	20	MS	0.29	No	No	No	No	No	No	No	No	No	No	(5)
332	Zhari Nam Co	1004	NS	11.64	9.6	2996	439	1.2	296	1723	5565	0	46.6	555.7	(7)
333	Zougangyou Caka	15	SC	27.04	10.0	11900	1486	3.6	123	4576	1984	4670	1836.2	408.9	(4)

APPENDIX 2

No.	Lake name	Area (km²)	Hydro-chemical type	Salinity (g/l)	pH	Na	K	Ca	Mg	Cl	SO₄	CO₃	HCO₃	B₂O₃	Note
334	Ning Co	8	SC	317.90	No	No	No	26.0	52	153915	14608	15091	3488.4	4080.2	(3)
335	Gomo Co	66	NS	313.36	No	No	No	10.9	581	183166	6457	478	431.8	829.2	(3)
336	Hee Co	4.5	MS	19.83	No	No	No	21.3	1059	2809	9166	289	1056.4	291.4	(3)
337	Wuming L.	9	NS	51.96	9.0	23100	500	10.9	614	22963	1981	141	2636.9	No	(4)
338	Laxong Co	59	MC	12.61	10.0	4620	263	3.3	229	2754	1982	1727	1026.2	No	(4)
339	Gangtang Co	11	C	72.43	9.3	24948	3969	12.0	61	38371	1457	1826	14.6	1549.8	(5)
340	Yibug Caka	88	NS	105.10	8.6	33862	1164	156.3	9002	43408	17123	85	74.2	219.0	(5)
341	Kangru Caka	9.5	MS	322.55	7.0	117195	2728	260.5	3847	185787	12266	42	26.2	286.0	(5)
342	Margog Caka	76	NS	323.02	7.7	100290	5630	497.9	13033	188887	13171	65	0.0	1009.0	(5)
343	Xo Caka	11	MS	320.20	7.3	122213	2051	34.7	1112	191691	2332	141	127.1	381.4	(5)
344	PiPa L.	30	NS	23.05	7.9	6380	324	411.5	873	10497	4337	0	211.4	No	(4)
345	Kungkung Caka	38	MS	350.61	7.4	117200	5600	287.6	3712	181270	6420	881	447.9	948.7	(3)
346	Longwei Co	50	C	2.10	8.0	86	3	4.5	7	No	64	0	32.3	3.9	(4)
347	Qagam Co	20	MS	184.28	No	66169	2706	340.0	1818	105730	6092	0	688.7	528.8	(1)
348	Niudu L.	10	NS	19.66	7.8	79	3	3.4	17	No	87	1	7.8	2.2	(4)

APPENDIX 2

Unit (mg/l)

No.	Lake name	Area (km²)	Hydro-chemical type	Salinity (g/l)	pH	Na	K	Ca	Mg	Cl	SO₄	CO₃	HCO₃	B₂O₃	Note
349	Darngo Co'ngoin L.	40	MS	136.16	No	44928	2335	710.0	3507	78820	5268	0	299.3	251.9	(1)
350	Amu Co	32	NS	218.84	8.1	61440	3750	942.4	12728	131862	7300	112	17.0	462.4	(1)
351	Linggo Co	140	NS	0.99	8.5	72	4	3.7	20	No	77	2	17.1	3.0	(4)
352	Ngoyêr Co	52	MS	111.18	No	24096	4669	495.0	9492	65050	6215	121	508.6	459.2	(1)
353	Bêlog Co	22	NS	331.07	8.3	116080	4850	21.2	5016	182178	21900	144	273.0	382.2	(1)
354	Cêdo Caka	36	NS	168.39	8.3	58192	1850	580.3	2585	86909	17700	12	166.9	296.9	(1)
355	Têrang Pun Co	20	NS	70.42	8.9	19119	1115	222.4	2843	18581	27652	196	196.4	406.2	(5)
356	Jangcang Co	80	NS	4.24	No	No	No	6.8	162	1478	877	129	241.3	22.2	(3)
357	Ogcom Co	4.5	NS	85.97	8.3	76320	2285	39.8	1463	34397	18086	401	416.9	897.2	(3)
358	Yaggain Can Co	92	MS	196.83	8.1	96800	12850	67.0	1847	108418	5786	266	375.6	1266.1	(4)
358-1	Yaggai Co *	92	MS	324.76	8.1	1059	564	287.6	14030	171992	23845	1322	447.9	2759.9	(3)
359	Meiriqie Co	75	NS	21.05	8.2	932	32	386.2	929	15814	2022	26	246.4	No	(4)
360	Laorite L.	25	NS	3.02	8.6	20	1	70.1	68	1462	305	17	135.3	No	(4)
361	Quedan Co	No	CL	0.22	8.7	9200	550	44.0	5	10	7	13	100.1	18.0	(7)
362	Bietang L.	78	MS	34.64	8.2	9200	550	529.5	1609	12696.6	9656.6	0.0	384.6	102.0	(14)

APPENDIX 2

No.	Lake name	Area (km²)	Hydro-chemical type	Salinity (g/l)	pH	Na	K	Ca	Mg	Cl	SO₄	CO₃	HCO₃	B₂O₃	Note
363	Lhabu Co	16	C	201.04	8.6	69960	9300	44.6	767.3	103128.5	9680	7555.9	572.2	3692.6	(14)
365	Xianaling Co	4	NS	59.68	9.3	21184	2575	27.9	486.8	199778	10951.8	3652	768.3	553.2	(14)
366	Cherkang Co	4	MS	5.42	9.2	1083	77.5	27.9	760.5	738	432.3	629.7	1664.6	101.2	(14)
367	Bawu Co	49	MS	53.36	8.2	7220	2000	125.4	10267.4	16702.2	13139.4	906.7	0.00	1884.6	(14)
370	Zuoqing Co	15	C	272.53	8.4	94200	5800	21.9	1504.2	160747	8837.2	1032.6	0.00	1222.7	(14)
371	Duolang Coguo	13	NS	64.93	9.7	24840	1475	50.2	1054.6	8144.2	1299	23926.9	4097.6	2606.3	(14)
372	Xiaxian Co	4	C	67.1	9.9	23704	750	27.9	371.8	5110.6	10372.2	26445.5	0.00	1135.1	(14)
373	Yanjan L.	12	NS	59.74	8.4	14400	1187.5	437.1	7637.6	24975.4	9783.6	251.9	1024.4	474.1	(14)
374	Niangring	1	C	129.99	9.2	70480	4487.5	41.8	451.2	56653.8	11934.8	33271	15622.1	3019.4	(14)
401	Lungmu Co	96	MS	172.50	7.6	35000	4255	1155.8	14959	109048	7697	79	90.7	45.5	(5)
402	Yanghu L.	76	CL	115.65	8.4	32550	886	950.4	8220	70633	2169	70	118.5	No	(4)
403	Tuzhong L.	12	MS	15.89	8.7	4000	429	148.6	1026	8060	1998	17	195.0	8.4	(4)
404	Wandao L.	2	MS	5.36	7.6	1250	71	284.3	439	813	2070	No	429.9	No	(4)
405	Qingwa L.	25	MS	8.01	No	1972	102	372.6	275	3770	1434	No	78.1	No	(4)
406	Yurba Co	54	MS	314.03	No	107000	5380	658.4	4926	187319	7908	No	664.2	No	(4)

Unit (mg/l)

APPENDIX 2

Unit (mg/l)

No.	Lake name	Area (km²)	Hydro-chemical type	Salinity (g/l)	pH	Na	K	Ca	Mg	Cl	SO₄	CO₃	HCO₃	B₂O₃	Note
407	Margai Caka	76	MS	78.72	No	28339	606	18.6	767	44678	2536	290	0.0	1309.1	(5)
408	Co Nyi L.	60	MS	62.73	8.5	18834	839	707.5	2133	30954	8366	165	193.4	445.9	(5)
409	Shuangduan L.	0.5	MS	1.76	8.2	456	11	49.4	71	822	215	3	127.0	No	(4)
410	Tuoba L.	30	CL	271.33	6.8	93200	4580	5431.8	2955	163312	1588	0	140.6	No	(4)
411	Xianhe L.	22.5	CL	8.22	8.3	2080	74	70.0	574	4695	677	5	43.5	55.7	(4)
412	Yuping L.	3.5	MS	1.08	7.7	220	16	38.0	50	158	400	0	188.9	8.4	(4)
413	Taiku L.	8	CL	45.08	8.2	17320	21	74.1	70	27362	136	12	78.9	No	(4)
414	Rola Co	56	MS	215.34	7.5	63600	3260	1522.5	10637	131218	4638	No	443.7	No	(4)
415	Changjing L.	11	CL	93.09	7.1	30800	1154	2897.0	943	53814	2666	0	584.3	203.8	(4)
416	Yupan L.	9.5	MS	143.72	7.3	1204	1864	288.0	7233	No	7918	0	1105.1	4269.0	(4)
417	Dogai Coring	350	CL	226.35	6.5	90	4	2.5	4	98	2	No	0.1	No	(4)
418	Cheyue L.	28	CL	60.28	6.7	85	3	7.3	5	96	3	No	0.3	No	(4)
419	Wuming L.	0.5	MS	22.31	9.3	5648	383	1160.7	705	10000	4200	0	144.8	48.0	(4)
420	Botao L.	30	MS	2.75	8.6	838	22	57.7	62	1339	270	14	146.2	No	(4)
421	84 Yan L (I)	No	MS	97.18	No	26934	2040	362.0	5466	53050	8830	37	252.0	320.3	(1)

APPENDIX 2

No.	Lake name	Area (km²)	Hydro-chemical type	Salinity (g/l)	pH	Na	K	Ca	Mg	Cl	SO₄	CO₃	HCO₃	B₂O₃	Note
										Unit (mg/l)					
421–1	84 Yan L (II)	No	MS	23.05	No	5700	213	71.0	872	9480	2992	0	241.0	67.9	(1)
422	Rierlama L.	No	NS	7.19	No	1280	31	638.0	297	1860	2850	0	207.0	8.1	(1)
501	Doqên Co	60	NS	1.70	7.7	No	No	103.0	57	34	1010	No	154.7	No	(6)
502	Chêm Co	42	NS	1.02	7.6	No	No	81.8	53	15	567	No	157.2	No	(6)
503	Puma Yum Co	295	MS	0.41	7.9	No	No	25.1	51	4	23	No	73.1	0.1	(7)
504	Yamzho Yum Co	649	NS	0.21	9.3	99	2	9.0	6	35	33	No	25.7	No	(2)
505	Chigu Co	56	MC	0.39	7.7	45	1	28.6	20	2	37	No	257.7	No	(6)
506	Nara Yum Co	27	NS	4.67	8.9	No	No	12.3	404	23	2778	97	519.5	No	(6)
507	Pasang Co	20	NS	0.12	7.2	No	No	14.3	4	6	19	0	58.6	1.6	(7)
508	Yi'ong Co	10	NS	0.09	5.6	2	1	18.2	2	1	9	No	55.0	No	(6)
509	Buchong Co	9	NS	0.06	7.2	2	2	8.8	2	1	10	0	32.3	No	(8)
510	Butuo Coqong	7	NS	0.69	7.4	3	2	9.8	3	2	10	0	38.9	No	(8)
511	Ra'og Co	15	NS	0.10	No	No	No	15.4	3	5	12	No	58.6	No	(6)
512	Bum Co	16	C	0.14	No	8	3	66.3	22	3	1	5	91.2	No	(7)
513	Gyaring L.	540	C	0.48	8.5	71	5	27.0	31	94	23	18	210.4	No	(5)

APPENDIX 2

No.	Lake name	Area (km²)	Hydro-chemical type	Salinity (g/l)	pH	Na	K	Ca	Mg	Cl	SO₄	CO₃	HCO₃	B₂O₃	Note
						\multicolumn{9}{c	}{Unit (mg/l)}	
514	Ngoring L.	630	C	0.38	8.4	45	4	33.0	21	55	14	12	192.0	No	(5)
515	Hajiang L.	9	MS	381.99	7.9	No	No	No	No	No	No	No	No	No	(8)
516	Qinghai L.	4635	NS	12.76	8.8	3419	164	10.5	781	5374	2241	371	397.0	No	(10)
517	Jainhaizi L.	0.001	C	98.35	No	28700	4366	5.5	34	34720	3741	18594	6371.5	1174.5	(7)
518	Chumba Yum Co**	11	NS	0.12	7.4	No	No	66.1	9	5	21	No	73.8	No	(7)
519	Caka L.	104	MS	323.11	6.8	80230	4470	130.0	26510	187710	23630	170	26.0	150.6	(5)
601	Aqqikkol L.	320	MS	78.19	No	21800	1320	36.1	3210	40194	9066	930	610.2	852.8	(12)
602	Jingyu L.	165	MS	61.96	No	20940	840	83.8	1506	34498	2926	387	776.1	No	(9)
603	Gas Hure L.	103	MS	359.85	7.6	62850	6680	150.0	43080	179560	64540	480	1760.0	547.5	(5)
604*	Dalangtan P.	500	MS	382.43	6.8	11500	4600	0.0	84227	251192	30425	92	0.0	254.3	(11)
605	Mangnai L.	1.5	NS	21.21	8.0	2210	510	1258.0	2040	3400	11440	0	310.0	20.1	(5)
606*	Kunteyi P.	1667	CL	328.92	6.2	53765	12219	7283.0	37917	216859	844	6	68.1	301.7	(5)
607*	Niulang Zhinu L.	1	CL	555.00	5.2	0	680	99470.0	72360	382410	0	76	0.0	No	(5)
608	Da Suhai L.	97.5	NS	32.25	8.9	8060	420	72.0	1910	9790	10730	200	940.0	93.8	(5)
609	Xiao Suhai L.	11	NS	1.63	7.4	No	No	No	No	No	No	No	No	No	(11)

APPENDIX 2

No.	Lake name	Area (km²)	Hydro-chemical type	Salinity (g/l)	pH	Na	K	Ca	Mg	Cl	SO₄	CO₃	HCO₃	B₂O₃	Note
610 *	Yiliping P.	360	MS	333.97	7.3	81860	11730	280.0	24440	193320	19670	480	390.0	1040.0	(5)
611	Xi Taijnar L.	82	MS	338.26	7.7	103670	6810	290.0	13580	188380	23960	28	170.0	996.3	(5)
612	Dong Taijnar L.	116	MS	331.54	7.8	116950	3560	470.0	5280	187210	17040	0	110.0	653.0	(5)
613	Dezong Mahai L.	11	MS	39.49	7.4	1626	2184	No	7938	26492	1143	29	876.3	88.9	(13)
614	Baluan Mahai L.	4.5	CL	257.85	7.4	81400	2300	3000.0	9400	156700	2600	470	1700.0	250.0	(13)
615	Suli L.	69	MS	333.39	6.9	94500	7270	270.0	18210	189140	22570	0	290.0	804.4	(5)
616	Da Biele L.	7.4	MS	259.88	7.0	75686	2803	584.0	9928	154176	1672	0	No	175.2	(11)
617	Xiao Biele L.	6.3	MS	205.17	6.2	56693	6858	1143.0	12002	124358	4115	No	No	80.0	(11)
618	Dabsan L.	184	MS	318.56	7.3	62760	7830	730.0	34830	203870	5700	120	450.0	220.0	(5)
618–1	Xi Dabsan L.	30	MS	262.10	7.3	9460	3400	1256.7	9460	155982	4566	No	210.5	73.3	(7)
619 *	Da Qaidam L.	36	MS	340.57	8.0	102320	5110	500.0	15670	186200	27650	No	No	2580.0	(5)
620 *	Xiao Qaidam L.	35.9	MS	239.53	7.8	81210	1100	410.0	4410	108500	41040	280	1080.0	1273.7	(5)
621	Tuanjie L.	6	MS	426.28	5.4	5800	7220	0.0	98690	273400	39400	800	12.0	756.5	(5)
622	Xiezuo L.	17	CL	359.50	5.5	16000	7720	17660.0	64870	252020	200	2	0.0	467.2	(5)
623	Nan Hulsan L.	33.4		312.69	7.8	86806	3627	1692.6	19948	195737	4232	No	604.5	24.2	(7)

APPENDIX 2   271

No.	Lake name	Area (km^2)	Hydro-chemical type	Salinity (g/l)	pH	Na	K	Ca	Mg	Cl	SO$_4$	CO$_3$	HCO$_3$	B$_2$O$_3$	Note
624	Bei Hulsan L.	90.4	CL	243.07	7.5	77467	790	6783.8	6377	148764	2333	0	57.2	69.6	(11)
625	Qarhan P.	5856	CL	417.50	5.0	4700	3160	12890.8	93308	299389	82	No	No	223.5	(5)
626	Hurleg L.	110	NS	0.89	8.0	No	No	No	No	No	No	No	No	No	(7)
627	Toson L.	140	MS	15.86	8.9	No	No	No	No	No	No	No	No	No	(7)
628	Gahai L.	37.4	NS	90.59	8.3	27330	250	310.0	4180	45260	13090	140	34.0	No	(5)
629	Hoh Yanhu L.	90	MS	335.66	6.8	82560	4410	170.0	27230	183000	37250	0	760.0	167.6	(5)
630	Xiligou L.	19.5	NS	261.32	7.6	78980	1640	480.0	11110	121130	47250	250	480.0	No	(5)
632	Dajuba L.	5	MS	217.88	No	86400	1009	360.7	21905	189921	15157	0	594.9	1323.0	(12)
633	Dasaziquan L.	9	NS	0.74	No	125	8	42.7	34	169	61	0	286.8	8.7	(12)
634	Beileke L.	21	MS	2.57	No	744	9	19.0	116	1302	51	31	290.4	4.4	(12)
635	Chaikai L.	18	MS	324.13	6.8	57000	2513	160.3	32007	155908	39054	363	1452.0	No	(5)
636	Ayakkum L.	535.5	MS	145.90	No	42400	936	148.3	7781	88918	4649	0	149.5	800.6	(12)
637*	Chahanshilatu P.	2000	MS	351.31	No	115600	6380	270.0	9820	183460	35400	48	95.0	177.6	(11)
638*	Potash L.	1.5	CL	321.20	No	59600	4320	6260.3	35015	209278	1103	No	0.0	57.8	(7)
639	Dongling L.	7.2	CL	356.90	6.9	17575	2812	24729.4	58394	246500	170	No	No	334.2	(7)

APPENDIX 2

No.	Lake name	Area (km²)	Hydro-chemical type	Salinity (g/l)	pH	Na	K	Ca	Mg	Cl	SO₄	CO₃	HCO₃	B₂O₃	Note
640	Gansenqnan L.	16	NS	51.35	7.9	13255	38	1030.1	1114	29016	6779	0	213.6	6.5	(7)
641	Nabei L.	15		29.00	No	No	No	No	No	No	No	No	No	No	(7)
642	Tart L.	4	MS	11.57	No	2708	39	No	1145	7150	310	79	125.0	13.0	(1)

Explanation

I. Location of the numbers in this table please look at Fig. 1.13.

II. Caption of the column "Hydrochemical types" as follows:

C: Carbonate type; SC: Strong–Carbonate subtype; MC: Medium–Carbonate subtype; WC: Weak–Carbonate subtype; NS: Na–sulfate subtype; MS: Mg–sulfate subtype; CL: Chloride type.

III. The others explained as follows:

* : means results of intercrystaline brine analysis;

* * : means unit use equivalent percentage.

IV. Data sources: 1. Zheng Mianping et al. (1989); 2. Zheng Mianping et al. (1980); 3. Geological Party of Northern Tibet (1963; 1964); 4. Regional Geological Party of Tibet (1980–1984); 5. Institute of Salt Lakes in Qinghai, C.A.S. (1976; 1978; 1984); 6. Nanjing Institute of Soil, C.A.S. (1970′s); 7. Committee of Comprehensive Survey, C.A.S. (1984); 8. Nanjing Institute of Geography, C.A.S. (1970′s); 9. Regional Geological Party of Xinjiang (1970′s); 10. Lanzhou Institute of Geology, C.A.S. 1980; 11. The First Geological Party of Qinghai, (1960; 1967; 1969); 12. Regional Geological Party of Xinjiang and Institute of Mineral Deposits (1984–1985); 13. Institute of Chemistry, C.A.S. (1957; 1960); 14. Tibet Bureau of Geology and Mineral Resources (1988–1991).

Appendix 3 Index of Place Names

English name	Chinese name	Original name	N. lat.	E. long.	No. in the app. map I	Note
Argog Co	阿果错	郭阿错	30°59′	82°14′	278	2
Aksayqin L.	阿克塞钦湖	阿克萨依湖	35°12′	79°50′		1
Altun Mts.	阿尔金山		38°36′	89°00′		1
Alxa block	阿拉善地块		40°30′	103°00′		1
Amdo	安多	安东买马	32°12′	91°36′		1
Amjog Co	安觉错		29°12′	88°19′	016	2
Amu Co	阿木错		33°27′	88°40′	350	1
Angdar L.	昂达尔湖	安德尔湖	32°43′	89°35′	254	1
A'ong Co	阿翁错	阿工湖	32°46′	81°44′	207	1
Aqqikkol L.	阿其克库勒湖		37°05′	88°28′	601	1
Aru Co	阿鲁错	阿如错	34°00′	82°20′	307	1
Ayakkum L.	阿牙克库木湖		37°17′	89°25′	636	1
Baidoi Co	帕度错	拔度错	32°48′	87°50′	233	1
Baingoin	班戈		31°18′	90°00′		1
Bairab Co	拜惹布错	麻克哈湖	35°02′	83°10′		1
Baluan mahai L.	巴伦马海		38°02′	94°13′	614	3
Bam Co	巴木错		31°15′	90°35′	265	1
Bangdag Co	邦达错	雅尔错	34°58′	81°32′		1
Bangkog Co	班戈错		31°44′	89°29′	251	1
Bangong Co	班公错	错木昂拉仁波	33°44′	79°15′	301	1
Bawu Co	八乌错		32°14′	85°33′	367	3
Bayan Har Mts.	巴颜喀拉山		35°18′	95°00′		1
Bayanheqian Mt.	巴颜和欠山		34°58′	96°53′		3
Bei Hulsan L.	北霍鲁逊湖		36°54′	95°57′	624	1
Beileke L.	贝勒克湖		36°22′	90°40′	634	3
Ben Co	苯错		30°56′	89°40′	253	3

Bêro zêco	别若则错		32°26′	82°56′	312	2
Bietang L.	别塘湖		33°02′	83°20′	362	3
Bêlog Co	比隆错	毕洛错,折罗茶卡	32°54′	88°35′	353	2
Bobsêr Co	瀑塞尔错		32°20′	89°27′	249	2
Bom	波密		29°48′	95°42′		1
Bong Co	崩错	孟错	31°13′	91°09′	272	1
Botao L.	波涛湖		34°00′	90°00′	420	3
Brackish L.	咸湖		37°09′	90°08′		4
Buchong Co	布冲错		39°38′	95°34′	509	3
Buerga Co	布尔嘎错		33°40′	84°23′	318	3
Buga Co	补嘎错	布尔卡湖	31°06′	89°33′	252	3
Bum Co	本错		29°32′	96°51″	512	2
Burang	普兰		30°18′	81°06′		1
Burhan Budai Mts.	布尔汗布达山		35°48′	96°00′		1
Burog Co	布若错		34°22′	85°45′		1
Butuo Coqong	布托错穷		31°37′	95°40′	510	3
Caiji Co	蔡几错		31°13′	85°26′	330	3
Caka L.	茶卡湖		36°42′	99°07′	519	1
Cam Co	茶错		31°59′	83°42′		1
Cam Co	仓木错	麻米错: Marmê Co	32°07′	83°32′	215	1
Cêdo Caka	才多茶卡	切多茶卡	33°10′	89°00′	354	2
Cêmar Co	才玛尔错	次玛错	32°33′	84°36′	322	2
Chabuluo Co	查布罗错		31°57′	86°59′	284	3
Chagbo Co	查波错		33°22′	84°13′	315	2
Chagcam Caka	扎仑茶卡	张藏茶卡,恰果错,克穷错	32°34′	82°16′	308	2
Chagmar Co	昌玛错	茶莫尔湖	32°16′	87°54′	231	2
Chagnag Co	查那错	查治那泊	33°18′	84°56′	314	2
Chahanshilatu P.	察汗斯拉图		38°39′	92°27′	637	3
Chaikai L.	柴凯湖		37°57′	98°08′	635	3

Chainjoin Co	振泉愤		35° 55'	86° 58'		1
Chalaka II L.	茶拉卡II湖	喳勒错	31° 45'	82° 23'	211	3
Chali Co	茶里错	鲁尔帕湖,内湖,错那错,册肉湖	31° 38'	82° 24'	212	3
Changjing L.	长颈湖		35° 05'	88° 20'	415	3
Changmu Co	常木错	扎木茶卡	33° 29'	80° 19'	303	3
Chaxia L.	茶夏湖		31° 56'	87° 53'	228	3
Chêm Co	泽错		34° 09'	79° 47		1
Chêm Co	沉错		28° 55'	90° 30'	502	2
Chengdu	成都		30° 36'	104° 06'		1
Cherkang Co	查尔康错		32° 47'	83° 54'	366	3
Cheyue L.	车月湖		34° 21'	89° 11'	418	3
Chigu Co	哲古错		28° 40'	91° 40'	505	1
Chumba Yumco	冲巴雍错		28° 15'	89° 38'	518	2
Cocholung L.	错戳龙		29° 12'	84° 45'	017	1
Cokeza L.	错克扎湖		32° 40'	81° 29'	206	3
Como Chamling L.	错姆折林	特索莫特里敦湖	28° 24'	88° 16'	009	1
Co Nag L.	错那	黑海湖,错那格湖	32° 03'	91° 29'	275	1
Co Ngoin L.	错鄂		31° 28'	91° 30'	277	1
Co Ngoin	触安姆错	错鄂,错俄木	31° 38'	88° 40'		2
Co Nyi L.	错尼	双湖	34° 35'	87° 16'	408	1
Coqên	措勤		31° 00'	85° 06'		1
Coqên Co	措勤错		30° 59'	85° 09'	329	2
Cungu L.	寸古湖		32° 00'	91° 53'	274	3
Cuowa L.	错哇		31° 58'	87° 02'	283	3
Da Biele L.	大别勒湖		36° 59'	94° 31'	616	3
Da Qaidam	大柴旦		37° 50'	95° 13'		1
Da Qaidam L.	大柴旦湖	伊克柴达木湖	37° 50'	95° 13'	619	1
Da Suhai L.	大苏干湖		38° 53'	93° 50'	608	1
Dabsan L.	达布逊湖		37° 00'	95° 07'	618	1

Daggyai Co	打加错		29°50′	85°44′	018	1
Darlag	达日		33°44′	99°37′		1
Dagzê Co	达则错	达格济湖	31°54′	87°32′	227	1
Dajamang Co	打加芒错		29°45′	85°40′	019	3
Dajiuba L.	大九坝湖		37°10′	90°09′	632	3
Dalangtan P.	大浪滩		34°14′	91°30′	604	3
Damxung Co	当雄错		31°37′	86°45′	223	1
Darngo Congoin L.	达尔沃错温		33°35′	88°45′	349	2
Daru Co	达如错	达仑湖	31°42′	90°45′	267	1
Dasaziquan L.	大沙子泉湖		37°19′	90°16′	633	3
Dawa Co	达瓦错		31°15′	84°58′	326	1
Delingha	德令哈		37°18′	93°18′		1
Dêngqên	丁青	甲木塘	31°24′	95°30′		1
Dezong mahai L.	德宗马海湖		38°14′	94°19′	613	3
Dinggyê	定结		28°18′	87°42′		1
Dinggyê Co	定结错		28°18′	88°07′	008	1
Dingzi L.	顶兹湖		32°29′	81°18′	204	3
Dogai Coring	多格错仁	曼特喀木湖	34°36′	88°49′	417	1
Dogaicoring Qangco	多格错仁强错		35°20′	89°15′		1
Domar Co	多玛错		32°57′	84°27′	319	2
Dong Co	洞错	洞卡错,洞嘎湖	32°11′	84°44′	323	1
Dong Taijnar L.	东台吉乃尔湖		37°30′	93°55′	612	1
Donggicona L.	冬给措纳湖	托索湖	37°08′	98°33′		1
Donggu Co	洞古错		31°59′	82°10′	281	3
Dongling L.	东陵湖		37°16′	94°50′	639	3
Doqên Co	多庆错		28°10′	89°22′	501	1
Dorqê Co	多尔改错	错仁德加	35°14′	92°07′		1
Dorsoidong Co	多尔索洞错	晒冷腾错	33°24′	89°50′		1
Dubuke I L.	都布克工湖	俄布错	31°57′	82°11′	209–1	3

Dubuke II L.	都布克 II 湖	俄布错	31° 57′	82° 11′	209	3
DuJiali L.	杜佳里湖	郭加林 Koggyaling L.	32° 00′	88° 38′	240	2
Dulan	都兰	察汗乌苏	36° 18′	98° 06′		1
Dulishi L.	独立石湖		34° 45′	81° 55′		3
Dung Co	懂错	蕫湖,洞湖	31° 42′	91° 10′	273	1
Dungqag Co	东恰错	东下湖	31° 47′	90° 25′	263	2
Duolang Coguo	夺廊错果		32° 14′	85° 50′	371	3
Duqu Co	都曲错		32° 24′	82° 51′	311	3
Eqong Co	恶穷错		32° 41′	82° 23′	376	3
Gahai L.	尕海		37° 08′	97° 33′	628	1
Gamba	岗巴		28° 12′	88° 30′		1
Gangdisê Mts.	冈底斯山		30° 36′	82° 48′		1
Gangmar Co	冈玛错		33° 50′	84° 21′	317	1
Gangtang Co	冈塘错		33° 11′	86° 40′	339	2
Gansenquan L.	甘森泉湖		37° 17′	92° 46′	640	1
Gar Zangbo R.	噶尔藏布		32° 00′	80° 00′		1
Garing Co	嘎仁错		30° 46′	84° 58′	328	2
Garkung Caka	嘎尔孔茶卡		33° 56′	86° 30′		1
Gas Hure L.	尕斯库勒湖		38° 07′	90° 48′	603	1
Gê'gyai	革吉		32° 24′	81° 06′		1
Gêrzê	改则		32° 18′	84° 00′		1
Goigain Co	戈梗错	比罗湖	32° 23′	89° 12′	247	2
Golmud	格尔木		36° 24′	94° 54′		1
Gomang Co	戈芒错		31° 34′	87° 17′	224	2
Gomang Co	戈昂错	果忙错,戈孟错 戈玛错	32° 14′	89° 12′	246	1
Gomo Co	戈木错	戈穆茶卡	33° 40′	85° 49′	335	1
Gonghe	共和	恰卜恰 (Qabqa)	36° 12′	100° 36′		1
Gopug Co	果普错		31° 54′	83° 10′		2
Gorong Co	戈荣错		30° 26′	85° 13′	287	3

Gozha Co	郭扎湖	里田错,明亮湖	35°02'	81°05'		1
Gyaco	加措		33°24'	83°18'		1
Gyarab Punco	甲热布错	加拉本湖	32°12'	87°46'	229	2
Gyaring Co	格仁错	加仁错,扎林湖	31°08'	88°20'	239	1
Gyaring L.	扎陵湖		34°55'	97°15'	513	1
Gyêsar Co	杰萨错		30°10'	84°48'		1
Gyêzê Caka	结则茶卡		33°57'	80°56'		1
Gyirong	吉隆		28°54'	85°12'		1
Hajiang L.	哈江盐池		35°03'	97°54'	515	3
Har L.	哈拉湖		38°18'	97°36'		1
Hee Co	合俄错		32°52'	85°46'	336	3
Heishi Bei L.	黑石北湖		35°37'	82°45'		1
Heishui R.	黑水河		37°44'	96°18'		3
Himalaya Mts.	喜马拉雅山		27°54'	86°54'		1
Hoh Sai L.	库赛湖		35°44'	92°50'		1
Hoh Xil L.	可可西里湖		35°34'	91°07'		1
Hoh Xil Mts.	可可西里山		35°18'	90°06'		1
Hoh Yanhu L.	柯柯盐湖		36°58'	98°14'	629	1
Hurleg L.	可鲁克湖		37°17'	96°55'	626	1
Jang Co	江错		31°33'	90°49'	268	1
Jangcang Co	江沧错		32°56'	89°14'	356	2
Jiabo Co	加波错		31°09'	84°32'	220	3
Jiaduo Co	加多错		34°03'	85°37'		3
Jiangdongrurui Co	浆东如瑞错		31°48'	89°18'	245	3
Jianhaizi L.	碱海子		29°41'	96°40'	517	3
Jianshui L.	碱水湖		35°18'	83°08'		1
Jibucaka Co	基布茶卡错		32°03'	83°58'	217	3
Jingyu L.	鲸鱼湖		36°20'	89°25'	602	3
Jinsha R.	金沙江		29°42'	98°54'		1

Juhongtu	居红土		37° 48′	96° 18′		1
Jurhen Ul Mt.	祖尔肯乌拉山		33° 57′	90° 45′		1
Kaketu R.	卡克图河		37° 42′	96° 14′		3
Kalong L.	嘎弄	尕龙湖	31° 54′	91° 32′	276	2
Kangru Caka	康如茶卡	坎米如茶卡	33° 32′	86° 58′	341	2
Karakorum Mts.	喀喇昆仑山		36° 00′	76° 00′		1
Karamiran R.	喀拉米兰河		37° 30′	84° 48′		1
Karamiran pass	卡拉米兰山口		36° 12′	87° 00′		1
Kayi Co	卡易错	扎西茶卡	33° 33′	80° 08′	302	3
Kong Co	控错		31° 30′	87° 31′	226	3
Kumkol	库木库里		37° 30′	89° 00′		1
Kumkol Basin	库木库里盆地		37° 30′	89° 00′		1
Kunggyu Co	公珠错		30° 40′	81° 30′	003	1
Kungkung Caka	孔孔茶卡	加利茶卡	33° 09′	88° 08′	345	2
Kunlun Mts.	昆仑山		36° 30′	86° 00′		1
Kunlun pass	昆仑山口		35° 36′	93° 48′		1
Kunming	昆明		25° 00′	102° 42′		1
Kunteyi P.	昆特依		38° 52′	93° 05′	606	3
Kushui L.	苦水湖		35° 42′	79° 22′		3
Kyêbxang Co	其香错	吉波江错, 气相错 色哇错	32° 27′	89° 58′	261	1
Lagkor Co	拉果错	拉格尔错	32° 03′	84° 13′	218	1
Lancang R.	澜沧江		31° 06′	97° 06′		1
Lang Co	浪错		29° 12′	87° 24′	007	3
Langa Co	拉昂错	兰加克湖	30° 45′	81° 15′	001	1
Langjang Co	浪强错		28° 43′	85° 53′	005	1
Langmari L.	浪马日湖	噶尔昆沙湖 Gargünsa L.	32° 10′	80° 00′	202	3
Lanzhou	兰州		36° 00′	103° 42′		1
Laorite L.	劳日特湖		33° 39′	90° 00′	360	3
Laxong Co	拉雄错		34° 21′	85° 14′	338	1

Leen L.	勒恩湖		31°19′	84°11′	214-3	3
Lenghu	冷湖		38°48′	93°24′		1
Lhabu Co	拉布错		32°58′	83°49′	363	3
Lhari	嘉黎		30°36′	93°30′		1
Lhasa	拉萨		29°36′	91°06′		1
Lhazê	拉孜		29°04′	87°36′		1
Lhünzê	隆子		28°24′	92°18′		1
Linggo Co	令戈错	东湖	33°45′	88°40′	351	1
Lixi'oidain Co	勒斜武担错		35°46′	90°12′		1
Longgar	隆格尔		31°06′	84°00′		1
Longmen	龙门		30°48′	106°06′		1
Longwei Co	龙尾错		33°50′	88°20′	346	1
Lumajangdong Co	鲁玛江东错	措作错, 查罗尔错	34°00′	81°40′		1
Lungmu Co	龙木错		34°37′	80°28′	401	1
Lunpola	伦坡拉		32°12′	90°10′		3
Luotuo L.	骆驼湖		34°26′	81°56′	305	3
Lushitage Mt.	吕什塔格山		36°12′	82°02′		3
Maindong Co	曼冬错	斯潘古尔湖	33°32′	78°55′	201	2
Mazhangcuoqin L.	玛章错钦		34°20′	91°34′		3
Malashan L.	马拉山湖		28°39′	85°44′	012	1
Mangnai	茫崖		38°18′	90°06′		1
Mangnai L.	茫崖湖		37°46′	91°50′	605	1
Mani Co	玛尼错		32°39′	81°21′	205	1
Mapam Yumco	玛旁雍错	玛法木湖	30°40′	81°30′	002	1
Margai Caka	玛尔盖茶卡	约基台错	35°08′	86°46′	407	1
Margog Caka	玛尔果茶卡		33°53′	87°00′	342	1
Markam	芒康		29°36′	98°30′		1
Mêdog	墨脱		29°12′	95°18′		1
Mêmar Co	美马错	明马茶卡	34°13′	82°19′	306	1

Mêrqung Co	麦穷错	曲依错	31°00′	84°34′		1
Migriggyangzham Co	米提江占木错		31°33′	83°05′		1
Meiriqie Co	美日切错		33°39′	89°43′	359	3
Moincêr	门士		31°06′	80°42′		1
Mugdê Co	木地错		32°11′	89°22′	248	2
Nabei L.	那北湖		37°07′	92°57′	641	3
Nagqu	那曲	墨河	31°24′	92°03′		1
Naiqam Co	纳江错	内江错	32°20′	88°42′	242	2
Nake Caka	纳克茶卡		34°20′	87°42′		3
Nam Co	纳木错	腾格里湖	30°40′	90°35′	266	1
Namka Co	纳卡错		31°52′	89°47′	258	2
Nan Hulsan L.	南霍鲁逊湖		36°44′	95°47′	623	1
Nangxian	朗县		29°00′	93°06′		1
Nara Yumco	拿日雍错		28°18′	91°57′	506	2
Narin Gol R.	那梭格勒河		37°00′	93°00′		1
Nau Co	纳屋错	春布茶卡	32°52′	82°15′	375	1
Ngamring Co	昂仁错		29°18′	89°11′	006	1
Ngang Co	昂错	玛日湖 Marri Co	32°12′	89°29′	250	2
Ngangla Ringco	昂拉仁错	昂拉陵湖	31°33′	83°05′	213	1
Ngangzê Co	昂孜错	则错	31°02′	87°05′		1
Ngoring L.	鄂陵湖		34°55′	97°43′	514	1
Ngoyêr Co	鄂雅错	鄂雅错琼,夏姑错	32°59′	88°42′	352	2
Niangring Co	娘荣错		32°48′	84°04′	374	3
Nieniexiongla	聂聂雄拉		28°24′	86°06′		3
Ning Co	宁错	宁日错,冬扎湖	33°19′	85°49′	334	3
Niudu L.	牛肚湖		33°30′	88°35′	348	3
Niulang Zhinu L.	牛郎织女湖		38°30′	93°40′	607	3
Niwumu	尼乌木		28°00′	94°05′		1
Norma Co	罗尔错	诺尔玛错,脑日错, 诺各错	32°23′	88°05′	236	1

Nujiang R.	怒江		24° 06′	98° 42′	1
Nyainqêntanglha Mts.	念青唐古拉山		30° 24′	90° 36′	1
Nyalam	聂拉木		28° 06′	85° 54′	1
Nyang R.	尼洋河		29° 48′	93° 30′	1
Nyêr Co	聂尔错	内尔错	32° 17′	82° 13′	210	2
Nyingchi	林芝		29° 30′	94° 18′	1
Ogcom Co	鄂纵错	鄂葱错, 切尔塞湖	32° 15′	89° 28′	357	2
Oma Co	物玛错		32° 26′	83° 11′	313	1
Orba Co	窝尔巴错	霍尔巴错	34° 32′	81° 03′	1
Pagri	帕里		27° 42′	89° 06′	1
Pasang Co	帕桑错	八松错 Pagsum Co	30° 01′	94° 00′	507	2
Paiku Co	佩枯错	拉错新错, 泊古湖	28° 54′	85° 36′	004	1
Palung Co	帕龙错	帕鲁错	30° 56′	83° 35′	1
Pamir	帕米尔		37° 54′	75° 00′	1
Pêli Co	北雷错	恰阿姆湖, 折乌茶卡	32° 54′	88° 25′	235	2
Pipa L.	琵琶湖		34° 10′	87° 45′	344	3
Pongcê Co	崩则错	傍则湖	32° 04′	88° 40′	241	2
Pongyin Co	朋彦错	榜于茶卡	32° 55′	88° 10′	234	1
Potash L.	钾湖		38° 24″	93° 05′	638	4
Puma Yumco	普莫雍错	博磨湖	28° 35′	90° 20′	503	1
Pung Co	蓬错	棒错, 硼错	31° 40′	90° 58′	271	1
Puoji Co	婆基错		31° 53′	87° 24′	286	3
Qagam Co	恰岗错	恰尔嘎木错	33° 13′	88° 23′	347	2
Qagar Co	吓嘎错		32° 20′	83° 50′	216	2
Qagoi Co	恰规错	察尔骨特湖, 夏尔骨湖	31° 49′	88° 15′	238	2
Qaidam Basin	柴达木盆地		37° 12′	94° 24′	1
Qamdo	昌都		31° 06′	97° 06′	1
Qarhan P.	察尔汗		37° 08′	95° 59′	625	1
Qêgê punco	吉给普错		32° 27′	89° 41′	256	2

Qieli Co	切里错	册律湖	31°41′	90°58′	270	3
Qigai Co	纪格错	齐格错	31°12′	85°33′	331	2
Qilian	祁连		38°06′	100°12′		1
Qilian Mts.	祁连山		39°06′	98°48′		1
Qimantag Mts.	祁漫塔格山		37°12′	91°06′		1
Qinghai L.	青海湖		36°54′	100°11′	516	1
Qingmuke L.	清木柯		31°15′	85°05′	327	3
Qingwa L.	青蛙湖		34°43′	86°20′	405	3
Qitaidaban L.	奇台大坂湖		35°44′	79°21′		3
Qoima Co	雀莫错		33°53′	91°12′		1
Qomolangma Peak	珠穆朗玛峰		27°54′	86°54′		1
Qonj	全吉	泉吉	37°12′	99°48′		1
Quanshuigou	泉水沟		35°38′	79°34′		1
Quedan Co	确旦错		34°20′	87°30′	361	3
Qumar R.	楚玛尔河		35°12′	94°00′		1
Qumarlêb	曲麻莱		34°30′	95°24′		1
Qungdo′gyang	邛多江		28°48′	92°06′		1
Rabang Co	热帮错		33°02′	80°34′	203	1
Rabgyai Co	热布杰错		30°25′	82°18′	015	2
Rabka	热布喀	多瓦	30°36′	87°30′		1
Raggryor Caka	热觉茶卡		33°41′	86°51′		2
Ra′og Co	然乌错		29°26′	96°47′	511	2
Rena Co	热那错		32°44′	84°16′	316	3
Rencogongma Ⅲ L.	仁错贡马Ⅲ湖	仑湖	30°56′	89°50′	259	3
Rierlama L.	日尔拉马湖		32°28′	92°49′	422	3
Rigain Pünco	日干配错		32°35′	86°15′		1
Rinqin Xubco	仁青休布错	休沃错	31°18′	83°29′		1
Riwan Caka	日弯茶卡		33°49′	84°33′	320	3
Rola Co	若拉错		35°25′	88°23′	414	1

Rutog	日土		33°24′	79°36′		1
Sêkazhig Co	色卡执错		32°00′	82°03′	280	2
Sên Co	申错		31°07′	90°29′	264	1
Sênggêzangbo R.	森格藏布	狮泉河 Shiquan R.	32°30′	80°30′		1
Sêngli Co	森里错	桑里错	30°35′	84°04′		1
Sêrbug Co	赛布错	狮湖错	32°00′	88°13′	237	1
Sêrxong	色雄		31°24′	92°57′		2
Sêrxü	石渠		32°59′	98°05′		1
Shiquanhe	狮泉河		32°30′	80°00′		1
Shuangduan L.	双端湖		35°18′	87°40′	409	3
Shuanghu	双湖		33°12′	88°50′		3
Siling Co	色林错	奇林错	31°48′	89°00′	243	1
Suojia	索加		34°14′	94°34′		3
Sogxian	索县		31°48′	93°42′		1
Subanxili	苏班西里		27°40′	94°16′		3
Suli L.	涩聂湖		37°04′	94°18′	615	1
Sumxi	松西		34°24′	79°54′		1
Tagchagpu Mt.	塔查普山		33°30′	82°12′		1
Tai Co	台错		33°43′	80°43′	304	3
Taiku L.	太苦湖		34°55′	88°15′	413	3
Taiyang L.	太阳湖		35°56′	90°38′		3
Talabu Co	塔拉布错		32°33′	83°33′	368	3
Tanggula Mts.	唐古拉山		32°54′	91°42′		1
Tanggula Pass	唐古拉山口		32°48′	91°54′		1
Tangra Yumco	当惹雍错	唐古拉攸木错	31°04′	86°37′	222	1
Tarim Block	塔里木地块		39°18′	83°00′		1
Taro Co	塔若错	塔罗克错	31°10′	84°10′	219	1
Tart L.	塔尔丁湖		36°53′	92°41′	642	1
Tatalin R.	塔塔梭河		37°42′	96°24′		1

Telashi L.	特拉什湖		34°49′	92°12′		1
Tèrang Punco	蒂让碧错		33°03′	89°04′	355	2
Tianjun	天峻	新源	37°18′	99°00′		1
Tongtian R.	通天河	直曲	34°36′	94°12′		1
Toson L.	托素湖		37°08′	96°56′	627	1
Tuanjie L.	团结湖		36°48′	95°21′	621	3
Tungpu Co	懂布错		31°20′	87°13′	285	2
Tuoba L.	托把湖		34°30′	88°10′	410	3
Tuoheping Co	托和平错		34°11′	83°09′		1
Tuotuo R.	沱沱河		34°12′	91°30′		1
Tuotuoheyan	沱沱河沿		34°12′	92°24′		1
Tuzhong L.	图中湖		34°32′	84°42′	403	3
Ulan	乌兰	希里沟	36°54′	98°24′		1
Ulan Ul # L.	乌兰乌拉湖		34°48′	90°30′		1
Urru Co	吴如错	阿达湖	31°43′	87°59′	232	1
Wandao L.	弯岛湖		34°51′	84°47′	404	3
Wanquan L.	万泉湖		34°15′	83°49′		1
Wudaoliang	五道梁		35°06′	93°00′		1
Wuguo Co	巫戈错		32°00′	86°38′	282	3
Wulanbaomu	乌兰保姆		37°42′	96°20′		3
Wulukele L.	乌鲁克勒湖		35°40′	81°37′		3
Wuming L.	无名湖		34°07′	85°05′	337	3
Wuming L.	无名湖		34°50′	89°20′	419	3
Xagbai Co	吓别错		32°13′	87°12′	225	2
Xianaling Co	夏那令错		33°02′	83°57′	365	3
Xagquka	下秋卡		31°48′	92°48′		1
Xainza	申扎		30°54′	88°42′		1
Xi Taijnar L.	西台吉乃尔湖		37°43′	93°29′	611	1
Xia Caka	夏茶卡	怡茶卡,叉茶卡	32°38′	81°52′	208	3

Xianhe L.	仙鹤湖		36°00′	88°10′	411	3
Xiangtao L.	香桃湖		34°08′	85°00′		3
Xiangyang L.	向阳湖		35°51′	89°32′		1
Xiao Biele L.	小别勒湖		37°02′	94°39′	617	3
Xiao Qaidam L.	小柴旦湖	巴嘎柴旦木湖	37°10′	95°30′	620	1
Xiaosuhai L.	小苏干湖		39°04′	94°13′	609	1
Xiaxian Co	吓先错		32°38′	86°33′	372	3
Xiezuo L.	协作湖		37°00′	95°39′	622	3
Xifeng L.	希峰冰成湖		27°55′	85°54′	014	3
Xigazê	日喀则		29°12′	88°48′		1
Xijir Ulan L.	西金乌兰湖		35°14′	90°22′		1
Xiligou L.	希里沟湖		36°50′	98°27′	630	1
Ximen Co	西门错		30°04′	85°54′	013	3
Xinhu L.	心湖		34°23′	84°15′		1
Xining	西宁		36°36′	101°48′		1
Xo Caka	肖茶卡		33°04′	87°47′	343	2
Xogor Co	徐果错	雪卧湖	31°57′	90°21′	262	2
Xuru Co	许如错	霞如错	30°20′	86°27′		1
Xuxu L.	徐旭		32°34′	82°42′	364	3
Yadong	亚东		27°24′	88°54′		1
Yagedong Co	雅个冬错	小触安姆湖	31°35′	89°01′	244	3
Yaggain Canco	雅根错	雅根查错,亚根亚姆茶卡	33°01′	89°48′	358	1
Yagra	亚热		31°30′	82°18′		1
Yalong R.	雅砻江		29°18′	101°06′		1
Yamatu R.	雅马图河		37°44′	96°10′		3
Yamzho Yumco	羊卓雍错		29°00′	90°45′	504	1
84 Yan L.	84 盐湖		32°10′	92°17′	421	3
Yangbajain	羊八井		30°00′	90°30′		1
Yanghu L.	羊湖		35°26′	84°39′	402	1

Yangnapen Co	洋纳朋错		32°19′	89°46′	257	3
Yanhuqu	盐湖区		32°30′	84°26′		1
Yanjian L.	盐碱湖		32°44′	86°30′	373	3
Yarlung Zangbo R.	雅鲁藏布江		29°18′	88°12′		1
Yarlungqu R.	雅鲁曲		28°16′	88°08′		2
Yashatu	雅沙图		37°42′	96°10′		3
Yejihai L.	野鸡海	年吉错	33°04′	96°18′		1
Yellow R.	黄河		34°30′	98°18′		
Yeniugou	野牛沟		38°24′	99°36′		1
Yibug Caka	依布茶卡	腰布茶卡, 巴尔替古错	32°59′	86°44′	340	1
Yiliping P.	一里坪		37°58′	93°09′	610	1
Yingshan L.	映山湖		35°21′	86°52′	423	3
Yinhu I L.	荫湖 I 湖	岗岗湖	28°19′	88°23′	010	3
Yinhu II L.	荫湖 II 湖	岗岗湖	28°00′	88°20′	010-1	3
Yinhu III L.	荫湖 III 湖	岗岗湖	27°54′	88°20′	011	3
Yi'ong Co	易贡错		30°13′	94°58′	508	1
Yiran Co	依然错	尼日阿改错	33°06′	93°14′		1
Yoigilangleb Mt.	约古宗列山		34°55′	96°30′		1
Yoqag Co	越怡错		30°30′	88°37′		2
Youbu Co	攸布错	敌布错, 杰马湖	30°48′	84°48′	325	3
Yupan L.	玉盘湖		34°55′	88°23′	416	3
Yuping L.	玉瓶湖		34°13′	88°15′	412	3
Yurba Co	涌波错		35°44′	86°42′	406	1
Zabuye Caka	扎布耶茶卡	查木错, Chabyêr Caka	31°26′	84°03′	214	1
Zadoi	杂多		33°00′	95°18′		1
Zainzong Caka	占宗茶卡	珍宗茶卡, 赞宗错	32°15′	89°36′	255	2
Zanda	扎达		31°24′	79°48′		1
Zaqu R.	扎曲		32°50′	95°30′		1
Zayü	察隅		28°36′	97°24′		1

Zhag'yab	察雅		30° 36'	97° 30'		1
Zhajieyi Co	扎杰意错		33° 06'	84° 34'	321	3
Zhangye	张掖		38° 54'	100° 24'		1
Zhari Namco	扎日南木错	塔热错	30° 57'	85° 37'	332	1
Zhaxi Co	扎西错		32° 12'	85° 07'	324	2
Zhidoi	治多		33° 48'	95° 36'		1
Zhongba	仲巴		29° 36'	84° 06'		1
Zhuqu Co	祝曲错		32° 02'	87° 45'	230	2
Zigê Tang Co	兹格塘错	孜格丹湖	32° 05'	90° 51'	269	1
Ziri L.	自日湖		31° 30'	91° 36'	279	2
Zogang	左贡		29° 36'	97° 48'		1
Zonag Co	卓乃错		35° 33'	91° 57'		1
Zongmugen Co	宗木根错		32° 46'	89° 56'	260	3
Zougangyou Caka	走岗由茶卡		33° 07'	85° 05'	333	3
Zuoqing Co	座倾错		33° 01'	85° 22'	370	3

Notes: 1. from Gazetteer of China, 1983.

2. from Gazetter of Xizang (Tibet) Autonomous Region, 1978.

3. spelled according to the Chinese phonetic alphabet with some phonetic annotations made by referring to Gazetter of Xizang Autonomous Region.

4. from the meaning

OUTSTANDING PROBLEMS RELATING TO THE INDEX OF PLACE NAMES

1. Place names in this index are arranged in the order of the Latin alphabet, but a few place names are changed somewhat during revision; so the revised place names are arranged in a new location in alphabetic order.
2. In view of the change of some place names, corresponding changes are also made in the Map I "Hydrochemical Zones of the Lakes of the Qinghai–Tibet Plateau". The place names changed are as follows:

APPENDIX 3

(1) 阿果错,	Original	Aguo Co	Now	Argog Co
(2) 比隆错		Bilong Co		Bêlog Co
(3) 波密		Bomê		Bomi
(4) 达日		Dagri		Darlag
(5) 岗巴		Ganba		Gamba
(6) 杜佳里		Dujiali L.		Koggyaling L.
(7) 拉果错		Laggor Co		Lagkor Co
(8) 美日切错		Miriqie Co		Meiriqie Co
(9) 木地错		Mudi		Mugdê
(10) 色卡执错		Sekazi		Sêkazhig Co
(11) 索加		Soga		Suojia
(12) 夏那令错		Xagnaglin		Xianaling

APPENDIX 4 EXPLANATION OF PLATES

Plate I

1. The medical use of borax recorded in the "Sibu Yidian", a pharmacopoeia written in Tibetan in 720 A.D.
2. Potala Palace in Lhasa that began to be constructed in the sixth century.
3. Jokhang Temple that began to be constructed in 647.
4. Ball granite in the rear mountain of the Drepung Monastery.
5. A chain bridge across the Yarlung Zangbo River in the Dobê district constructed more than 500 years ago.
6. A flock of sheep carrying rock salt on the back ready and waiting besides the South Zabuye Lake—an age–old way of carrying rock salt.

Plate II

1. The saline lake expedition headed by the author encamps on the bank of the South Zabuye Lake in 1982.
2. The members of the expedition cross a salt marsh of the South Zabuye Lake in 1984.
3. The expedition makes an investigation in Taro Co (a recharge lake of the South Zabuye Lake) in 1984.
4. Drilling in the South Zabuye Lake in 1984.
5. The author and others make an on–the–spot observation of the ecology of halophilic algae growing on a travertine mound in 1984.
6. The expedition vehicle sinks into a salt marsh by the Zabuye Lake (1993).
7. The expedition passes through a pass of the Gangdise Mountains (1990).
8. Dr. Wu Ande (a Chinese American) and the family of Pomco, a local Tibetan herdsman, on the shore of the Zabuye Lake (1994).

Plate III

1. View of Chagcam Caka.
2. Lake–worn line (L) in the northeast of Chagcam Caka, about 170 m higher than the modern lake surface.
3. Magnesium borate terrace–I of Chagcam Caka: the upper part consists of a magnesium borate ore beds and marl beds folded due to deglaciation, while the lower part a thin carbonate–clay bed.

4. Swell of a deglaciation fold of boron—bearing sediments.
5. The expedition carrying out an investigation on the shore of Chagcam Caka II (1982).
6. The expedition making topographic measurements with the infrared range finder in Chagcam Caka.
7. View of the open—pit of the saline magnesium borate deposit in Chagcam Caka.
8. The boron—tolerant *Polyonum sibiricum* community growing on terrace—I of boron—bearing lime—mud deposits on the shore of Chagcam Caka.

Plate IV

1. View of a playa of Bangkog Lake II. Distant view is a Cretaceous cuesta.
2. The site of an abandoned borax mine in Bangkog Lake II; the dired lake surface 4522 m above sea level.
3. View of the Qarhan Playa; the playa surface 2675—2680 m above sea level (photo by Yang Qian).
4. Mining ditch of potassium salts in the Qarhan Playa (photo by Yang Qian).
5. Mining boat for potassium salts in Qarhan (photo by Yang Qian).
6. Shallow brine district in the east part of the Da Qaidam Lake. Halite deposited in waters. The lake surface 3110 m above sea level.
7. Salt solution cave in the Da Qaidam Lake.
8. Mining district of ulexite—pinnoite in the Xiao Qaidam Lake; the Lake surface 3118 m above sea level.

Plate V

1. Nam Co, the largest lake of the Tibet Plateau (4718 m above sea level; area 1920 km^2).
2. Siling Co, the second largest lake of the Tibet Plateau (4530 m above sea level; area 1640 km^2).
3. Zhan Namco, the third largest lake of the Tibet Plateau (4613 m above sea level; area 1020 km^2).
4. Lagkor Co (4470 m above sea level) yielding abundant *Artemia salina*.
5. Jibu Caka (yielding magnesium borates) separated from Lagkor Co by a sand levee (I).
6. Jiabo Co yielding borax.
7. Marmê Co yielding abundant ulexite.
8. Marmê Co. An ulexite bed (white) lying below a thin carbonate— clay bed (brown).

Plate VI

1. The highest ancient lake water line (L) at the foot of the Jiadonglongba Mountains northeast of the Zabuye Lake is about 200 m higher than the modern lake surface. The distant mountains are the Lunggar Mountains.
2. The low planation surface (P) and multi–step high lake terrace (T) southeast of the Zabuye Lake.
3. The low planation surface (P), multi–step lake terrace (T) and salt flat (S) at Jiu'ershan south of the Zabuye Lake.
4. Thick calcareous gravel bed on the 8th step of high sand levee of the Zabuye Lake.
5. Hydromagnesite marl bed— chemical deltaic deposits— in the upper part of the 5th step of sand levee of the Zabuye Lake.
6. A lake–worn trough (whose bed rock is a Lower Cretaceous limestone bed) about 100 m higher than the modern lake surface in the west of the Zabuye Lake.
7. A modern calcium carbonate cofferdam formed by ascending springs in front of Chabuya Island.
8. Travertine mound in the North Zabuye Lake with a halite patch on the surface. The distant view is the Lunggar glacier.

Plate VII

1. Second electron image of Li–bearing dolomite from Zabuye Lake. 25KV, ×270.
2. Zabuyelite crystal.
3. Zabuyelite crystal (diameter 2.66 mm).
4. Zabuyelite twin crystal.
5. Inderite crystal, ×90.
6. Euhedral pinnoite (Pi) replaces the slaty kurnakovite (Ku) crystal (diameter 4.33 mm).
7. Carbonate clay pellets (c–c) surrounded by radiating pinnoite (Pi); K–quartz (diameter 1.8 mm).

Plate VIII

1. Hair–like ulexite (U) replaces pinnoite (Pi) viewed under the optical microscope (diameter 2.5 mm).
2. Irreqular quadrilateral inyoite (in) distributed in kurnakovite (Ku) viewed under the optical microscope (diameter 2.6 mm).
3. Syngenite with radial texture fromed by the euhedral, thin–platy syngenite aggregates viewed under the optical microscope (diameter 4.3 mm).
4. Salt shell on Qarhan Salt Lake.

APPENDIX 4

5. Rhombic dipyramid carnallite crystal × 3 (from the man–made dam, East Qarhan Lake, photographed in 1959).
6. Light grey carnallite (B) filled intermitently in halite bed (A) (photographed in 1957, from the southeast side of the salt highway, Qarhan Salt Lake).

Plate IX

1. Vermiform efflorescence formed by cappilary action on the saline lake surface (Da Qaidam Lake).
2. Cross section of sub–terrace I–3 northeast of Bangkog Lake I. Hydromagnesite gravels (white) mixed with round limestone pebbles (dark), roughly bedded.
3. Hydromagnesite crumble structure in Bangkog Co, × 60.
      a: Hydromagnesite pellets; b: muddy hydromagnesite; c: pore.
4. Cylindrical Li–bearing magnesite, × 30.
5. Li–bearing magnesite monocrystal.
6. Electron diffraction pattern of Li–bearing magnesite monocrystal.

Plate X

1. The sand–gravel bar (sub–terrace I–3) formed by hydromagnesite bed on the southern shore of Bangkog Co II. Below the hydromagnesite bed is carbonate clay.
2. Coarse–grained–platy borax from Bangkog Co.
3. Illite with interference stripes, from carbonate clay at the lake bottom of Bangkog Co II, × 17000.
      ESCA: Mg 1%; Al 29.69%; Si 47.79%; K 15.67%; Fe 5.12%.
4. Schistose chlorite in the bottom mud of Bangkog Co III, × 22000.
5. Fine–flaky illite, flocculent montmorillonite and tubular endellite in the bottom carbonate clay of Bangkog Co I, × 13000.
6. Flaky illite, chlorite (with blurred edge) and a small amount of hexagonal kaolinite in the lake bottom sandy clay (hole CK4) of Bangkog Co III, × 13000.
7. Fine and thin flaky illite, chlorite and a small amount of tubular endellinte in the bottom blue clay of Bangkog Co III, × 13000.

Plate XI

1. Borax crystal occurring in the Zabuye Lake, × 7.
2. Wedge–shaped gaylussite occurring in Zabuye Lake, × 10.
3. Short–prismatic and "envelope"–shaped gaylussite occurring in the Zabuye Lake, × 10.
4. Trona occurring in the Zabuye Lake. × 10.
5. Zabuyelite crystal found in the Zabuye Lake. × 10.

6. Glaserite occurring in the Zabuye Lake. × 10.
7. Zabuyeite (Z) intergrown with gaylussite (G) and Li-bearing dolomite (D). × 30.
8. Northupite (occurring in the Bangkog Lake) enclosing Li-bearing magnesite (white dot). × 20.

Plate XII

1. Octahedral and dodecahedral mirabilite (Bangkog Lake). × 20.
2. Pseudo-octahedral tincakonite (Bangkog Lake). × 20.
3. Kurnakovite (Chagcam Caka). × 10.
4. Inderite (I) intergrown with kurnakovite (K) (Chagcam Caka). × 20.
5. Platy syngenite. × 10.
6. Eggs of natural *Artemia salina* (Lagkor Co).
7. Larvas of *Artemia salina* pushed to and accumulated between gravels on the Lakeshore by winds and waves (Lagkor Co; on 23 September 1993).

Plate XIII

1. The lower jawbone with teeth of *Cervus albirostris*; ^{14}C age 9430 ± 290 a.
2. A long-term observation station set up at 4424 m above sea level in the Zabuye Lake on the Tibet Plateau.
3. Meteorological apparatus of the station.
4. Experiment of natural evaporation of brines at the station.
5. Self recorder of the brine water level at the station.
6. Stage rod of the station.

I

II

III

IV

V

VI

VII

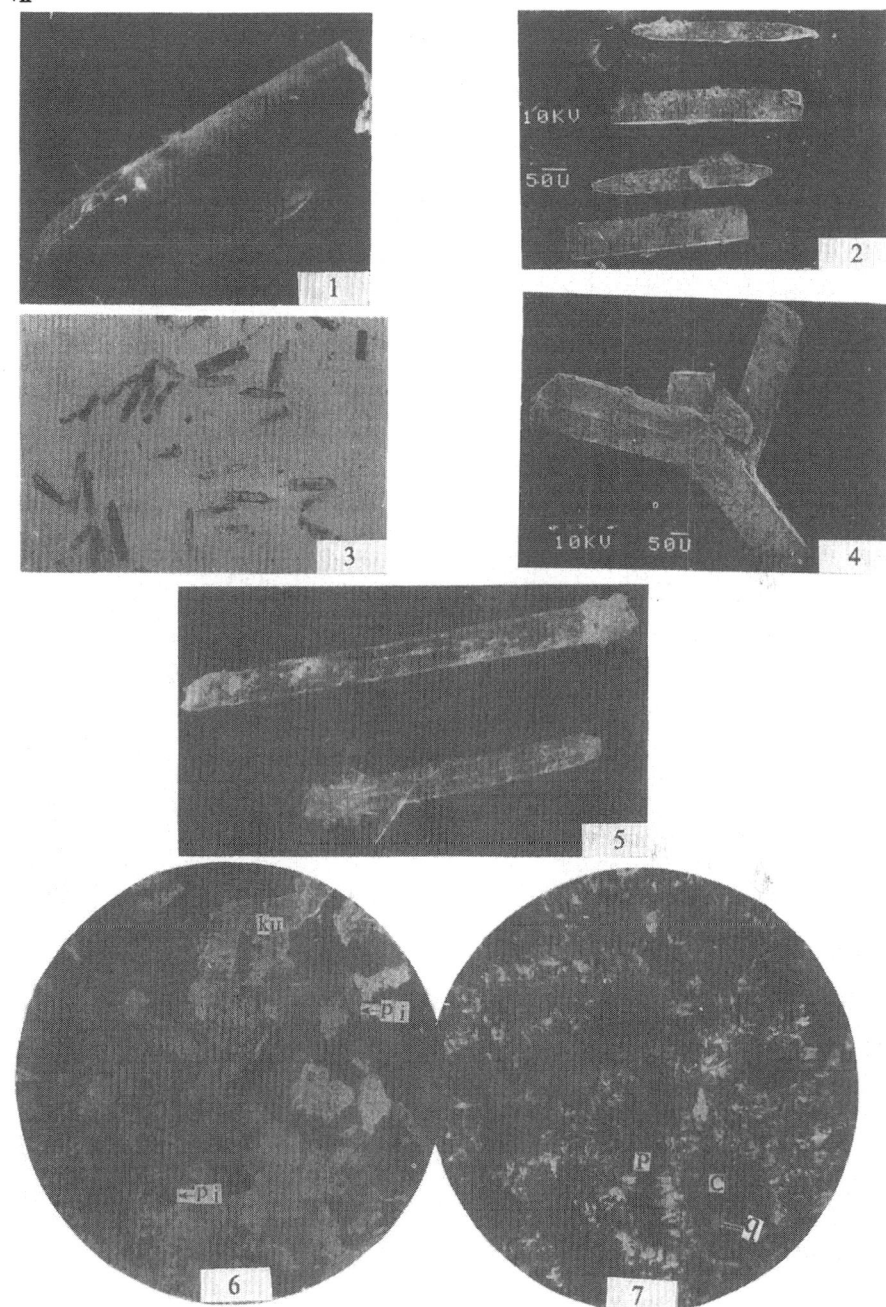

VIII

IX

X

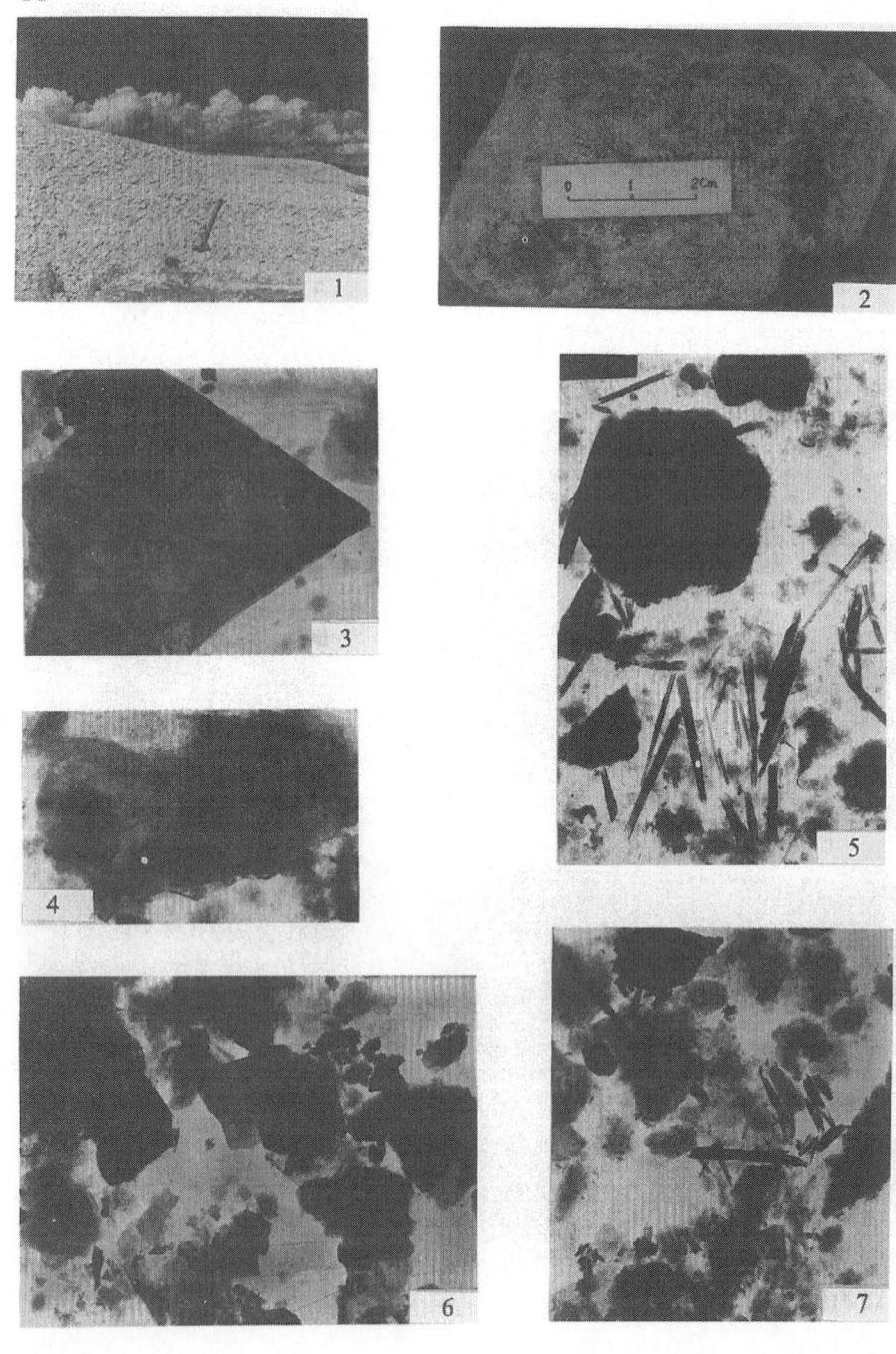

1	2
3	5
4	
6	7

XI

XII

XIII

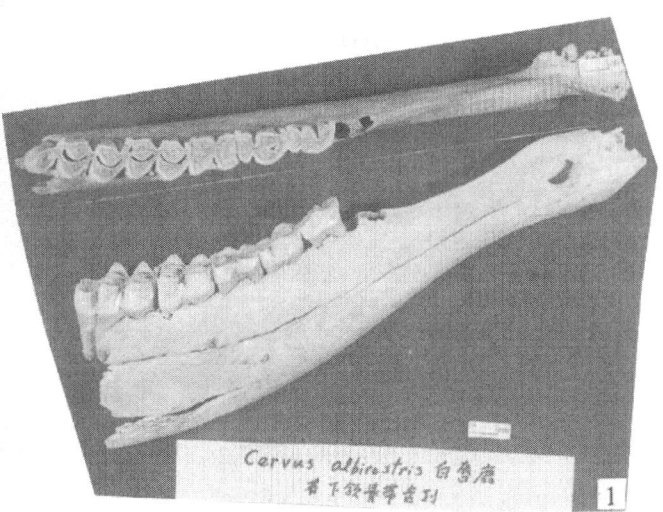

Cervus albirostris 白唇鹿
若下钦兽等古到

XIV

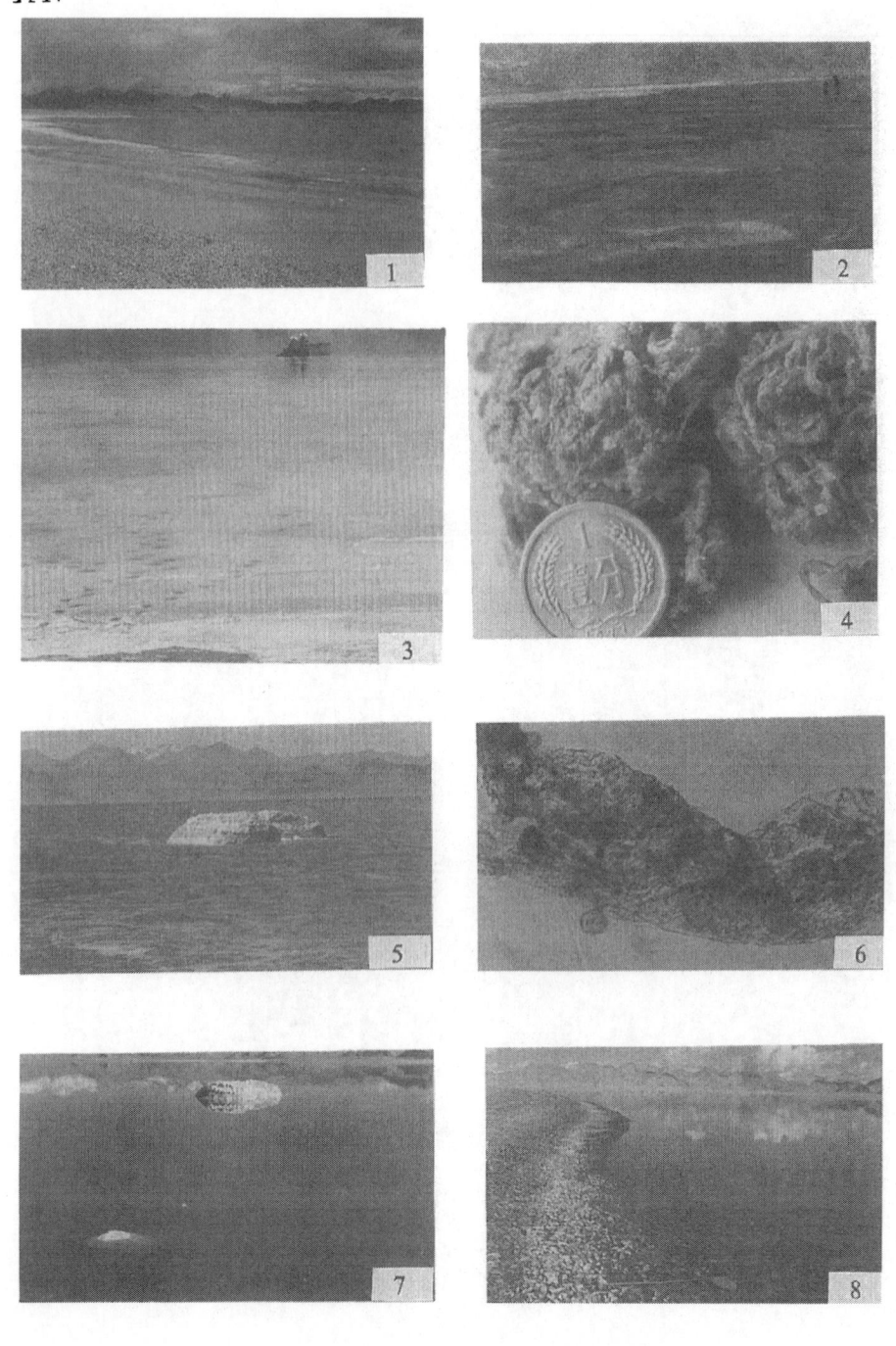

Previous published books

Biogeography and Ecology of Turkmenistan, Victor Fet and Khabibulla Atamuradov. 1994   ISBN 0-7923-2738-1

Lake Baikal and Its Life, M. Kozhov. 1963   ISBN 90-6193-064-2

Lake Kariba. A Man-Made Tropical Ecosystem in Central Africa, Eugene K. Balon and A.G. Coche. 1974   ISBN 90-6193-076-6

Lake Kinneret. Lake of Tiberias, Sea of Galilee, C. Serruya. 1978   ISBN 90-6193-085-5

Lake Chilwa. Studies of Change in a Tropical Ecosystem, M. Kalk, C. Howard-Williams and A.J. McLachlan. 1979   ISBN 90-6193-087-1

Lake Sibaya, B.R. Allanson. 1979   ISBN 90-6193-088-X

Lake Biwa, S. Horie. 1984   ISBN 90-6193-095-2

Saline Lake Ecosystems of the World, U.T. Hammer. 1986   ISBN 90-6193-535-0

Limnology in Australia, P. de Deckker and W.D. Williams. 1986   ISBN 90-6193-578-4

Printed by Publishers' Graphics LLC USA
JCIMO131106.15.17.40